T0140439

The Information Retrieval Series

Volume 42

Information Retrieval (IR) deals with access to and search in mostly unstructured information, in text, audio, and/or video, either from one large file or spread over separate and diverse sources, in static storage devices as well as on streaming data. It is part of both computer and information science, and uses techniques from e.g. mathematics, statistics, machine learning, database management, or computational linguistics. Information Retrieval is often at the core of networked applications, web-based data management, or large-scale data analysis.

The Information Retrieval Series presents monographs, edited collections, and advanced text books on topics of interest for researchers in academia and industry alike. Its focus is on the timely publication of state-of-the-art results at the forefront of research and on theoretical foundations necessary to develop a deeper understanding of methods and approaches.

More information about this series at http://www.springer.com/series/6128

Deepak P • Tanmoy Chakraborty • Cheng Long •
Santhosh Kumar G

Data Science for Fake News

Surveys and Perspectives

 Springer

Deepak P
Queen's University Belfast
Belfast, UK

Cheng Long
School of Computer Science
and Engineering
Nanyang Technological University
Singapore, Singapore

Tanmoy Chakraborty
Indraprastha Institute of Information
Technology
Delhi, India

Santhosh Kumar G
Department of Computer Science
Cochin University of Science
and Technology
Cochin, Kerala, India

ISSN 1871-7500   ISSN 2730-6836   (electronic)
The Information Retrieval Series
ISBN 978-3-030-62698-3   ISBN 978-3-030-62696-9   (eBook)
https://doi.org/10.1007/978-3-030-62696-9

This Springer imprint is published by the registered company Springer Nature Switzerland AG.
The registered company address is: Gewerbestrasse 11, 6330 Cham, Switzerland

Preface

Fake news is among the major challenges facing global society in the twenty-first century. There has been a steadily growing awareness about its importance. In a recent study by the Pew Research Center,[1] it was found that fake news is being perceived as more important than climate change. The study also reports that a majority of respondents observed that it has an impact on confidence in government and confidence in each other. While the study was conducted in the USA, the nature and size of challenges that it poses elsewhere are along similar lines. Fake news has destabilized governments, undermined trust in institutions all over the world, and sowed seeds of disharmony in the global society. Social and online media form a major vehicle for fake news, and this causes fake news to leave a vivid digital footprint. The digital footprint of (fake) news consumption forms a rich resource for online platforms to profile fake news effectiveness and develop better micro-targeting mechanisms that will aid the spread of more (fake) news. On the other hand, the digital footprint also provides a rich data resource, which allows us to understand fake news and check its growth and spread. Consequently, most work on combating fake news has been within data science. A search with the term *fake news* on DBLP,[2] a computing publication repository, shows that there were 100+ publications with *fake news* in the title in 2018, with that roughly doubling to 200+ in 2019. As of this writing (September 2020), there have been another 150 publications in 2020 with *fake news* in the title. While data science work on fake news is likely to significantly influence the way society responds to fake news in the future, combating fake news is not a pursuit that should be nestled within one discipline. Fake news propagation exploits cognitive biases and societal stereotypes and causes serious concern about the nature of politics and healthcare. The predominantly textual nature of fake news also needs to be seen within the broader context of linguistic inquiry. Knowledge from such domains

---

[1] https://www.journalism.org/2019/06/05/many-americans-say-made-up-news-is-a-critical-problem-that-needs-to-be-fixed/.

[2] https://dblp.org/.

would inevitably enhance the breadth, depth, and impact of efforts targeted at combating fake news.

This book presents what we believe is a first-of-a-kind effort toward an inter-disciplinary take on data science for fake news intended to open up conversations with non-computing disciplines in tackling this grand challenge of the twenty-first century. This book is organized into two parts. First, it covers a number of surveys on data science advances toward fake news detection. While we do have chapters covering the "mainstream" work on fake news detection (e.g., deep learning), we have taken care to include underexplored directions of scholarly inquiry, namely, unsupervised methods, graph-based methods, and knowledge-oriented directions. Second, we have a set of chapters on interdisciplinary perspectives on data science for fake news authored by scholars who are located within those disciplines (except for two chapters, which are authored by data scientists with expertise in those disciplines). The second part is structured in a way that is accessible to the layperson or a generalist with an interest in fake news. Given the diversity of the content in the book, we envisage the following ways in which it could be used:

– *Advanced Computing Course on Data Science for Fake News:* The survey chapters form the material for an advanced computing course and can perhaps be augmented with more material on deep learning methodologies that are quite popular these days. The chapter on ethics (in the perspectives section of the book) could be included as supplementary reading for the course as well.
– *Researchers' Reference:* The survey chapters will provide a ready reference for researchers working on data science efforts toward fake news.
– *Reading Material for Non-computing Courses:* The perspective chapters may be treated as additional reading material for courses within those disciplines. For example, a course on linguistics for digital media could use the linguistics perspective chapter as additional reading on how linguistics applies to fake news. The same would hold for other perspective chapters.
– *Popular Science Reading:* The introduction followed by the perspective chapters together can be treated as general reading material or a short popular science book on fake news.

There could be other ways of using the book and the individual chapters, so the above may be treated as a non-comprehensive list of possibilities. We expect that this book will help in deepening the debate on combating fake news.

Acknowledgments

The authors would like to express their sincere gratitude to the authors of the perspective chapters: Ninu Poulose, Muiris MacCarthaigh, Girish Palshikar, Lyric Jain, Anil Bandhakavi, Renny Thomas, and Jane Lugea. Without their contribution, the essence of this book project would not have materialized. The authors would also like to thank their collaborators for various insightful discussions that enriched the book. They would also like to acknowledge the wholehearted support from their families, without which these efforts would undoubtedly have been impossible.

Belfast, UK   Deepak P
Delhi, India   Tanmoy Chakraborty
Singapore, Singapore   Cheng Long
Cochin, India   Santhosh Kumar G
September 2020

Contents

A Multifaceted Approach to Fake News

Deepak P

Abstract This introductory chapter provides the contextual basis and inspiration for this volume/book, as well a summary of the chapter contents. We believe that the multifaceted approach to fake news we have used in this book is quite unique and hopefully will open up fresh perspectives and questions for the interested reader. In this chapter, we start by motivating the need for such a multifaceted approach toward fake news, followed by introducing the two-part structure of this book: surveys and perspectives. Further, we briefly describe the various surveys and perspectives that we cover in the subsequent chapters in this book.

Keywords Fake news · Data science · Multidisciplinary perspectives

1 Introduction

Research into *fake news* (across various flavors ranging from *rumors* to *misinformation* and *disinformation*) has been on the rise in the computer science literature, especially within data analytics and machine learning. DBLP,[1] a popular computer science publication indexing service, has 500+ publications with *fake news* in the title, whereas the corresponding numbers for *misinformation* and *disinformation* are 250+ and 90+, respectively; most of these have been from very recent years. While *fake news* has been well understood as an interdisciplinary topic, the quantum of research within computing and data sciences on the topic is easily much more than the attention it has attracted within other fields, including pertinent social science disciplines. However, the machine learning perspective that drives virtually all computing advances toward fake news is arguably parochial in outlook. With several computing conferences being increasingly attended by experts from across disciplines (e.g., some even encourage work from other disciplines to be

---

[1] https://dblp.org/.

© Springer Nature Switzerland AG 2021
Deepak P et al., *Data Science for Fake News*, The Information
Retrieval Series 42, https://doi.org/10.1007/978-3-030-62696-9_1

presented, such as FAccT, Web Science, HyperText, The Web Conf, and so on), the shortcomings of the machine learning perspective are increasingly revealed. I have had the chance to attend many presentations of *data science for fake news* work across several conferences, and a slightly dramatized mash-up of several conversation fragments across Q&A sessions below reveals some of the shortcomings of the machine learning approach:

- *Question:* Thanks for your presentation of an interesting approach against fake news. How does your technique generalize to other domains?
- *Presenter:* We have not come across such large datasets outside the political domain in which we have performed our empirical evaluation. Thus, we have not tested. However, we believe that the high-level assumptions of our method would generalize well.
- *Question:* But, well, you make an assumption of temporal coherence; it doesn't seem to me that this would generalize to domains such as health and science where the fake news does not necessarily relate to an "ongoing" event or debate.
- *Presenter:* I see your point; perhaps it won't. But we are yet to find large-scale timestamped datasets to test and validate them for health and science.
- *Question:* Populist and right-leaning governments within several nations have repeatedly used the term *fake news* to discredit legitimate news that goes against them. How does the conception of fake news within your method relate to such tendencies?
- *Presenter:* Our method is robust to such illegitimate usage of the term *fake news*. Our method itself does not use a dictionary and makes use of state-of-the-art deep learning methodologies and is thus guided by the correlations between the news and the *fake–real* labelling within the training data, and, as you can see, it works well on the held-out test set too.
- *Question:* But, well, have you considered that the labelling may be polluted by labellers being exposed to such ambiguous usage of the term?
- *Presenter:* We did not create the labelled dataset ourselves, so we are not aware of the specific details. However, the dataset creators have reported a good inter-annotator agreement.
- *Question:* Given that your dataset covers recent political news, I would be interested in knowing how successful your method has been in handling the fake news about the wall in the US-Mexican border.
- *Presenter:* Unfortunately, we have not performed any qualitative analysis on specific topics. Overall, across tens of thousands of articles, we achieve an accuracy of 90%; thus we would naturally expect that it performed well on the separate topics in the corpus.
- *Question:* How would you think your method could be incrementally retrained for continued usage, given that the dynamics of fake news could change quite dramatically with time?
- *Presenter:* We have not considered incremental learning aspects. We hope to do so in future work. Thanks for your question.

I do not know what you would think about how well the presenter answered the questions posed, but one could rate them as quite revealing of the narrow perspective toward fake news in the work undertaken. First, machine learning researchers quite routinely have taken the data as given, so any deeper questions about data collection or labelling are considered as *beyond scope*. Second, there is an impetus toward justifying an approach by the sophistication of methods employed and the results of the quantitative analyses performed, and any qualitative analysis of the workings is often considered superfluous. Third, there is little consideration given to how such a technique addressing a societally important problem would be used in a real application scenario where the character of fake news continuously changes and the standard training-test apparatus is deemed sufficient. Fourth, researchers engaged in such work are *disinterested in the details of the domain* and treat it as yet another accuracy maximization challenge.

In fact, such considerations are hardly unique to fake news detection. The widening of the ambit of data science to various societally important domains manifests in the form of similar issues in such domains. As an example, the raw application of data science to criminality prediction from face images, recidivism prediction based on demographic details, and the like have resulted in similar critiques.

This book draws inspiration from such critiques of data science efforts in societally important domains, and out of a belief that attention to the *softer* and *domain-specific* parts of the space of fake news, in particular, and digital media, in general, is necessary to develop the next generation of AI techniques for fake news detection. Accordingly, this book adopts a two-part structure. The first part surveys the current state-of-the-art data science approaches to fake news, with a focus on facets that have not been used much before in surveying data science work on fake news detection. The second part, in a very unique and first-of-a-kind effort, comprises several chapters from experts in various disciplinary domains who provide key disciplinary inputs that would help inform the development of the next generation of data science techniques for fake news.

The following sections describe each of the survey and perspective chapters in the book, with specific attention to their positioning in the context of the multifaceted approach to data science for fake news that the book attempts to advance.

2 Surveys

Machine learning, data science, and AI efforts toward any interdisciplinary problem are usually spread across multiple subdisciplines that the data analytics community has organized itself into. For the case of fake news detection, the subdisciplines of relevance would be the following:

- *Natural language processing (NLP):* This relates to the text-oriented approach toward the task, with a focus on using NLP methods. While NLP methods could span traditional parsing-based approaches as well as statistical approaches, statistical approaches (and neural methods in particular) have been popular over the last two decades.

- *Computer vision and image processing:* The discipline of computer vision and image processing targets to focus on the image and video aspects of the task, which is often quite abundant in the domain of digital media and is thus critical for the task of fake news detection. Much like the case of NLP, this area has also seen abundant uptake of deep learning methods.
- *Information retrieval (IR):* The information retrieval community has roots in the 1960s, and while they largely rely on text data like the NLP community, they adopt a more statistical and less linguistic approach toward text processing. The core task of interest for the community is retrieval and allied tasks that can improve the accuracy or efficiency of retrieval. There has been limited work on fake news within the information retrieval community.
- *Web science:* The web science community, spearheaded by ACM SIGWEB, has considered computational and non-computational tasks that are societally important, are interdisciplinary, and relate in some way to the World Wide Web. This community, and the more recent community around AI ethics, has a strong focus on humanities and social science aspects.
- *Social media:* The social media analytics community blends techniques from network science, NLP, and temporal dynamics modeling in order to attack several analytics tasks that are central to social media. Social media is often construed in a broad manner, covering conventional social media (e.g., Facebook, Twitter) along with other forms of online user-created content such as blogs. Work on social media, due to the primacy of the social component, also tends to be quite interdisciplinary.
- *Databases and data mining:* The databases and data mining community have seen attention to fake news from a fact-checking perspective, especially focusing on the usage of external knowledge sources such as authoritative databases and ontologies for the task of fake news detection.
- *Classical AI:* Classical AI methods (sometimes called *Good Old-Fashioned AI* or *GOFAI*) could adopt a reasoning approach to fake news detection, whereby the focus could be on the soundness of the reasoning employed in a news narrative, where shallow reasoning or leaps of imagination could be identified as indicative of lack of veracity.

While there are significant overlaps across the above communities (e.g., NLP and IR), they still tend to be separate, with publication avenues (e.g., conferences and journals) clearly aligning within specific communities. In such an ecosystem, it is natural for surveys to be focused on particular aspects relevant to each subcommunity. For example, an NLP survey on fake news appears at [2], whereas an image- and media-focused survey appears at [3] and a social media-oriented survey appears elsewhere [4]. Our approach is to survey fake news work from viewpoints that are not very deep-seated within such subdisciplines above. We now describe the various surveys we incorporate in this book, explaining the nature of the survey at a high level herein.

2.1 Unsupervised Methods for Fake News Detection

While supervised learning has been the dominant stream for the task of fake news detection, unsupervised learning has seen some emerging interest in recent times. Deepak's chapter on unsupervised learning considers reviewing this emerging body of work. The chapter starts by observing the distinctions between supervised and unsupervised approaches with specific reference to fake news, particularly noting the increased implicit responsibility placed within unsupervised methods on heuristics/assumptions to offset the unavailability of labelled data. It further outlines the flavor and main families of assumptions made across unsupervised learning methods and describes the main techniques. The chapter concludes with some observations on the road ahead for unsupervised fake news detection.

2.2 Multimodal Fake News Detection

While there has been an emerging interest in effectively leveraging the complementarities and synergies across multiple modalities of data in accomplishing machine learning tasks (particularly within the multimedia community), the domain of fake news poses unique challenges in multimodal analytics. Chakraborty's chapter on multimodal fake news detection notes that the fakeness of multimodal fake news often rests "between" the modalities. An image from one event being used along with text from another event is among the most powerful forms of fake news generated through political propaganda. Chakraborty considers three main streams of approaches used in multimodal fake news detection: early fusion, late fusion, and hybrid fusion. These differ in the stage at which data from across modalities are fused in order to perform fake news detection. The author surveys the state-of-the-art approaches for multimodal fake news detection across the above three categories as well as across emerging streams such as those that use adversarial training. The chapter also includes a listing of several kinds of multimodal fake news data that would undoubtedly be a useful resource for researchers aspiring to work in this area.

2.3 Deep Learning Methods for Fake News Detection

Deep learning techniques have revolutionized the field of machine learning and AI over the last decade. As in other realms of machine learning, deep learning methods have been used heavily for fake news detection as well. Kumar's chapter on deep learning for fake news detection targets to provide an overview of the usage of deep learning frameworks among the state-of-the-art models for fake news detection. Kumar starts with a gentle introduction to popular deep learning building blocks and illustrates how they have been used to build complex deep learning methods for fake news detection. The author illustrates how complex frameworks can be understood by decomposing them into several component modules and their learning "responsibilities" within the framework. The chapter describes how image-oriented (e.g., CNN) and text-oriented (e.g., LSTM) layers have been integrated into methods targeted at multimodal fake news detection. Kumar then provides an overview of various datasets and evaluation metrics that have been popular for deep learning-based fake news detection to aid researchers who are interested in the task. The chapter concludes by summarizing a few emerging directions in deep learning-based fake news detection such as exploitation of graph structures and explainable decision-making.

2.4 Dynamics of Fake News Diffusion

Diffusion is usually used to refer to the nature of propagation or spreading, and efforts in modeling diffusion mathematically have ranged across several disciplines including epidemiology. Chakraborty's chapter on fake news diffusion focuses on the ways in which the propagation of fake news has been leveraged in fake news analysis and detection, within computational efforts. The author notes that there have been interesting and sometimes conflicting analyses on the speed of fake news vis-à-vis real news, as well as the importance of the role played by bots vis-à-vis humans in spreading fake news. The chapter surveys a number of propagation models that have been employed for modeling the spread of fake news. While some of them are able to work backward and identify the source of misinformation, others are more suited toward predicting the nature of fake news spread and characterizing user trustworthiness. Despite the variety of computational modeling methods employed in diffusion analysis of fake news, the author notes that there are still several avenues for furthering research in the area. The chapter ends with a review of how an understanding of fake news propagation could be used in order to curb fake news by actively intervening within the user network.

2.5 Neural Text Generation

This is the age of *deepfakes*[2], when data, predominantly images, is synthetically generated to much perfection. An upcoming frontier in fake news is that of deep text fakes, where fake news text would be generated automatically by machine learning models. While this is still not quite a reality, there are enough indicators that it would be so. Kumar, in a chapter on neural text generation, surveys the current

---

[2]https://en.wikipedia.org/wiki/Deepfake.

state of the art in text generation using neural language models. The author starts by considering the various kinds of neural architectures and how they have evolved into yielding (pre-trained and non-pre-trained) language models that are quite effective in generating real-sounding synthetic text. The author goes on to illustrate how these could take the next step and yield a suite of methods that would generate fake news automatically. Would computational methods be able to identify such clever computationally generated fake news? We would need to wait for the future to unfold.

2.6 Fact Checking on Knowledge Graph

Knowledge graphs from sources such as Wikipedia have been quite popular in various areas of data science. However, they have been used relatively lightly in the context of fake news detection. With knowledge graphs widely regarded as among the most systematic resources of semantic information on how humans understand the world, these are invaluable resources for fake news detection. Luo and Long, in a unique endeavor, identify ways in which knowledge graphs could be used for the task of fact checking. The authors outline several concepts involving paths between entities and knowledge streams, all the time relating to how they could be of use for debunking examples of highly popular fake news. They illustrate how such concepts could be blended together to assess the veracity of claims involving mentions of concepts (e.g., people, religions, locations, institutions) and illustrate the empirical effectiveness of such methods for fact checking. The authors also outline potential future directions for research on leveraging knowledge graphs for fact checking.

2.7 Graph Mining Meets Fake News Detection

Graphs, comprising nodes that are interconnected through edges, form a very popular form of data representation used in data analytics. In a dedicated chapter, Yu and Long consider how graph mining may be used within fake news detection. They first consider the various challenges, followed by elucidating the different kinds of data and how they may be modeled within varying types of graphs. Within the unimodal data scenario, they illustrate how graph mining tasks aligned with fake news detection (e.g., anomalies) can be addressed through statistical analysis of graphs and dense graph mining. This is followed by a review of methods for fake news detection over multimodal graphs, with a focus on the usage of temporal information in identifying structures of interest for the task of fake news detection. The authors conclude by pointing out the plethora of work that could be done in fake news detection using graph analytics methods.

3   Perspectives

Given the earlier discussion, it is superfluous to reemphasize that fake news detection is an interdisciplinary topic. We now outline some of the several connections it has to various disciplines.

- *Politics:* This is the realm in which the impact of fake news has been most felt. The abundance of fake news within the 2016 US presidential election is what brought this topic to the forefront. Fake news is often used by political parties to sway the debate to topics of their convenience. They are often effective at that since it takes very little effort to create fake news, following which the affected party would need to undertake much time and effort to debunk it. This undermines effective public reason and, thus, the functioning of democracy.
- *Media:* Various forms of new media have been the vehicles of fake news, and in an era of constant exposure to digital media (contrast this with the newspaper-only era), the race to capture user attention has caused a race toward the sensationalist end of the news spectrum, one that is plagued with fake news. This has caused significant issues in journalism as a discipline of practice and scholarly inquiry.
- *Health:* Fake news related to health and well-being has been rampant in our times. This focuses on topics that are not necessarily of real-time importance, and its consequences are not often visible. For example, a person who tried a magic cure suggested by fake news and suffered due to it is not likely to admit it in the public domain. Admission in social media could even result in people pointing fingers at the sufferer for lack of due diligence and aggravate the suffering. It is even more of a consequence that fake news often targets deadly diseases such as cancer, making the problem worse.
- *Sociology:* The prevalence of fake news within society moderates and modulates the interpersonal behavior and social interactions in population. The abundance of fake news narratives, and people buying into it, is likely to cause people to move toward extreme opinions, deepening societal divisions. Further, resilience to fake news requires a deep appreciation of the nuanced nature of various topics, and not mere adherence to the expectation of a single absolute truth.
- *Psychology and cognition:* A huge fraction of fake news is designed to exploit cognitive biases that have evolved over millennia and are deeply entrenched within humans. The oft-quoted example is that of confirmation bias and fake news [1], where popular societal stereotypes are used within stories in order to enhance their believability. A deeper understanding of the various cognitive biases and how they function in fake news is absolutely necessary in order to build techniques to effectively counter them.
- *Law and order:* Fake news has led to serious law and order issues, and a case in point is the abundant usage of anti-minority fake news in India[3] to instigate

---

[3]https://asiatimes.com/2019/10/indias-fake-news-problem-is-killing-real-people/.

localized violence, as well as wider clashes and riots. This requires on-the-ground enforcement officers to be on constant vigil and adopt well-thought-out measures to identify and curb such violence at early stages.

We have only illustrated a small cross section of the interdisciplinary dimensions of fake news. Fake news has additionally created several problems for brands and has caused the shallowing of trust in reputation systems such as crowdsourced reviews (e.g., e-commerce platforms and map services). The usage of emotions in fake news may be interesting to analyze from an anthropological perspective, and a historical context of fake news could help understand and inform strategies to counter them within particular societies. The nature of writing and style used within fake news would be undoubtedly interesting for experts in linguistics. The massive burden of debunking fake news often means that asymmetric power structures within societies are intensified in the wake of fake news; for example, a small retail store may be doomed with just one fake news, whereas a multinational chain store can potentially overcome multiple attacks against it. These effects also work across class, caste, and ethnicity lines, with fake news often impacting lower classes and minorities much more than others.

Given the limited contents that we can cover in the book, we feature an array of solicited chapters from experts from various disciplines. We briefly outline the scope of the various chapters herein.

3.1 Fake News in Health Sciences

Poulose, a cancer researcher, provides perspectives and recent studies on the topic of health-related fake news. The author highlights the importance of health fake news as a significant topic for our times, when social media is being increasingly used to look up health-related information. Poulose highlights the preeminence of anti-vax and cancer fake news, the latter being quite apparent even by simple studies over Google Trends. In particular, the author notes that the popularity of unproven and fake cancer cures among digital search trends has increased much more than the general interest in cancer within search trends. Poulose observes that the main streams of cancer fake news have been magic cures, natural remedies, political criticism of big pharma touted as evidence of ineffectiveness of modern cures, and immunity boosters. The author notes a crucial difference between pandemic-time and other fake news; the large majority of pandemic-time fake news is shared by common users out of altruism (if it saves a life, let it) rather than through organized campaigns. Poulose also provides a detailed overview of the various kinds of consequences of health misinformation, ranging from the hurdles it poses to effective medical care to the false hope it promotes. The chapter concludes by outlining possible strategies to curb health fake news, such as avoiding oversimplification in presenting scientific results and users exercising abundant caution before forwarding such news.

3.2   Ethical Considerations in Data-Driven Fake News Detection

In a first-of-a-kind effort, Deepak considers ethical questions in a chapter titled "Ethical Considerations in Data-Driven Fake News Detection." The chapter starts by noting the immense importance of ethical issues in AI-driven fake news detection, given the domain of operation is in the space of digital media, one that has a central role within democratic societies. AI, the author observes, is still guided by automation-era metrics aligned with utilitarianism, and this mismatch of values needs to be addressed as it starts to play a more central role in societies. It may even be argued, as the author points out, that the fundamental assumptions behind using historical data for making future decisions are unsuitable for operating in a space of high societal significance. The chapter points out that the properties of the digital media domain make certain kinds of decision-making impossible. The chapter does a deep dive into fairness as well as alignment of fake news detection uptake modalities with democratic values and outlines a set of recommendations that could potentially help develop the next generation of responsible AI for fake news detection.

3.3   A Political Science Perspective on Fake News

MacCarthaigh and McKeown, both political science researchers, observe the origins of fake news within the historical context of parliamentary democracy and the role played by media reporting in determining the success or otherwise of the government's political agenda. They outline the modalities of historical flows of fake news along class lines, often characterized as horizontal and vertical fake news, and the connections between tabloid/yellow journalism and fake news. They provide an insightful contrast between conventional media (fourth estate) and online media (fifth estate) in their relationship to fake news, especially referencing the echo chamber effect that the latter is often accused of. The authors observe that, in contrast to the fourth estate being often regarded as fundamentally important to democracy, the fifth estate might be a threat to it. The experience of the twenty-first century so far, the authors analyze, has been that fake news has worked as an important catalyst for populism and led to debates on geopolitical questions. The chapter also raises interesting questions on the nuanced relationship between fake news and trust in government. It concludes by noting the distinctive role that political scientists need to play in understanding the effects of fake news on political institutions and processes, as well as moderating the relationship between citizens and government.

3.4 Fake News and Social Processes

Palshikar, an industry researcher, considers the topic of fake news and its impact on social and political processes within society. The author rightly argues that fake news has had much impact on society and compares and contrasts the two streams of fake news research, viz., computational and sociological. Using vaccine hesitancy and political propaganda as case studies, the chapter analyzes how fake news systematically undermines scientific outlook and democratic processes, with specific reference to the explicit and implicit intents of fake news peddlers. Palshikar observes that fake news has far-reaching effects in a variety of social scenarios and has even led to the loss of lives, motivating the urgent need to curb it within society.

3.5 Misinformation and the Indian Election

In a unique chapter for a book on fake news, Jain and Bandhakavi, from Logically, a company in the space of tackling misinformation, describe their experiences from a wide deployment of fake news detection capability in the context of India's federal elections in 2019. India's elections, held once in 5 years, are by far the largest electoral exercise in the world. Jain and Bandhakavi outline the historical context of India, especially the ethnic, linguistic, and religious diversity among its people, one that has been exploited lately by right-wing forces for electoral gains. They observe that digital misinformation campaign has been aided by significant penetration of smartphones and messaging-based social media such as WhatsApp. Based on their analysis, they have identified as many as 133k stories that could be classified as "unreliable" around the time of the elections and that these have been shared many millions of times, likely influencing people's opinion in the run-up to the election. They describe their unique approach to misinformation, motivated by the context of Mexico's recent elections. In what is a very insightful experience narration, they outline the anti-misinformation strategies that worked well, and how they were designed. Their main strategy, WhatsApp-based user reporting of news stories followed by verified shares, is likely a model that could hold much promise for upcoming elections globally. They also note the specific challenges, such as the encrypted nature of WhatsApp, making it difficult to source content from closed groups where none may be willing to share information. It has also been noted that the eventual victor in the elections, the Hindu nationalist right-wing political formation led by the Bharatiya Janata Party, has been responsible for sharing a huge quantum of election-time junk news, a strategy whose continued usage is likely to pose significant challenges to the continuation of India's democratic tradition, unless checked.

3.6 Science and Technology Studies (STS) and Fake News

Renny Thomas considers the avenues where science and technology studies (STS) could aid data science in its pursuit against fake news. Thomas outlines the importance of viewing data science as a player in society, and not as a laboratory science. This property of data science makes it well suited to engage in conversations with STS and other related disciplines. Thomas warns against making the assumption of the "single objective truth" in the case of fake news detection; this assumption has been challenged by STS in other scholarly disciplines. The author makes a compelling case for STS-oriented explorations and data science technologies working hand in hand in combating the menace of fake news.

3.7 Linguistic Approaches to Fake News Detection

In a chapter on linguistics and fake news detection, Jane Lugea, a linguist, analyzes the linguistics aspects of fake news. Starting off with a categorization of the broad disciplinary subareas of linguistics, the author considers the linguistic characterization of news as well as that of deception styles in text. The author discusses the complexities in *characterizing* fake news and outlines how a data-centric approach could risk brushing a lot of nuances under the carpet. Further, Lugea looks at the various streams of linguistic approaches to fake news detection, organized according to the language approach adopted. The streams discussed include corpus linguistics, sentiment lexicons, readability, deep syntax, as well as rhetorical structure and discourse analysis. The chapter ends with thoughts on positioning linguistics-based inquiry into an integrated approach for fake news detection that combines intrinsic and extrinsic cues. Directions for potential further inquiry such as leveraging the nature of responses to identify fake news are also highlighted.

4   Concluding Remarks

We hope that this book, with its multifaceted approach to the topic of fake news, will provide data scientists as well as researchers within other disciplines an insightful overview of fresh perspectives on fake news. We are ever thankful to each and every one of our guest authors, especially interdisciplinary researchers who kindly agreed to step outside their core research interests in order to contribute chapters to enrich this book. We look forward to receiving feedback on this book, especially critical feedback, which will help us shape similar endeavors in the future.

References

1. Fisch, A.: Trump, JK Rowling, and confirmation bias: an experiential lesson in fake news. Radic. Teach. **111**, 103–108 (2018)
2. Oshikawa, R., Qian, J., Wang, W.Y.: A survey on natural language processing for fake news detection. arXiv preprint:1811.00770 (2018)
3. Parikh, S.B., Atrey, P.K.: Media-rich fake news detection: a survey. In: Proceedings of the 2018 IEEE Conference on Multimedia Information Processing and Retrieval (MIPR), pp. 436–441. IEEE, New York (2018)
4. Shu, K., Sliva, A., Wang, S., Tang, J., Liu, H.: Fake news detection on social media: a data mining perspective. ACM SIGKDD Explor. Newsl. **19**(1), 22–36 (2017)

Part I
Survey

On Unsupervised Methods for Fake News Detection

Deepak P

Abstract In this chapter, we consider a reasonably underexplored area in fake news analytics, that of unsupervised learning. We intend to keep the narrative accessible to a broader audience than machine learning specialists and accordingly start with outlining the structure of different learning paradigms vis-à-vis supervision. This is followed by an analysis of the challenges that are particularly pertinent for unsupervised fake news detection. Third, we provide an overview of unsupervised learning methods with a focus on their conceptual foundations. We analyze the conceptual bases with a critical eye and outline other kinds of conceptual building blocks that could be used in devising unsupervised fake news detection methods. Fourth, we survey the limited work in unsupervised fake news detection in detail with a methodological focus, outlining their relative strengths and weaknesses. Lastly, we discuss various possible directions in unsupervised fake news detection and consider the challenges and opportunities in the space.

Keywords Unsupervised learning · Fake news detection

1 Introduction

Fake news, the topic of this book, is a phenomenon of increasing concern over the last many years. Unlike the vast majority of machine learning tasks that seek to automate tasks that humans are quite adept at, such as image segmentation [7], action recognition [10], and emotion analysis [30], fake news identification [25] is a task of a different nature. Humans often find it hard to assess the veracity of news they come across due to a plurality of factors. First, in certain cases such as those of magic cures and anti-vaccination news, laypersons do not have enough knowledge of the domain to assess the veracity of a given news piece. Second, the news may pertain to real-time events that have not had time to gain enough of a footprint in public discourse, so there is no reference point to judge its veracity. Third, much fake news is carefully tailored to exploit human cognitive biases such as confirmation

© Springer Nature Switzerland AG 2021
Deepak P et al., *Data Science for Fake News*, The Information
Retrieval Series 42, https://doi.org/10.1007/978-3-030-62696-9_2

bias, echo chamber effects, and negativity bias; some discussions appear in the literature (e.g., [6, 26]). There are various other challenges that undermine the lay-person's ability to fact check for herself without the aid of additional technology or knowledge, but for the purposes of this chapter, it is enough to emphasize that humans could legitimately find the task difficult. In a way, the machine learning (ML) models for fake news detection seek to surpass the accuracy levels achieved by humans within reasonable time, effort, and knowledge limits.

1.1 Paradigms of Machine Learning vis-à-vis Supervision

The two broad streams of machine learning, viz., supervised and unsupervised, differ in terms of whether they assume the availability of historical labelled data to enable learning a statistical model that would then be used to label new data. Supervised learning, broadly construed, can be thought of as a mechanism of taking a training dataset of input–output pairs $\mathcal{T} = \{\ldots, [I, O], \ldots\}$ and producing a statistical model that embodies a mapping from the domain of inputs to outputs, $\mathcal{F} : D(I) \rightarrow D(O)$. For the task of fake news detection, the target domain is a veracity label, which could be one of $\{Fake, Legitimate, Doubtful\}$ or a number in a $[0, 1]$ range with the ends indicating *fake* and *legitimate*, respectively. The shape and form of the statistical model is *guided* by the labels in the training data but is constrained in ways to ensure its generalizability and/or conformance to knowledge about how the domain functions. On the other hand, the raw material for unsupervised methods is simply a set of unlabelled data objects, $\mathcal{T} = \{\ldots, I, \ldots\}$, from which the statistical models should learn to differentiate fake news from legitimate news. In contrast to supervised learning, the unsupervised methods may not necessarily produce a mapping from an input data object to a veracity label but could instead provide a grouping or representation whose subspaces are homogeneous with respect to veracity. For example, a clustering that is able to group a set of articles into two unlabelled groups, one of which is all fake articles and the other one all legitimate ones, could be considered successful from the perspective of fake news detection despite not being able to indicate which cluster is fake and legitimate. That said, producing a label along with output clusters only enhances the usefulness of the clustering with respect to the task.

There are other paradigms of machine learning that can make use of different flavors of supervision rather than the all-or-nothing cases discussed above. These include semi-supervised learning [39], active learning [24], and reinforcement learning [12]. Our focus in this chapter will be on unsupervised approaches to the task.

1.2 Challenges for Unsupervised Learning in Fake News Detection

When considering any analytics task, it may be observed that addressing the task in the unsupervised setting is obviously much more challenging than addressing it in a supervised setting. The former does not have the luxury of *label guidance* to complement or supplement domain knowledge-based directions in searching for effective statistical modeling. Thus, unsurprisingly, the effectiveness of unsupervised learning often falls well-short of that of supervised models.

We now consider some challenges for unsupervised learning for fake news detection. To offset the unavailability of labelled data, a natural pathway would be to develop a deeper understanding of the nature of fake news. This could be along dimensions such as *author*, *metadata* (e.g., article category, time, location), *propagation*, and *content*. For example, we may want to identify authors who regularly post content of limited veracity and categories (e.g., magic cures) that regularly get populated with disinformation. Similarly, if the news propagation is deeply dichotomous on the emotional aspect (e.g., either extremes of love or hate, without much in the middle ground), it may suggest correlation with disinformation or other aspects such as highly opinionated or divisive content. Some patterns in the content could itself be highly revealing; examples include clickbait-ish contents where the title and the article are highly divergent, or a sensationalist image placed strategically. Broadly speaking, the unavailability of label guidance could be offset by identifying some high-level patterns that correlate with veracity, which could then be folded into an unsupervised method. It may, however, be noted that such high-level patterns are unlikely to generalize across domains. For example, a fake news that deals with celebrity gossip may have a different structure than disinformation that deals with a COVID-19 cure (the fake news around COVID-19 has been called an infodemic [38]). Thus, the unsupervised methods that embed deeper domain knowledge could implicitly be very specific to the domain given that the deeper domain knowledge would itself be domain-specific. This may be contrasted with supervised learning where the label-guided learning framework may be generalizable across domains; concretely, it may learn *different models* for different domains using the *same* learning strategy since the labels in different domains could *pull* the learner in different directions that are suited for those domains.

The discussion suggests that efforts toward crafting unsupervised learning algorithms for fake news detection would entail the following:

– **Deeper Efforts at Understanding the Domain:** It would be useful, if not necessary, to understand the dynamics of the target domain through extensive studies. These may involve other scholarly realms beyond computer science; for example, the usage of confirmation bias as a tool may be more prevalent among xenophobic, anti-minority, and far-right rhetoric in political fake news (e.g., [16]) and thus could naturally be an effective factor in fake news identification too. On the other hand, an authoritative or assertive linguistic style with an abundance of anecdotes may characterize medical fake news. Such explorations may be situated within other disciplines such as psychology and linguistics or at their intersection with computing. This would likely make the body of literature around unsupervised fake news detection (UFND) more interdisciplinary than its supervised counterpart.

- **Empirical Generalizability:** While making use of insights from across disciplines as well as through extensive data analysis, there should also be an unrelenting focus toward empirical generalizability. If we focus on a single dataset and try out various combinations from a vocabulary of insight-driven heuristics, it is possible to be able to arrive at a spurious technique that performs very well for that dataset. This is often due to the well-understood mechanism of spurious pattern discovery called *data dredging* [29]. The vocabulary of fake news patterns that come from a deep understanding of specific domains may not be amenable to manual audit due to vocabulary size, complexity, and the deep expertise required for such analyses. Thus, there should be a particular focus on empirical generalizability to ensure that the developed methods are practically usable as well as legitimate. This may be achieved through verification over a large number of datasets from the target domain or by vetting for the validity of patterns with scholarly expertise in the target domain. This is particularly crucial when there is reliance on patterns identified through extensive empirical experimentation.

- **Ethical Considerations:** Machine learning methods more often rely on empirical than analytical analyses to make their point. Crudely put, it considers that the past is predictive of the future and develops techniques that project historical patterns for usage in unseen data from the future. This makes it systematically less capable of identifying novel and emerging patterns, something which has been very well understood in machine learning, with phenomena such as *concept drift* [32] and methods such as *transfer learning* [18] being well explored. When machine learning is used for tasks such as fake news detection, there is a chance that its widespread adoption would itself skew the data. For example, a novel pattern of legitimate news may be mistaken for fake news and may never be shown to users, leading to it never being labelled by humans anymore. Thus, the next generation of algorithms that work on the data would not be able to correct for it, given the lack of feedback. Such data bias and how they are exacerbated through algorithms have been well studied in the law enforcement domain [21]. Furthermore, the patterns embodied in the method could possibly be differentially equipped to identify fake news in subspaces; for example, a model incompetent at detecting fake remedies for tuberculosis, a predominant disease in some parts of Africa, may still fare well on the overall accuracy when tested over a dataset procured from the Western world where tuberculosis is rare. Unlike the case of supervised learning, there is an increased likelihood of biased high-level heuristics, over and above biased data, to be embedded in unsupervised learning algorithms.

- **Continuous Refinement:** Supervised learning systems can be retrained with new and updated datasets to some extent despite issues such as algorithms affecting the dataset, as described above; however, the analogous refinement of unsupervised learning algorithms requires updates to the algorithm design itself. Such refinements with changing data and societal discourse would require, as in the case of the algorithm design process, identifying and updating high-level heuristics with continuous vetting with domain expertise. We will return to this issue later on in this chapter.

2   Unsupervised Fake News Detection: A Conceptual Analysis

We now consider a conceptual positioning of the various research efforts on unsupervised fake news detection (UFND). As outlined in the discussion above, each unsupervised fake news detection method is invariably driven by high-level assumptions about patterns in the data that correlate with the veracity of news. In this section, we target to position the methods at a conceptual level, without getting into technical details; the technical and methodological details would form the topic of a subsequent section.

2.1   Conceptual Basis for UFND Methods

Given the paucity of UFND methods in the literature, we are able to consider the conceptual basis of each work separately. We have come across four research papers proposing UFND methods, which we use in our discussion as state-of-the-art methods. We have italicized the high-level heuristics employed, as and when discussed, for convenient reference. These are as follows:

- **Truth Discovery:** Truth discovery is the task that deals with estimating the veracity of an information nugget when it is reported by multiple sources (e.g., multiple websites), with conflicts existing across the multiple reports; a survey appears here [14]. An early work, perhaps the first UFND technique [37], makes use of truth discovery heuristics in estimating the veracity of information. It makes use of the high-level heuristic that *a piece of news is likely to be true when it is provided by many trustworthy websites*. Trustworthiness is not assumed as given a priori but estimated in an iterative fashion along with veracity estimation of various news pieces.
- **Differentiating User Types:** Many social media websites such as Twitter provide a way for users to be labelled as *verified*. This label is regarded as broadly honorific and could be interpreted as indicating a higher status or trustworthiness. UFND [36] exploits this user verification process in fake news detection. In particular, it models news veracity as being determined by

user opinion, modeling the way user opinion is factored into veracity analyses differently for verified and unverified users. Their heuristic, as quoted verbatim from Sect. 3.1, is the following: "an implicit assumption is imposed that verified users, who may have large influences and high social status, may have higher credibility in differentiating between fake news and real news." They make use of a generative framework to employ the above assumption into a veracity detection framework.

- **Propagandist Patterns:** The first unsupervised method to make use of behavioral analyses of user groups is a work that targets identifying propagandist misinformation in social media [17]. Their task is motivated by the increasing prevalence of orchestrated political propaganda and misinformation in social media, possibly facilitated by authoritarian governments and usually driven by large groups of users who work collectively to enhance acceptability of the official version. The proposed method for detecting propagandist misinformation relies on identifying *groups of users who write political posts that are textually and temporally synchronized, and aligned with the "official" vision or "party line."* Their method makes use of repeated invocations of clustering and frequent itemset mining [11], both of which are popular unsupervised learning methods.

- **Inter-user Dynamics:** GTUT [8], Graph Mining over Textual, User and Temporal Data, a recently proposed graph-based method for fake news detection, makes use of a phased approach that relies on heuristics that exploit assumptions on user dynamics, in what may be seen as a generalization of the user dynamics approach in [17] to cover a broader spectrum of fake news. In the first phase, they assume that *a set of articles posted by the same users at similar times through textually similar posts are fake.* This assumption follows, as they point out, from orchestrated behavior that is often observed in sharing fake news. Once such a core set of fake news articles are identified, the labels are propagated to other articles based on both *user correlation* and *textual similarity.* Thus, the heuristic beyond the first phase can be summarized as *articles that are similar to core fake articles based on posting users and textual similarity are likely to be fake.* The above heuristics are also analogous (i.e., as vice versa) to identifying a core set of trustworthy/legitimate articles and propagating trustworthiness labels.

Any single pattern or a single cocktail of patterns embedded in an algorithm being used in a widespread manner to counter fake news has high potential risks. This is best understood when fake news detection is seen from the perspective of gamification. When a single technique becomes widespread, the heuristics used by it would become well understood, and the authors of fake news would consequently *game* it by identifying ways to circumvent being caught by this. User dynamics heuristics could be circumvented by automated or semiautomated staggered posting of messages, while majority-oriented heuristics can be circumvented by organizing an orchestrated posting of messages aided by blackmarket services [5]. This makes any single static solution infeasible for effective fake news debunking in the long run. The existence of multiple methodologies for fake news detection that are continuously refined to be in tune with the current realities of the social media ecosystem is likely the best way to tackle the disinformation menace.

2.2   Critical Analysis of UFND Conceptual Bases

In the following discussion, we consider the relative merits and demerits of the conceptual basis of the techniques discussed above. This is not to undermine their value in being part of a mix of effective methodologies for UFND, but just to ensure a more nuanced understanding. We consider each of the techniques discussed above, in turn.

Truth Discovery

The truth discovery approach has a distinctly majoritarian flavor, whereby a more widespread opinion is likely to be regarded as truer than a narrowly shared one. While the authors in [37] explicitly clarify their assumption that they expect *a higher divergence of false facts* (Heuristic 3 in Sect. 2.2), the validity of their assumption may be challenged if multiple sources may be persuaded, with the aid of a mushrooming market around blackmarket services (e.g., [5]), to post the same fake content. This is plausible especially in narrow-domain topics such as fake news intended to malign a particular local enterprise. Such a situation could persuade the algorithm to consider the fake version as true and vice versa. However, this possibility is somewhat limited by the fact that trustworthy services are less likely to engage in such blackmarket orchestration, which places their trustworthiness at stake in the long run.

Differentiating User Types

The user type differentiation and the assumption of *enhanced credibility of verified users* employed by Yang et al. [36] are an interesting heuristic to analyze. Account verification in social media, according to Wikipedia,[1] was initially a feature for public figures and accounts of public interest, individuals in music, acting, fashion, government, politics, religion, journalism, media, sports, business, and other key interest areas. It was introduced by Twitter in June 2009, followed by Google Plus in 2011, Facebook in 2012, Instagram in 2014, and Pinterest in 2015. On YouTube, users are able to submit a request for a verification badge once they obtain 100,000 or more subscribers. In July 2016, Twitter announced that, beyond public figures, any individual would be able to apply for account verification. With the

---

[1] https://en.wikipedia.org/wiki/Account_verification—accessed 28 June, 2020.

focus of [36] on Twitter, we will consider Twitter verified users more carefully. Twitter's *request verification* service was temporarily suspended in February 2018, following a backlash over the verification of one of the organizers of the far-right *Unite the Right* rally due to a perception that verification conveys "credibility" or "importance." As of June 2020, Twitter is reportedly still working on bringing back the *request verification* feature.[2] Given this background, the usage of verified accounts as those with enhanced credibility raises some concerns. First, the authors in [36] say: "... in preparing our data, we only consider the tweets created by verified users and the related social engagements (like, retweet, and reply) of the unverified users." This data preparation principle severely limits the ability of their method to detect fake news within narrow domains that may involve very few or no verified users. While the techniques proposed are generalizable, in principle, to any kind of classification of higher-status users, it is yet to be empirically verified for the general case. Second, given that verified users were intended to involve public figures in areas such as politics, religion, music, acting, fashion, journalism, media, etc., the definition could exclude domain experts who may be best positioned to provide credible and well-studied opinions. For example, academics who may be able to provide credible analyses of science fake news, or doctors who may be able to identify health fake news, are kept out of the ambit of verified users. This also likely renders the method to be of limited utility even for many broad domains. Third, while we have not found any analyses of verified user distribution across geographies, it may be reasonably assumed that it is skewed in favor of areas of deep social media penetration such as the developed world. This geographic skew would reflect in the method and could dent its applicability for pressing issues in the global south, such as Africa and South Asia.

Propagandist Patterns

The paper that considers identifying propagandist patterns [17] is quite friendly for analysis in that it explicitly lays down the assumptions. We re-produce them below: *We assume that propaganda is disseminated by professionals who are centrally managed and who have the following characteristics:*

1. *They work in groups.*
2. *Disseminators from the same group write very similar (or even identical) posts within a short timeframe.*
3. *Each disseminator writes very frequently (within short intervals between posts and/or replies).*
4. *One disseminator may have multiple accounts; as such, a group of accounts with strikingly similar content may represent the same person.*
5. *We assume that propaganda posts are primarily political.*

---

[2]https://www.theverge.com/2020/6/8/21284406/twitter-verified-back-badges-blue-check.

Singh Is King 👑 ▶
@Singh996606

One of my distant relative in Delhi who
voted for AAP because of low
electricity bill was discharged
yesterday from a #COVID19 hospital
after paying a bill of Rs 13.5 lakh
@ArvindKejriwal
@AamAadmiParty

8:10 AM · 27/06/2020 · Twitter for Android

sumairabaloch
@sumairabaloch19

One of my distant relative in Delhi who
voted for AAP because of low
electricity bill was discharged
yesterday from a #COVID19 hospital
after paying a bill of Rs 13.5 lakh
#pakistan

9:43 AM · 27/06/2020 · Twitter for Android

Rishi Bagree 🇮🇳
@rishibagree

One of my distant relative in Delhi
who voted for AAP because of
low electricity bill was discharged
yesterday from a #COVID19 hospital
after paying a bill of Rs 13.5 lakh

9:36 AM · 27 Jun 20 · Twitter for Android

Fig. 1 A propaganda-based misinformation from the Indian context

6. *The content of tweets from one particular disseminator may vary according to the subject of an "assignment," and as such, each subject is discussed in disseminator's accounts during some temporal frame of its relevance.*
7. *Propaganda carries content similar to an official governance "vision" depicted in mass media.*

The above observations, partly motivated in the paper through examples from the Russian social network *VK*,[3] are likely to hold true for most regimes with shallow democracies and autocratic tendencies. Figure 1 shows a political fake news from the Indian context, which illustrates agreement to most of the assumptions above. The easiest way to game the system that works using the above assumptions would be to make the posts textually dissimilar; however, this would require much work and could undermine the ability of such fake news armies to mass produce fake tweets with high throughput. This makes the assumption fairly robust, at least in the short run. The limitations of the approach are largely engrained in the assumptions themselves, in that these apply only to fake news in the political domain produced in favor of the authoritarian regimes. In particular, in a federal governance system

---

[3]https://vk.com/.

such as those in the USA, India, or Spain, with different political parties leading different provincial governments, there may not be a *coherent official governance vision*, undermining assumption #7 above to some extent. It is likely that a subset of such assumptions above also apply to some other domains, such as religion-based fake news, but more studies may be needed to evaluate those aspects.

Inter-user Dynamics

The recent work on using inter-group dynamics in UFND, called GTUT [8], makes use of three phases, with the core assumption embedded in the first phase of identifying a core set of fake news and legitimate news. Their key assumption is that a core set of fake news articles can be identified as *a set of news articles shared by across a set of users using tweets that are temporally and textually similar*. This resembles some parts of the behavioral identification assumptions used in [17]; however, by relaxing the assumptions of *official vision adherence* and certain others, this is likely applicable to a broader set of scenarios. Analogous to the above, they use a curiously analogous assumption for identifying a core set of legitimate articles. In essence, *a set of news articles shared across a set of users using tweets that are temporally and textually dissimilar* are identified as a core set of legitimate news articles. While a reasoning for this is not adequately described, it is unclear as to the nature of legitimate news articles that would be shared in a temporally and textually dissimilar fashion. Clearly, this heuristic would have limited applicability in the political realm where legitimate news and fake news are often shared synchronously, when the event is in public memory. However, it is notable that these heuristics are only used in order to identify a core set of fake and legitimate articles (around 5% of the dataset, as mentioned in Sect. 4.1). In the subsequent phases, the fake and legitimate news labels are propagated using similarity between articles estimated as a mix of commonality between users and textual content of tweets. Another aspect of the method that may limit the applicability is the reliance on textual similarity. The method assumes that there is accompanying text along with an article over which textual similarity is assessed in the core set finding phase. It is not uncommon to simply share articles without posting any comment in social media; the applicability of GTUT over such posts would be evidently limited. On the positive side, much like observed in the case of [17], inter-user behavioral heuristics are harder to circumvent, making that a strong point of this method.

2.3   Building Blocks for UFND

While end-to-end techniques for UFND have evidently been limited in the literature, empirical analyses that could provide some building blocks for UFND have been explored lately. These are generally one of two types: (1) *computational social science studies* that seek to computationally verify a hypothesis rather than building a technology for a particular task or (2) work on supervised learning methods that establish the utility of certain features that implicitly indicate fertile directions for UFND research. Work of the latter kind typically is limited in making an observation that a particular feature is useful without indicating the nature of difference between fake news and legitimate news along that feature. For example, if *punctuation* is found to be a useful feature, it does not tell us whether fake news is better or worse in punctuation vis-à-vis real news (though one may be able to guess easily, in this case, as to which is more likely). We consider a few such works below, without claiming to provide a comprehensive overview:

- **Satirical Cues:** Rubin et al. [22] study the usage of satirical cues in supervised fake news detection and provide evidence that *absurdity*, *grammar*, and *punctuation* are useful features.
- **Propagation Patterns:** Vosoughi et al. [34] present evidence that "Falsehood diffused significantly farther, faster, deeper, and more broadly than the truth in all categories of information, and the effects were more pronounced for false political news than for false news about terrorism, natural disasters, science, urban legends, or financial information."
- **Topical Novelty:** Vosoughi et al. [34], in the same study as above, illustrate the utility of topical novelty against recent history as a useful way of identifying fake news, with fake news expected to be more novel topically.
- **Political Orientation and Age:** In a study based on Facebook, Guess et al. [9] say: "Conservatives were more likely to share articles from fake news domains, which in 2016 were largely pro-Trump in orientation, than liberals or moderates. We also find a strong age effect that persists after controlling for partisanship and ideology: On average, users over 65 shared nearly seven times as many articles from fake news domains as the youngest age group."
- **Effect of Fake News Based on Behavioral Traits of the Reader:** In a recent work, Pennycook and Rand [19] identify personality traits with respect to fake news vulnerability and say: "individuals who overclaim their level of knowledge also judge fake news to be more accurate." While this does not necessarily form a building block for UFND, it potentially indicates who may benefit more from the methods.
- **Psychological Appeal:** Acerbi [1] analyzes the cognitive appeal of online misinformation and suggests that misinformation may be correlated with *psychological appeal* in that it aims to exploit various cognitive inclinations of humans.
- **Language Style:** Rashkin et al. [20] illustrate that language style modeled through lexical features can help differentiate fake news from legitimate ones in a supervised task. Linguistic cues were also explored in [4].
- **Network Patterns:** An analysis [27] of dissemination patterns of news through the network indicates that the type of network formed through propagation can be revealing of the veracity of news.
- **Emotions:** Anoop et al. [2] report a computational social science study providing empirical evidence that the emotion profile of fake news differs from legitimate news, through an innovative mechanism that illustrates that emotion-amplified

fake news is farther away from their legitimate counterparts. Emotions and sentiments were also found to be useful in detecting fake reviews in another study [15].

- **Users Who Like:** In a large-scale study of Facebook likes, Tacchini et al. [31] suggest that *users who like* a post is a reasonable predictor of post veracity. This likely points to the existence of some consistent patterns of *liking* activity across the veracity dimension, which may be of use in UFND.
- **Lexical Coherence:** A recent computational social science study [28] considers the various ways of quantifying lexical coherence, and observes that word embedding based on coherence analyses is best suited to tease out the differences between fake and legitimate news.

The above is by no means an exhaustive list but serves to indicate the diversity of directions to explore toward building effective UFND methods. While several minor building blocks, even when packaged into a UFND method, may not have the muscle to compete with the state of the art in UFND, such efforts nevertheless contribute to building a diversity of UFND methods, diversity being an important factor as pointed out earlier. We may also add here that such research efforts are likely more suited to avenues focused on computational social science, such as the many avenues that have been instituted recently, viz., *Journal of Computational Social Science,*[4] *ACM Transactions on Social Computing,*[5] and *IEEE Transactions on Computational Social Systems.*[6]

3   Unsupervised Fake News Detection: A Methodological Analysis

Having introduced the various methods for UFND at the conceptual level in the previous section, we now endeavor to provide a tutorial overview of their methodological details. As in the previous case, we cover each method in turn.

3.1   Truth Discovery

The approach proposed in [37] makes use of an iterative approach toward veracity identification. The approach attacks two estimation problems concurrently:

- *Trustworthiness Estimation of Websites:* Estimating a non-negative trustworthiness score for each website as $T'(w)$.

---

[4]https://www.springer.com/journal/42001.

[5]https://dl.acm.org/journal/tsc.

[6]https://ieeexplore.ieee.org/xpl/RecentIssue.jsp?punumber=6570650.

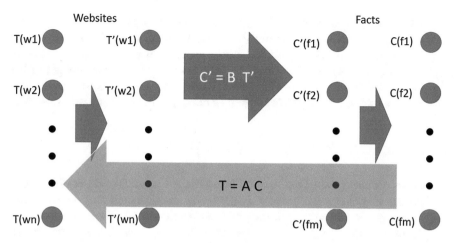

Fig. 2 Truth discovery approach from [37]. Figure adapted from across illustration in the paper

– *Confidence Estimation for Facts:* Estimating a confidence score for each fact as $C'(f)$.

The scores are directly related to trustworthiness and confidence, respectively; that is, higher scores indicate higher trustworthiness and higher confidence. There is also an additional construct, the *objects* associated with each fact, that is also used in the estimation. While the estimation process bears resemblance to the hub-authority score estimation in Hyperlink-induced Topic Search [13], the actual estimation process, as the authors say, is quite different in mathematical character. We provide an overview of the methodology employed in [37], to aid understanding of the spirit of the approach. The exact details are in the paper.

Figure 2 depicts an overview of the method. The set of websites are $W = \{w_1, w_2, \ldots, w_n\}$, across which a number of facts are mentioned, $F = \{f_1, f_2, \ldots, f_m\}$. For each website, there are two trustworthiness scores, $T(w)$ and $T(w')$; these are easily convertible across each other and serve to simplify the iterative computation process only. Analogously, there are two confidence scores, $C(f)$ and $C'(f)$, for facts that are also similarly inter-convertible.

The method starts with initializing all websites to be of equal trustworthiness, say 0.9, for $T(w)$. This is used to estimate $T'(w)$, which is then followed by two key matrix multiplication operations that form the key steps within each iteration:

– *Confidence from Trustworthiness:* Consider the $\{\ldots, T'(w), \ldots\}$ as an $n \times 1$ vector. This vector is transformed using an $m \times n$ matrix B that is structured as follows:

$$
B_{ij} = \begin{cases} 1 & \text{if } f_i \text{ is provided by } w_j \\ \rho \times imp(f_k \to f_i) & \text{if } w_j \text{ provides } f_k \text{ and } o(f_k) = o(f_i) \\ 0 & \text{otherwise} \end{cases}
$$

As obvious, B_{ij} quantifies the support from website w_j toward the fact f_i. The second case above takes care of the scenario where w_j does not directly provide the fact f_i but provides a related fact f_k that relates to the same object as f_i ($o(f)$ denotes the object the fact f relates to). In that case, the strength of the implication from f_k to f_i (which could be negative when f_k conflicts with f_i), denoted by $imp(f_k \to f_i)$, is scaled by a factor ρ. The transformation operation is

$$\overrightarrow{C'} = B \, \overrightarrow{T'}$$

– *Trustworthiness from Confidence:* The estimation of trustworthiness from confidence is quite straightforward. In particular, the trustworthiness of a website is simply the average confidence of facts provided by it. In terms of matrix operations, this is modeled as a matrix A that is $n \times m$ whose entries are as follows:

$$A_{ij} = \begin{cases} \frac{1}{|F(w_i)|} & if \ f_j \in F(w_i) \\ 0 & otherwise \end{cases}$$

where $F(w)$ is the set of facts provided by the website w. The transformation is then

$$\overrightarrow{T} = A \, \overrightarrow{C}$$

The iterative process is stopped when the trustworthiness scores do not change much, and the confidence scores are returned as an estimation of veracity for each fact.

The empirical analysis of this method has been predominantly performed over datasets involving books and movies, and it is not clear about the applicability of this method for social media fake news debunking. One way to use this, however, would be to treat each profile as the equivalent of a website, and the *facts* contained with each post as similar to the facts provided by websites. A particular notable aspect of this method is that it provides a trustworthiness estimate along with confidence scores; thus, this could be used in order to assess the trustworthiness of social media profiles, when considering profiles as the equivalent of websites, as outlined above.

3.2 Differentiating User Types

We now consider the approach proposed in [36] and describe the methodological framework. The cornerstone of this work, as outlined earlier, is the differentiation between the *verified* and *unverified* users. They limit their remit to assessing the veracity of news stories that have been tweeted by at least one verified user. Each

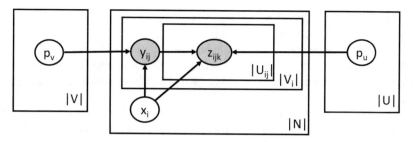

Fig. 3 Simplified graphical model from [36]

tweet of a news story by a verified user could be commented on or reacted to by one
or more unverified users.

We introduce some notation to make the ensuing narration easier. Let $N = \{\ldots, n_i, \ldots\}$ be a set of news stories. Each news story n_i has an associated truth value $x_i \in \{0, 1\}$, estimating which forms the core target of the learning process in UFND. Let the opinion made by a verified user v_j on n_i be $y_{ij} \in \{0, 1\}$. The technique considers this opinion as an observed variable, since y_{ij} can be identified using sentiment or opinion analysis techniques. When a verified user v_j expresses an opinion on n_i, it is by means of a *tweet* or a *post* onto which *unverified* users can then engage and express their own opinion. Let z_{ijk} be the opinion expressed by the unverified user u_k on the v_j's post with n_i. This $z_{ijk} \in \{0, 1\}$ is also an observed variable estimated using sentiment or opinion analysis methods. The task is now to estimate x_is given the various y_{ij}s and z_{ijk}s. The authors use a probabilistic graphical model for this purpose.

Figure 3 depicts a simplified version of the graphical model omitting the details as well as hyperparameters for narrative simplicity. Each verified user is represented by a set of parameters p_v, and each unverified user by a different set p_u. The observed opinion y_{ij} is modeled as being influenced by both the truth value of the news x_i and the personal parameters of the user v_j. Similarly, the opinion z_{ijk} is influenced by all of (1) the truth value of x_i, (2) the opinion of the verified user y_{ij}, and (3) the parameters of the unverified user u_k. The parameters for verified and unverified users are modeled differently. The verified users are modeled using their *true positive rate* and *false positive rate*. Given that unverified users can only interact with a news within the context of a verified user's post, the unverified user has four parameters: the positivity rate for each combination of truth value of the article and opinion polarity of the verified user. For example, $p_u(z_{ijk} = 1 | x_i = 0, y_{ij} = 0)$ indicates the likelihood of the unverified user expressing a positive opinion on a fake article (fake article since $x_i = 0$) to which the verified user has expressed a negative opinion (since $y_{ij} = 0$). The authors use a Gibbs sampling approach to estimate the latent parameters in the model, details of which are in the paper.

We had indicated in an earlier section that the authors of [36] had opined that "an implicit assumption is imposed that verified users, who may have large influences and high social status, may have higher credibility in differentiating between fake news and real news." However, nothing in the methodology, as far as we understand, prevents verified users from having lower true positive rates (or higher false positive rates) than unverified users. There is evidently differentiated modeling of verified and unverified users, which may be implicitly pushing toward configurations that confer higher credibility to verified users, though it is far to reason analytically as to how such configurations are favored.

They evaluate the method against the truthfinder method as well as other baselines over two public datasets, LIAR [35] and BuzzFeed News data, and report accuracies of around 70% or higher.

3.3 Propagandist Patterns

The third work we describe, from [17], looks at using propagandist patterns in order to tackle misinformation that is aligned with the *official version*, probably inspired by scenarios in shallow democracies around the world. The technique itself is structured as a human-in-the-loop method that targets to identify patterns that need to be vetted by humans in order to complete the misinformation detection pipeline.

The automated part of the process follows the illustration in Fig. 4. We trace the process in reference to the seven assumptions outlined in Sect. 2.2. The target domain is Twitter, with the tweets ordered in temporal order indicated on the left-hand side. Tweets are split into temporal buckets to align with assumption #2. The tweets inside each time window are then clustered to ensure the textual similarity

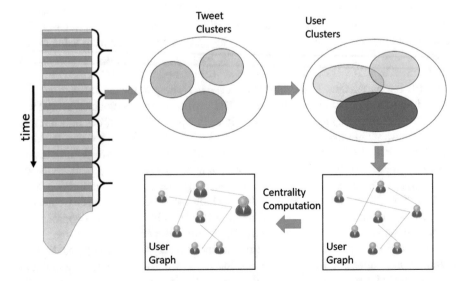

Fig. 4 Propagandist misinformation detection pipeline [17]

part of assumption #2. The tweets within the clusters are replaced with the userids of the authors, thus converting a disjoint clustering of tweets to an overlapping clustering of users; this is so since a user may have authored tweets that fall into disjoint clusters under the tweet clustering. This is inspired by both assumptions #1 and #2. The user clusters are converted into a user graph with cluster colocation being the criterion for edge induction. This user graph is then subjected to centrality detection to identify key users. In parallel, not shown in the diagram, there are two additional steps:

- A topic analysis over tweets to identify political topics in accordance with assumptions #5 and #7.
- Identification of user groups by application of a priori algorithm over user clusters. This addresses mostly assumption #4 and aligns with certain others.

The other assumptions, among the seven listed, are used by the human process. The authors do not perform a large empirical evaluation in the absence of labelled information but indicate the validity of the results from the method through manual vetting.

It may be seen that the manual steps in the process severely limit the applicability of the method in a large-scale manner. Additionally, given the lack of empirical validation over a labelled dataset, the recall (i.e., quantifying what has been missed) is not clear either. However, this presents a first effort in using inter-user behavioral dynamics within misinformation detection pipeline.

3.4   Inter-user Dynamics

We now come to the most recent work [8], one that uses inter-user behavioral dynamics in fake news detection using graph-based methods. GTUT, the method, relies on identifying temporally and textually synchronous behavior among users, as the key bootstrapping heuristic for identifying misinformation. This is enabled through a graph-based approach outlined below.

The graph employed by GTUT is a biclique, containing two kinds of nodes, *users* and *articles*. There exists an edge between a user and an article if the user has tweeted mentioning the article. In fact, a specific user may have tweeted about an article multiple times, leading to multiple tweets. Thus, an edge may *contain* multiple tweets. The first phase in the three-phase GTUT starts by identifying bicliques, a combination of a set of users and a set of articles such that each user–article pair in the combination is connected. One such biclique is illustrated in Fig. 5. Once such bicliques are identified, they are scored based on their *temporal* and *textual* coherence.

$$TTScore(B) = \lambda \times Temporal(B) + (1 - \lambda) \times Textual(B) \qquad (1)$$

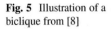
Fig. 5 Illustration of a
biclique from [8]

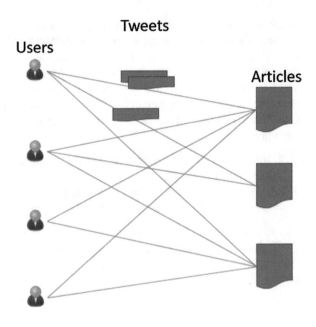

These biclique level scores are transferred to each article A as follows:

$$TTScore(A) = \frac{\sum_{B \in BiCliques(A)} TTScore(B)}{|BiCliques(A)|} \tag{2}$$

where $BiCliques(A)$ indicates the bicliques that article A is part of. In other words, the score of an article is simply the average of the scores of bicliques that contain it. The 5% of articles with the highest coherence scores (indicating highly synchronous posting activity) are labelled as a core set of *fake* articles, with the analogous set at the other end being labelled as a core set of *legitimate* articles. This completes the first phase in GTUT.

The second phase propagates the fake and legitimate labels from the core set to all articles contained across the bicliques. The label propagation uses a graph structure with nodes being articles and edges being weighted as a weighted sum of *biclique similarity, user similarity*, and *textual similarity*.

$$E(A, A') = \alpha \times Jacc(BiCliques(A), BiCliques(A')) \tag{3}$$
$$+\beta \times Jacc(Users(A), Users(A')) + (1 - \alpha - \beta) \times Sim(A, A')$$

where $Jacc(., .)$ indicates the Jaccard similarity and $Users(A)$ are the set of users who shared the article A, and $Sim(., .)$ is a textual similarity measure. At the end of this phase, each article contained in a biclique is labelled as either *fake* or *legitimate*.

The third phase propagates the labelling from within the bicliques to articles outside the bicliques; this uses the same structure as in the second phase, employing label spreading. However, being outside the bicliques, there are only two factors in determining edge weights, which are user similarity and textual similarity. This completes the labelling process for all articles.

The methodology outlined above starts with identifying a core set of fake and legitimate articles and spreads the labels progressively outward to eventually cover all articles. This serial order of labelling imposes a high dependency on the initial core set finding; inaccurate finding of core sets of fake and legitimate articles could potentially lead the next two phases wayward. While the authors illustrate good empirical accuracies over two large-scale datasets, more studies could be used to assert the generalizability of the initialization heuristic.

4   The Road Ahead for Unsupervised Fake News Detection

We now outline some pathways in which research on unsupervised fake news detection could progress, in order to advance the state of the art. This is purely based on opinions that are in turn based on observations in the field, and an understanding of the fake news domain developed through engaging in research in the field and need to be taken with abundant caution.

4.1   Specialist Domains and Authoritative Sources

Of particular concern in 2020, as this chapter is being written, is that of COVID-19[7] fake news. These have been peddled by authoritative sources such as heads of state.[8] The debunking of such news, in the offline world, often happens through specialists considering the claim in the light of scholarly evidence and assessing whether the claim is tenable. A natural approach to automate fake news detection in such specialist domains is to similarly make use of authoritative knowledge sources. This would require a significant effort in developing bespoke techniques depending on the structure and nature of reliable knowledge sources in each domain. For example, while NHS[9] and CDC[10] provide information for the layperson in the form of semi-structured articles, other sources such as PubMed[11] provide access to scholarly articles. Other sources, such as TRIP,[12] reside somewhere midway in the spectrum, while providing reliable and trustworthy information. We presume

---

[7]https://www.who.int/emergencies/diseases/novel-coronavirus-2019.

[8]https://www.bbc.co.uk/news/technology-52632909.

[9]https://www.nhs.uk/.

[10]https://www.cdc.gov/.

[11]https://pubmed.ncbi.nlm.nih.gov/.

[12]https://www.tripdatabase.com/.

a similar landscape might characterize other domains such as scientific domains (e.g., fake news around climate change) and history (e.g., painting a nonfactual picture of historical events). We have found scanty usage of authoritative knowledge sources even among supervised methods for fake news detection; a notable work is MedFact [23], which targets to adopt principles from evidence-based medicine, albeit superficially, using information retrieval methods, in the process of medical fake news verification.

4.2   Statistical Data for Fake News Detection

Consider a particular fake claim that was made by Gerard Batten, a British politician, in the context around Brexit. The claim, illustrated in Fig. 6, says that there are only approximately 100 lorries that cross the border between the UK and Ireland in the island of Ireland. This was promptly debunked by various fact-checking agencies and media in the UK and Ireland, including TheJournal[13] and FactCheckNI.[14]   Both of them pointed to a reference from a UK parliamentary report[15] that indicated that there were 177k *heavy goods crossings* across the Irish border each month, which equated to 5.9k such crossings each day on an average. There are two key aspects to this fact-checking effort: identifying that lorries refer to heavy goods vehicles and that the daily crossings can be computed from aggregate numbers. While the former is a task that relies on NLP and domain knowledge, the latter involves mathematical calculations, an elementary one, that of division, in this case. Such statistical claims appear all the time in the political domain, and those may involve population statistics of religious groups (heavily employed by the right wing in India) among others. Debunking these often involves the following steps:

- **Identification:** Identifying the pertinent statistic from an authoritative source, along with information about it.
- **Data Processing:** Normalization, interpolation, or extrapolation to enable direct comparison with the statistic in the claim.
- **Domain Conditioning:** Conditioning the processed statistic on well-understood patterns in the domain. For example, a high population growth rate is often correlated with low economic conditions and thus needs to be conditioned on the latter, to enable comparison across different cohorts.
- **Comparison and Veracity Assessment:** Once the data is processed and the comparable statistic identified, it may then be compared with the statistic in the claim and the veracity assessed in the backdrop of the knowledge of the domain patterns.

---

[13]https://www.thejournal.ie/factcheck-lorries-4469494-Feb2019/.

[14]https://factcheckni.org/fact-checks/is-border-trade-0-5-of-uk-eu-trade/.

[15]https://publications.parliament.uk/pa/cm201719/cmselect/cmniaf/329/32906.htm.

Fig. 6 Brexit fake news example

Gerard Batten MEP ✓
@GerardBattenMEP

Here's a fact about the Irish border. There are approx 100 lorries per day crossing the RoI & NI border. This accounts for 0.5% of UK/EU trade. One company accounts for 50% of these trips - that company is Guinness. And this is the false premise that is holding up Brexit.

LBC ✓ @LBC
The backstop is only there until we find a solution to the Irish border and Brexiters say a solution is easy. So it shouldn't matter, says James O'Brien.

...

12:46 · 30 Jan 19 · Twitter for iPhone

390 Retweets **640** Likes

Based on informal conversations with a UK-based fact-checking agency, we learnt that a significant number of fake claims that they perform manual fact checking on involve statistical analysis and number crunching. This might also be seen as a fake news detection problem that is hard to be analyzed from within the supervised learning framework due to the very nature of the task, making bespoke UFND likely the best mode of attack for the task.

4.3   Early Detection

A number of existing supervised learning methods for fake news detection make abundant use of propagation information in order to identify fake news. These, due to their design, are incapable of addressing emerging fake news accurately, since the dense feature footprint would need to accumulate before accurate veracity computation can be performed. Thus, unsupervised methods may be the only resort until the news has time to pass through the network enough to amass a significant digital footprint. This demarcates a niche space for unsupervised fake news detection.

4.4   Miscellaneous

We now outline a few other promising directions for unsupervised fake news detection.

Maligning Brands Through Fake Information

Within the space of e-commerce, there has been an increasing trend of using fake information to malign particular brands [33] or particular stores. These could differ on the basis of the kind of narrative used, and one in which historical labelled information may be of limited utility, making this a fertile area for unsupervised fake news detection. These also include reviews about brands posted on trading websites as well as maps; recent studies have established the utility of emotion and sentiment information in fake review detection [15].

Explainability in UFND

There is an increasing appreciation that any ML algorithm should not just provide its decision but also a rationale supporting the decision. Facilitating user engagement was also highlighted in the EU High-Level Expert Group report on disinformation [3], in the interest of ensuring that democratic practices be upheld and the diversity of the media ecosystem be preserved. This makes explainability or other forms of enhancing interpretability an interesting area for fake news detection in general. In fact, unlike supervised methods, unsupervised methods (and their designers) cannot relegate the decision to historical labelled data and hold more liability for the decision.

5   Conclusions

In this chapter, we provided a bird's-eye view of work in unsupervised fake news detection. In what we designed as a unique perspective, we endeavored to provide a critical analysis that is accessible to an informed layperson (rather than just the machine learning specialist). We started off by situating unsupervised methods among the plethora of paradigms in machine learning and outlined the specific challenges that are of high importance in unsupervised learning for fake news detection. This was followed by a conceptual analysis of UFND methods, a critical analysis of such conceptual foundations, and a listing of possible conceptual building blocks that may enhance both the existing UFND methods as well as provide a platform to design newer UFND methods. This was followed by a methodological analysis of UFND methods, along with a critical perspective outlining their limitations and strengths. We then concluded the chapter with a set of possible interesting directions to advance the frontier in unsupervised fake news detection.

References

1. Acerbi, A.: Cognitive attraction and online misinformation. Palgrave Commun. **5**(1), 1–7 (2019)
2. Anoop, K., Deepak, P., Lajish, L.V.: Emotion cognizance improves fake news identification. CoRR, abs/1906.10365 (2019). http://arxiv.org/abs/1906.10365
3. Buning, M.d.C., et al.: A multidimensional approach to disinformation. In: EU Expert Group Reports (2018)
4. Conroy, N.K., Rubin, V.L., Chen, Y.: Automatic deception detection: methods for finding fake news. Proc. Assoc. Inf. Sci. Technol. **52**(1), 1–4 (2015)
5. Dutta, H.S., Chakraborty, T.: Blackmarket-driven collusion among retweeters—analysis, detection, and characterization. IEEE Trans. Inf. Forensics Secur. **15**, 1935–1944 (2019)
6. Fisch, A.: Trump, JK Rowling, and confirmation bias: an experiential lesson in fake news. Radical Teach. **111**, 103–108 (2018)
7. Fu, K.S., Mui, J.: A survey on image segmentation. Pattern Recogn. **13**(1), 3–16 (1981)
8. Gangireddy, S.C., Deepak, P., Long, C., Chakraborty, T.: Unsupervised fake news detection: a graph-based approach. In: ACM Hypertext and Social Media (2020)
9. Guess, A., Nagler, J., Tucker, J.: Less than you think: prevalence and predictors of fake news dissemination on Facebook. Sci. Adv. **5**(1), eaau4586 (2019)
10. Herath, S., Harandi, M., Porikli, F.: Going deeper into action recognition: a survey. Image Vis. Comput. **60**, 4–21 (2017)
11. Jamsheela, O., Raju, G.: Frequent itemset mining algorithms: a literature survey. In: Proceedings of the 2015 IEEE International Advance Computing Conference (IACC), pp. 1099–1104. IEEE, New York (2015)
12. Kaelbling, L.P., Littman, M.L., Moore, A.W.: Reinforcement learning: a survey. J. Artif. Intell. Res. **4**, 237–285 (1996)
13. Kleinberg, J.M.: Authoritative sources in a hyperlinked environment. J. ACM **46**(5), 604–632 (1999)
14. Li, Y., Gao, J., Meng, C., Li, Q., Su, L., Zhao, B., Fan, W., Han, J.: A survey on truth discovery. ACM Sigkdd Explor. Newsl. **17**(2), 1–16 (2016)
15. Melleng, A., Jurek-Loughrey, A., Deepak, P.: Sentiment and emotion based representations for fake reviews detection. In: Mitkov, R., Angelova, G. (eds.) Proceedings of the International Conference on Recent Advances in Natural Language Processing (RANLP) 2019, Varna, Bulgaria, 2–4 September 2019, pp. 750–757. INCOMA Ltd., New York (2019). https://doi.org/10.26615/978-954-452-056-4_087
16. Murungi, D., Yates, D., Purao, S., Yu, J., Zhan, R.: Factual or believable? negotiating the boundaries of confirmation bias in online news stories. In: Proceedings of the 52nd Hawaii International Conference on System Sciences (2019)

17. Orlov, M., Litvak, M.: Using behavior and text analysis to detect propagandists and misinform-
    ers on twitter. In: Annual International Symposium on Information Management and Big Data,
    pp. 67–74. Springer, Berlin (2018)
18. Pan, S.J., Yang, Q.: A survey on transfer learning. IEEE Trans. Knowl. Data Eng. **22**(10),
    1345–1359 (2009)
19. Pennycook, G., Rand, D.G.: Who falls for fake news? the roles of bullshit receptivity,
    overclaiming, familiarity, and analytic thinking. J. Pers. **88**(2), 185–200 (2020)
20. Rashkin, H., Choi, E., Jang, J.Y., Volkova, S., Choi, Y.: Truth of varying shades: analyzing
    language in fake news and political fact-checking. In: Proceedings of the 2017 Conference on
    Empirical Methods in Natural Language Processing, pp. 2931–2937 (2017)
21. Richardson, R., Schultz, J.M., Crawford, K.: Dirty data, bad predictions: how civil rights
    violations impact police data, predictive policing systems, and justice. NYUL Rev. Online
    94, 15 (2019)
22. Rubin, V.L., Conroy, N., Yimin, C.: Towards news verification: deception detection methods
    for news discourse. In: Hawaii International Conference on System Sciences (2015)
23. Samuel, H., Zaiane, O.: Medfact: towards improving veracity of medical information in social
    media using applied machine learning. In: Canadian Conference on Artificial Intelligence, pp.
    108–120. Springer, Berlin (2018)
24. Settles, B.: Active learning literature survey. In: Technical Report University of Wisconsin-
    Madison Department of Computer Sciences (2009)
25. Sharma, K., Qian, F., Jiang, H., Ruchansky, N., Zhang, M., Liu, Y.: Combating fake news: a
    survey on identification and mitigation techniques. ACM Trans. Intell. Syst. Technol. (TIST)
    10(3), 1–42 (2019)
26. Shu, K., Wang, S., Liu, H.: Exploiting tri-relationship for fake news detection, vol. 8 (2017).
    arXiv preprint:1712.07709
27. Shu, K., Bernard, H.R., Liu, H.: Studying fake news via network analysis: detection and
    mitigation. In: Emerging Research Challenges and Opportunities in Computational Social
    Network Analysis and Mining, pp. 43–65. Springer, Berlin (2019)
28. Singh, I., Deepak, P., Anoop, K.: On the coherence of fake news articles. CoRR abs/1906.11126
    (2019). http://arxiv.org/abs/1906.11126
29. Smith, G.D., Ebrahim, S.: Data Dredging, Bias, or Confounding: they can all get you into the
    BMJ and the Friday Papers (2002)
30. Strapparava, C.: Emotions and NLP: future directions. In: Proceedings of the 7th Workshop on
    Computational Approaches to Subjectivity, Sentiment and Social Media Analysis (2016)
31. Tacchini, E., Ballarin, G., Della Vedova, M.L., Moret, S., de Alfaro, L.: Some like it HOAX:
    automated fake news detection in social networks. arXiv preprint:1704.07506 (2017)
32. Tsymbal, A.: The problem of concept drift: definitions and related work. Comput. Sci. Dep.
    Trinity Coll. Dublin **106**(2), 58 (2004)
33. Visentin, M., Pizzi, G., Pichierri, M.: Fake news, real problems for brands: the impact of
    content truthfulness and source credibility on consumers' behavioral intentions toward the
    advertised brands. J. Interact. Mark. **45**, 99–112 (2019)
34. Vosoughi, S., Roy, D., Aral, S.: The spread of true and false news online. Science **359**(6380),
    1146–1151 (2018)
35. Wang, W.Y.: "liar, liar pants on fire": a new benchmark dataset for fake news detection. arXiv
    preprint:1705.00648 (2017)
36. Yang, S., Shu, K., Wang, S., Gu, R., Wu, F., Liu, H.: Unsupervised fake news detection on
    social media: a generative approach. In: Proceedings of the AAAI Conference on Artificial
    Intelligence, vol. 33, pp. 5644–5651 (2019)
37. Yin, X., Han, J., Philip, S.Y.: Truth discovery with multiple conflicting information providers
    on the web. IEEE Trans. Knowl. Data Eng. **20**(6), 796–808 (2008)
38. Zarocostas, J.: How to fight an infodemic. Lancet **395**(10225), 676 (2020)
39. Zhu, X., Goldberg, A.B.: Introduction to semi-supervised learning. Synth. Lect. Artif. Intell.
    Mach. Learn. **3**(1), 1–130 (2009)

Multi-modal Fake News Detection

Tanmoy Chakraborty

Abstract The primary motivation behind the spread of fake news is to convince the readers to believe false information related to certain events or entities. Human cognition tends to consume news more when it is visually depicted through multimedia content than just plain text. Fake news spreaders leverage this cognitive state to prepare false information in such a way that it looks attractive in the first place. Therefore, multi-modal representation of fake news has become highly popular. This chapter presents a thorough survey of the recent approaches to detect multi-modal fake news spreading on various social media platforms. To this end, we present a list of challenges and opportunities in detecting multi-modal fake news. We further provide a set of publicly available datasets, which is often used to design multi-modal fake news detection models. We then describe the proposed methods by categorizing them through a taxonomy.

Keywords Multi-modal fake news · Multimedia · Microblogs · Supervised methods · Unsupervised methods

1 Introduction

A new article usually gains more visibility when it is accompanied by attractive visuals—images, videos, etc. Human psychology often relates the multi-modal content more to an individual's daily life than a textual content. Therefore, it is not surprising that fraudulent content creators often take advantage of such human cognition of biased multi-modal/multimedia content consumption to design catchy fake news in order to increase overall visibility and reach. Studies revealed that tweets with images receive 18% more clicks, 89% more likes, and 150% more retweets than those without images.[1] Moreover, visual component is frequently

---

[1]https://www.invid-project.eu/tools-and-services/invid-verification-plugin/.

© Springer Nature Switzerland AG 2021

Deepak P et al., *Data Science for Fake News*, The Information
Retrieval Series 42, https://doi.org/10.1007/978-3-030-62696-9_3

considered as a proof of the trustworthiness of the story in our common sense. This is another reason for a multimedia story to attract a large audience.[2] The dissemination of multi-modal fake news is thus even more detrimental than usual unimodal (text only or image only) fake news. Note that a fake image without any caption or description may not have the storytelling capability of a fake image with textual content. For example, an image depicting "a black person is beaten by several white persons" may not be that attractive if it is not accompanied by the associated story, such as where the incident happened (say, New York City) and what was the reason behind the incident (say, the black person challenged the state authorities). Such stories also lead to communal hatred, regional riot, etc. In this chapter, we will cover several recent research that deal with fake news detection by leveraging "multi-modal" or "multimedia" content.

Note that we refrain ourselves from discussing image/video forensics such as forgery, doctoring, or tampering detection [3] as well as fake news detection methods, which leverage only images or videos in isolation. Readers are encouraged to read notable studies in this direction, such as Gupta et al. [18], which made an effort to understand the temporal, social reputation, and influence patterns for the spreading of fake images on microblogs, and Angiani et al. [2], which proposed a supervised method for image-based hoax detection, etc.

Another body of research deals with image repurposing detection, where the task is to detect visual content that is real (not manipulated) but is published together with a false caption about the depicted event. These studies attempt to measure the semantic integrity of images and their corresponding captions using reference resources or knowledge bases [16, 20, 21, 52]. We also purposefully skip them in this chapter because they fall under the study of image caption generation. Captions are often not considered as equivalent news, tweets, or posts. Moreover, these models mostly look at the manipulation of image metadata such as image creation date, owner, location, etc., which are often not publicly available with social media content and online news articles.

We strictly confine our discussion to methods that consider *at least text and visual content* of an article/post for fake news detection. Also note that a "news" can be a social media post such as a "tweet" or an article in a newspaper or blog. Figure 1 shows an example of the type of fake news considered in this chapter.

In 2015, a workshop, called MediaEval,[3] was organized as a satellite event of Interspeech conference,[4] where one of the competitions was "Verifying Multimedia

---

[2]https://www.businesswire.com/news/home/20190204005613/en/Visual-SearchWins-Text-Consumers%E2%80%99-Trusted-Information.

[3]http://www.multimediaeval.org/mediaeval2015/.

[4]http://interspeech2015.org/.

This is a difficult time for everyone and I, for
one, am grateful for gifts such as these...

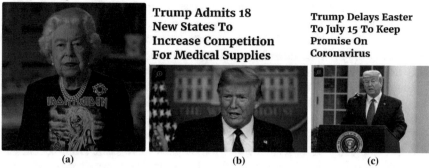

**Trump Admits 18
New States To
Increase Competition
For Medical Supplies**

**Trump Delays Easter
To July 15 To Keep
Promise On
Coronavirus**

(a)   (b)   (c)

Fig. 1 Three examples of multi-modal fake news. Example (**a**) was picked up from a recent video
of Queen Elizabeth regarding the current situation of coronavirus in the United Kingdom. The
Queen's dress in the photo was modified, and the caption was changed to create the fake news
[61]. Example (**b**) was picked up from a speech of Trump, and the above fake caption was attached
to it, which states that the number of states in the United States of America was increased by 18,
making the total number of states to 68 due to the current situation of coronavirus [46]. Example (**c**)
is another example where the coronavirus situation has been used. A photo of a speech of Trump
attached with the fake caption that states that Easter has been postponed to a future date. This is a
reference to the other events around the world, which are being postponed due to coronavirus to
avoid mass gatherings. Since Easter is a festival, its date cannot be changed [47]

Use (New in 2015!)." The organizers defined the following task:
   "Given a tweet and the accompanying multimedia item (image or video) from an
event that has the profile to be of interest in the international news, return a binary
decision representing verification of whether the multimedia item reflects the reality
of the event in the way purported by the tweet."
   As a part of the task, the organizers released the MediaEval dataset,[5]   which
contained ∼400 images used in about ∼20K different tweets in the context of
∼10 events (Hurricane Sandy, Boston Marathon bombings, etc.). This dataset is
considered as one of the first multi-modal fake news datasets and has been used
extensively for evaluating different models (see Table 3). Three competing teams
were shortlisted to present their systems [5]: Middleton [37], Jin et al. [22], and
Boididou et al. [6], which achieved 0.83, 0.92, and 0.91 F1-scores, respectively.
This was followed by another recent competition hosted jointly by the Institute of
Computing Technology, Chinese Academy of Sciences and the Beijing Academy
of Artificial Intelligence (BAAI) Research Institute, called MCG-FNews19,[6] where
three different tasks were given related to fake news: False News Text Detection,
False News Image Detection, and False Multi-modal News Detection.

---

[5]https://github.com/MKLab-ITI/image-verification-corpus.

[6]https://biendata.com/competition/falsenews/.

Two recent studies are worth mentioning: (i) Volkova et al. [63] explained the multi-modal deceptive news detection models by studying their behavior on a curated Twitter dataset. The authors categorized deceptive news into six classes and defined them: disinformation, propaganda, hoaxes, conspiracies, clickbait, and satire. They empirically showed that although text-only models outperform image-only models, combining both image and text modalities with lexical features performs even better. The authors also developed ErrFILTER,[7] an online interactive tool that allows users to explain model prediction by characterizing text and image traits of suspicious news content and analyzing patterns of errors made by the various models. (ii) Glenski et al. [14] performed fake news detection on a dataset comprising $7M$ posts in a variety of languages—Russian, English, Spanish, German, French, Arabic, Ukrainian, Portuguese, Italian, and unknown. Using a simple framework consisting of user network extractor and text and image feature extractors, they achieved 0.76 F1-score.

Li et al. [33] surveyed various datasets and methods for rumor detection. Cao et al. [10] defined fake news as follows and presented a survey on multi-modal approaches:

Definition 1 "A piece of fake news is a news post that shares multimedia content that does not faithfully represent the event that it refers to."

In this chapter, we start by discussing the major challenges faced by the multi-modal fake news detection models (Sect. 2). Section 3 introduces relevant multi-modal datasets that are often used for fake news detection. Section 4 presents the overview of the tools and techniques used for multi-modal fake news detection, which are further elaborated in Sects. 5–10. Section 11 concludes the chapter with possible future directions.

2   Challenges and Opportunities

The major challenges faced by multi-modal fake news detection methods can be divided into the following categories, based on which the existing methods can be differentiated:

– *Scarcity of Data*: Most of the publicly available datasets are small as human annotation is extremely costly and time consuming. Even if someone manages to employ multiple human annotators, it is extremely challenging to annotate a news as fake or real without knowing its context. For example, an expert in the social media domain may not be able to annotate news related to healthcare.

---

[7]https://github.com/pnnl/errfilter.

- *Class Imbalance*: The number of instances labelled as "fake" should be significantly smaller than that of the "real" category, thanks to the current online media that are mostly reliable and trustworthy. Therefore, most of the models face difficulties in handling highly skewed classes.
- *Capturing Multiple Modalities*: How to efficiently capture multiple modalities present in a news article is a challenge. Most of the methods extract features from different modalities independently and fuse them to obtain a combined representation of the article. Such methods usually fail to capture the dependency between modalities in the final representation.
- *Novel Fake News*: Fraudulent content creators are continuously adopting intelligent obfuscation strategies to evade quick detection of their story. Therefore, a model trained on an outdated dataset may not be able to spot the newly invented fake news articles.
- *Early Detection*: The effect of a highly damaging fake news may be detrimental to the society. Therefore, it is essential to adopt a strategy to detect fake stories immediately upon their publication. A model that takes into account time-dependent features, such as the number of shares/retweets and the underlying user network properties, may not be able to fulfill this requirement.
- *Explainability*: An additional challenge is to understand why a news is marked as "fake," explaining the root cause and answering the "why" and "how" of the method. Most of the existing methods fail to explain their results.
- *Generalizability*: A model may suffer from three types of problems: (i) *Domain Adaptation*: if it is trained on a healthcare-related fake news dataset, it may not perform equally well on social media posts; (ii) *Entity-Type Adaptation*: if it is trained on short texts such as tweets, it may not be able to generalize well on long news such as blogs or full-length news articles; and (iii) *Geo-location Adaptation*: if it is trained on a news dataset related to the US presidential election, it may not be able to perform well on Indian general elections (as the major sociological issue in the West is "black vs. white," on which the fake stories are often written, whereas in India, it is "Hindu vs. Muslim").

These challenges open up a tremendous opportunity to the research community to solve this problem in an efficient way in terms of both scalability and accuracy.

3 Multi-modal Fake News Datasets

In this section, we briefly describe some of the popular multi-modal fake news datasets. Table 1 presents a brief statistics of the datasets along with the link to obtain them. The datasets are broadly divided into two categories—datasets containing microblog posts (tweets, Weibo posts, Reddit posts, etc.) and datasets containing full-length news articles.

Table 1 Summary of the datasets used by various approaches for multi-modal fake news detection. Datasets are arranged in chronological order of the year of publication. The last column indicates the link where the source code is available

Dataset	Brief description	Entity	Size	Year	Link
MediaEval [5]	Tweets related to events or places along with images	Tweet	Training: ~5K genuine, ~7K fake, Test: ~1.2K genuine, ~2.5K fake	2015	[7]
Weibo-JIN [23]	Fake tweets from the official rumor busting system of Sina Weibo[a], genuine tweets verified by Xinhua News Agency	Tweet	50,287 tweets (19,762 of them have images attached), 25,953 images, 42,441 distinct users	2016	N.A.[b]
Weibo-att [24]	Human verified false rumor posts, genuine tweets verified by Xinhua News	Tweet	Training (rumor, 3749; real, 3783), Test (rumor, 1000; real, 996)	2017	[25]
Twitter [35]	Keywords were extracted from 530 rumors obtained from snopes.com, and tweets were scraped by queries using the keywords	Tweet	498 events each for rumor and non-rumors, 491,229 users, 1,101,985 tweets	2016	[36]
TI-News [66]	Real news obtained from authoritative news sites (*NYT, Washington Post*, etc.) and fake news collected by B.S. Detector[c]	News	8074 real and 11,941 fake news articles	2016	[51]
PHEME [71]	Tweets collected based on the newsworthy events identified by the journalists; news was marked as rumors/real by human annotators	Tweet	1972 rumor and 3830 real tweets	2017	[72]
PolitiFact [55]	A fact-checking website authenticating claims by elected officials. It contains news content, corresponding images, users' retweets/replies, and news profile (source, publisher, and keywords)	News, tweets, replies, users	624 real and 432 fake news, 558,937 users, and 552,698 replies	2018	[56]
Gossip Cop [55]	A fact-checking website for celebrity reporting investigating the credibility of entertainment stories	News, tweets, replies, users	16,817 real and 5323 fake news, 1,390,131 users, 379,996 replies	2018	[56]
Tampered News [40]	News articles written in English and published in 2014 across different domains (sports, politics, etc.); entities are tampered automatically using a tampering technique	News	72,561 news articles	2020	[42]

Name	Description	Type	Count	Year	Ref.
News400 [40]	News articles crawled from popular German news websites (faz.net, haz.de, and sued-deutsche.de); each news has at least one image and text	News	4000 news articles	2020	[41]
BuzzFeed News [58]	Facebook provided a dataset consisting of political orientation and portal label	News	181 fake and 757 real news articles	2016	[58]
NewsBag [26]	Real (fake) news collected from *The Wall Street Journal* (The Onion); human experts annotated the news	News	200,000 real and 15,000 fake news articles	2020	N.A.[d]
NewBag++ [26]	Augmented version of NewsBag where the size of two classes is balanced	News	200,000 real and 389,000 fake news articles	2020	N.A.[d]
NewsBag Test [26]	Real (fake) news articles scrapped from The Real News (The Poke)	News	11,000 real and 18,000 fake news articles	2020	N.A.[d]
Fakeddit [43]	1M submissions from 22 different subreddits were collected (March 19, 2008–October 24, 2019) and passed through several filtering steps; two-way, three-way, and six-way classification labels	Reddit posts and their metadata	628,501 fake and 527,049 real posts, 682,996 multi-modal posts	2020	[44]

N.A.: Not available

[a] http://service.account.weibo.com/

[b] Authors did not respond to our email

[c] https://github.com/selfagency/bs-detector

[d] Authors informed us that the dataset will be shared upon request

3.1 Fake Microblog Datasets

Each sample of these datasets is relatively small. Two widely used datasets in this category are MediaEval and Weibo-att. Along with them, we also describe a few other datasets that are often being used to detect fake news.

1. **MediaEval:** The dataset was collected as a part of the *Verifying Multimedia Use* task of MediaEval 2015 [5]. It contains tweets related to events or places along with images. A tweet was annotated as "genuine" if the associated image corresponds to the event that the text of the tweet points to; otherwise, it was marked as "fake." Overall, there are 400 images that are used in about ~20K different tweets in the context of ~10 events (Hurricane Sandy, Boston Marathon bombings, etc.).

2. **Weibo-JIN:** Jin et al. [23] collected tweets related to diverse events from Weibo. Instead of human annotation, the ground-truth was prepared based on the authenticity of the news sources. Specifically, fake news events were collected from the official rumor busting system of Sina Weibo, and real events were gathered from a hot news detection system of Xinhua News Agency, the official and most authoritative news agency in China, as the main source. From 146 event-related news articles, keywords were extracted based on which tweets were collected from Weibo. This dataset is larger than that of MediaEval.

3. **Weibo-att:** Jin et al. [24] collected false rumors posted from May 2012 to January 2016 from the official rumor debunking system of Weibo. The real tweets were collected from Xinhua News Agency, an authoritative news agency in China. This is one of the highly used datasets in multi-modal fake news detection.

4. **Twitter:** Ma et al. [35] collected 778 verified rumor and real events during March–December 2015 from www.snopes.com. Upon extracting the keywords and iteratively refining them, composite queries were fired on Twitter API. Non-rumor events were collected from some existing datasets [11, 30].

5. **PHEME**: Zubiaga et al. [71] collected this dataset by emulating the scenario in which a journalist is following a story. They hired few expert journalists and kept getting information about the new events. Upon receiving information about a new event, the crawler immediately started collecting tweets related to the event. After preprocessing, the remaining tweets were annotated by the experts based on whether there was any evidence about the trustworthiness of the fact expressed in the tweet or any authoritative source was found. The collected tweets were related to five events—Ferguson unrest, Ottawa shooting, Charlie Hebdo shooting, Sydney siege, and Germanwings plane crash.

6. **Fakeddit**: Nakamura et al. [43] collected 1M submissions from 22 different subreddits posted between March 19, 2008, and October 24, 2019. The dataset contains the title of the submission, images, comments made by the users, other user information, scores, upvote and downvote counts, etc. Around 64% of text comments have accompanying images. Initial quality assessment was done based on the metadata information such as the ratio of upvotes and downvotes, users'

karma score, etc. Second-level assessment was conducted by the experts. A series of preprocessing steps were followed to clean up the subreddit posts before entering the annotation process. The annotation was done in three levels—*two-way*, whether a sample is real or fake; *three-way*, whether a sample is completely real or it is fake and contains text that is true or the sample is fake with false text; and *six-way*, whether a sample is real, satire/parody, misleading content, imposter content, false connection, and manipulated content.

3.2 Fake News Datasets

Each sample of these datasets is relatively large and contains a full-length article. Two widely used datasets in this category are PolitiFact and Gossip Cop [55]. Along with these datasets, we also discuss some other datasets of this type that are often used for fake news detection.

1. **TI-News**: Yang et al. [66] created a collection of news from Megan Risdal and Kaggle, containing 11,941 fake and 8074 real news articles. We call this dataset TI-News. The real news articles were related to well-known authoritative sites such as *The New York Times*, *The Washington Post*, etc. Along with the text and image information, each sample contains the author of the news and the website where it was posted.

2. **PolitiFact and Gossip Cop**: Shu et al. [55] utilized two fact-checking websites, namely, PolitiFact[8] and Gossip Cop.[9] The former accommodates news related to politics, and the latter contains fact-checking stories related to films and entertainment. The ground-truth labels were provided by their expert teams. True news were collected from E! Online,[10] which is a well-known trusted media website for publishing entertainment news pieces. Social contexts were collected by searching Twitter API with the titles of the news articles. Users' responses were also collected for every post. Along with these, spatiotemporal information such as locations (if explicitly provided by the users), timestamps of user engagement, replies, likes, retweets, etc. enriched the dataset.[11]

3. **TamperedNews**: Müller-Budack et al. [40] collected an existing dataset, called BreakingNews [50], which covers $100K$ news related to different domains (sports, politics, healthcare, etc.). They further designed a tampering mechanism such as random replacement of named entities to synthetically generate fake news.

---

[8]https://www.politifact.com/.

[9]https://www.gossipcop.com/.

[10]https://www.eonline.com/.

[11]PolitiFact and Gossip Cop are combined in FakeNewsNet dataset [54, 55].

4. **News400**: In order to evaluate their model on cross-language datasets, Müller-
   Budack et al. [40] further created News400, a repository containing news articles
   from three popular German news websites (faz.net, haz.de, and sueddeutsche.de).
   The news were published during August 2018–January 2019 and were related to
   four topics—politics, economy, sports, and travel. Similar tampering mechanism
   was applied to obtain fake news.
5. **NewsBag**: Jindal et al. [26] created the largest dataset of multi-modal fake news
   articles. Real and fake news were collected from *The Wall Street Journal* and
   The Onion,[12] respectively. Several human experts were asked to verify 15,000
   articles as fake. However, the number of fake articles was much lesser than the
   real articles. To make a balanced dataset, the authors further created NewsBag++,
   comprising $200K$ real and $389K$ fake news by running a data augmentation
   method on NewsBag. They also created NewsBag Test, a separate dataset for
   testing the models. This dataset contains $11K$ real news collected from The Real
   News[13] and $18K$ fake news collected from The Poke.[14]

4 State-of-the-Art Models

Most of the existing models are supervised and follow *fusion technique*—low-level
features are extracted from different modalities (text, image, etc.) and combined
using various fusion mechanisms, based on which existing models can be divided
into three broad categories: early fusion, late fusion, and hybrid fusion [34]. Let $\mathbf{v_m}$
be the low-level feature representation of modality m, and there are M modalities in
a post. Semicolon (;) is used to indicate concatenation operation. The three fusion
techniques are defined below:

– *Early Fusion*: Low-level features from different modalities are combined (generally through concatenation), and a joint representation is created from the
  combined features. Next, a single model is trained to learn the correlation and
  interactions between low-level features of each modality. Let h be the single
  model and p be the final prediction. Then,

$$p = h([\mathbf{v}_1; \mathbf{v}_2; \cdots, \mathbf{v}_m; \cdots; \mathbf{v}_M])$$

– *Late Fusion*: From different modalities, unimodal decisions are obtained using
  other models. These decisions are then fused with some mechanism (such as
  averaging, voting, or a learned model). Let h_m be the model for mth modality,

---

[12]The Onion publishes satirical articles on both real and fictional events. Link: https://www.
theonion.com/.

[13]https://therealnews.com/.

[14]https://www.thepoke.co.uk/.

and F is the mechanism used to fuse the decisions as in the early fusion. Then, the final prediction will be

$$p = F([h_1(\mathbf{v}_1); h_2(\mathbf{v}_2); \cdots ; h_m(\mathbf{v}_m); \cdots ; h_M(\mathbf{v}_M))$$

– *Hybrid Fusion*: It is a combination of early and late fusion. A subset of features is passed through separate models to obtain the unimodal decisions as in the late fusion. These decisions are combined with the remaining features to obtain a combined representation, which is further passed through a single model for the final decision. Let $n, n+1, \cdots, m-1, m$ be the modalities that follow late fusion. Then, the final prediction will be

$$p = h([h_j(\mathbf{v}_j)]_{n \leq j \leq m}; [\mathbf{v}_i]_{1 \leq i \neq j \leq M}])$$

There are some methods that follow unsupervised techniques; some other methods follow advanced neural network techniques such as adversarial learning and variational autoencoder. Table 2 provides a brief summary of the state-of-the-art methods for multi-modal fake news detection, and Fig. 2 shows the taxonomy of the methods. Tables 3 and 4 show a comparative analysis of the methods on four widely used datasets. The following sections elaborate on these methods. The methods in each section are arranged in chronological order of the year of publication.

5 Unsupervised Approach

Müller-Budack et al. [40] introduced the task of cross-model consistency verification in real-world news. The idea is to quantify the coherence between image and text. They proposed the first unsupervised approach for multi-modal fake news detection, which we call CCVT (cross-model consistency verification tool).[15] CCVT links every named entity (person, location, and event) extracted from the text to its corresponding image using some reference image database. Then, the consistency between the texts and images present in the post is measured. CCVT is composed of three major components:

– *Extraction of Textual Entities*: CCVT utilizes spaCy [19] to extract the named entities and link them to the Wikidata [9] knowledge base. To extract the context of the text, sapCy is applied to obtain all nouns (general concepts such as politics, sports, actions, etc.). fastText [8] is used to obtain the embedding of each candidate.
– *Extraction of Visual Features*: Multi-task cascaded convolutional network [69] is used to detect faces from images. The feature vector of each face is extracted using DeepFace [53].

---

[15]Code is available at https://github.com/TIBHannover/cross-modal_entity_consistency.

Table 2 Summary of the methods used for multi-modal fake news detection. Methods are in chronological order of the year of publication

Method	Approach	Entity	Dataset	Year
JIN [23]	Five types of visual features are extracted and combined with textual features; concatenated feature set is fed to classifiers	Tweet	Weibo-JIN	2016
AGARWAL [1]	Augmentation of classification systems with a learning to rank scheme	Tweet	MediaEval	2017
att-RNN [24]	RNN with attention mechanism to fuse features from text, image, and social context	Tweet	Weibo-att, MediaEval	2017
EANN [64]	Event adversarial neural networks, composed of multi-modal feature extractor, event discriminator, and fake news detector	Tweet	Weibo-att, MediaEval	2018
TI-CNN [66]	Explicit text and image features are extracted and combined with the implicit features obtained from the CNNs and combined for the detection	News	TI-News	2018
MVAE [28]	Multi-modal variational autoencoder that uses a bimodal variational autoencoder coupled with a binary classifier	Tweet	Weibo-att, MediaEval	2019
MVNN [49]	An end-to-end neural network to learn representations of frequency and pixel domains simultaneously and effectively fuse them	Tweet	Weibo-att	2019
MKEMN [68]	Multi-modal knowledge-aware network to obtain text, visual, and external knowledge, and an event memory network to capture event-invariant feature	Tweet	Twitter, PHEME	2019
SAME [12]	Triplet (news publisher, user, and news) extraction followed by adversarial learning for detecting a semantic correlation between different modalities and finally incorporation of users' sentiment	News	PolitiFact, Gossip Cop	2019
SpotFake [59]	A concatenation of BERT-based text embedding and VGG-19-based image embedding	Tweet	Weibo-att, MediaEval	2019
SpotFake+ [60]	A transfer learning-based approach by combining XLNet and VGG-19 modules	News	PolitiFact, Gossip Cop	2020
CCVT [40]	An unsupervised approach that measures the consistency of image and text to detect fake news	News	Tampered News, News400	2020
MCE [27]	After obtaining the embedding from each modality, a combined representation is learned to score each news based on its magnitude and consistency	News	MediaEval, BuzzFeed News	2020
SAFE [70]	A fusion model is used to obtain a joint representation of news; two representations are compared to measure their similarity; both of them are combined to obtain final loss	News	PolitiFact, Gossip Cop	2020

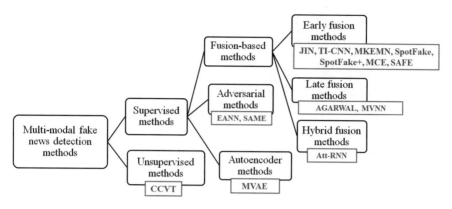

Fig. 2 Taxonomy of the multi-modal fake news detection models with respect to the techniques used for the detection

Table 3 Performance of the multi-modal fake news detection methods, which were evaluated on two popular microblog datasets—MediaEval and Weibo-att. The accuracy corresponding to the best setting of each model was taken from the original paper

Model	MediaEval				Weibo-att			
	Accuracy	Precision	Recall	F1-score	Accuracy	Precision	Recall	F1-score
JIN	0.898	–	0.835	–	–	–	–	–
att-RNN	0.682	0.78	0.615	0.689	0.788	0.862	0.686	0.764
EANN	0.715	0.822	0.638	0.719	0.827	0.847	0.812	0.829
MVAE	0.745	0.801	0.719	0.758	—	0.689	0.777	0.730
MVNN	–	–	–	–	0.846	0.809	0.857	0.832
MVNN+att-RNN	–	–	–	–	0.901	0.911	0.901	0.906
MVNN+EANN	–	–	–	–	0.897	0.930	0.872	0.900
MVNN+MVAE	–	–	–	–	0.891	0.896	0.898	0.897
SpotFake	0.777	0.751	0.900	0.820	0.892	0.902	0.964	0.932
MCE	0.967	0.875	0.976	0.923	–	–	–	–

Table 4 Performance of the multi-modal fake news detection methods, which were evaluated on two popular news datasets—PolitiFact and Gossip Cop. The accuracy corresponding to the best setting of each model was taken from the original paper

Model	PolitiFact				Gossip Cop			
	Accuracy	Precision	Recall	F1-score	Accuracy	Precision	Recall	F1-score
SAME	–	–	–	0.772	–	–	–	0.804
SpotFake+	0.846	–	–	–	0.856	–	–	–
SAFE	0.874	0.889	0.903	0.896	0.838	0.857	0.937	0.895

– *Verification of Shared Cross-Model Entities*: The scene contexts extracted from images and texts are compared. First, for each named entity, a set of k images is retrieved from Google/Bing search engine. Second, a denoising step is executed to remove irrelevant images from the set. It is followed by a clustering technique, and the mean feature vector corresponding to the majority cluster serves as the representative of the queried person. Finally, the feature vectors of all faces in the image are compared to the vector of each person in the text. Similarly, the consistency of locations and events is measured.

CCVT was evaluated on the TamperedNews and News400 datasets to show its efficacy compared to other baselines.

6 Early Fusion Approaches

6.1 JIN

Jin et al. [23] proposed JIN,[16] an early fusion approach to separate fake and real *events* (instead of detecting fake tweets/news). An event is composed of a set of tweets containing certain keywords, which indicate a real incident. The authors observed that given the same number of tweets in events, real events tend to contain more images than fake events. Their major contribution was to come up with five novel visual features:

- *Visual Clarity Score* (VCS): The intuition behind this score is that if a set of images (corresponding to an event) is distinct from the entire collection, then the event is likely to be genuine. First, the local descriptor of each image is extracted. Second, all descriptors are quantized to form a visual word vocabulary. Third, each image is represented by a bag-of-words model. Fourth, two language models are calculated—one from the event and the other from the entire collection. Finally, the "clarity score" is defined as the Kullback–Leibler divergence between two language models.
- *Visual Coherence Score* (VCoS): It measures how coherent images in a certain event are. GIST-based global image descriptor [45] is extracted from each image within an event, and an average similarity of all pairs of images within the event is computed.
- *Visual Similarity Distribution Histogram* (VSDH): For each event, the inter-image similarity is measured between all pairs of images based on VCoS. The similarity scores are divided into 10 bins. For each bin, the normalized number of elements indicates the entry of the feature. Ten features (corresponding to ten bins) are obtained after this step.
- *Visual Diversity Score* (VDS): For every event, images are ranked based on the popularity on social media. For each image, the average dissimilarity score (1-VCS) is then calculated between the image and all the other images ranked higher than the given image. The final VDS score is the average of the VDS scores of all the images in the event.

---

[16]If there is no explicit name of the method mentioned in the original paper, we use the name of the first author to denote the method.

Text Content	User
Count of Message,	Count of Distinct Users, Fraction of Popular Users,
Average Word/Character Length,	Average Followers/Followees/Posted Tweets,
Fraction of Question/Exclamation Mark,	Fraction of Verified User/Organization.
Fraction of Multi Question/Exclamation Mark Ratio,	
Fraction of First/Second/Third Pronouns,	
Fraction of URL/@/#,	**Propagation**
Count of Distinct URL/@/#,	Size of Max Subtree, Average Likes,
Fraction of Popular URL/@/#,	Average Degree/Non-zero Degree.
Count of Distinct People/Location/Organization,	
Fraction of People/Location/Organization,	
Fraction of Popular People/Location/Organization,	
Average Sentiment Score,	
Fraction of Positive/Negative Tweets.	

Fig. 3 Set of statistical features used by the JIN model

- *Visual Clustering Score* (VCIS): Each image is represented by the bag-of-word model as in VCS. For every event, images are placed in a Euclidean space, and a hierarchical agglomerative clustering (single-link strategy) is used to detect the number of clusters, which constitutes the feature of the event.

JIN also considers 42 statistical features, broadly divided into 3 categories—text content, user, and propagation based (as shown in Fig. 3).

Four classifiers, namely, SVM, Logistic Regression, KStar, and Random Forest, were run on the Weibo-JIN dataset, among which Random Forest was reported to be the best model considering both non-image- and image-based features, achieving 0.83 F1-score.

6.2 TI-CNN

Yang et al. [66] mentioned that the lexical diversity and cognition of the deceivers are totally different from true tellers. Images play a major role in fake news detection. For instance, a fake image is often of low resolution and not correlated with the text. The authors proposed TI-CNN (Text and Image information-based Convolutional Neural Network), which takes explicit user-defined features and implicit CNN-based features and gets trained on both texts and images.

TI-CNN is composed of two major components:

- *Text Feature Extractor*: Several features (such as the length of the news, number of question marks, exclamation, capital letters, etc.) are explicitly extracted from the text and passed through a Fully Connected Layer (FCL). Latent textual features are extracted using a Convolutional Neural Network (CNN). Both of them are concatenated to obtain a combined textual representation.

– *Image Feature Extractor*: Several image features (such as the number of faces, resolution of the image, etc.) are extracted and combined with the latent features obtained from another CNN.

Both these features are further combined and passed through a FCL for the final detection.

TI-CNN outperformed various unimodal classifiers with 0.921 F1-score on the TI-News dataset.

6.3   MKEMN

Zhang et al. [68] argued that along with the text and multimedia, one should also consider the rich knowledge information present in the existing rumor texts, which might often be used for rumor verification. Their proposed method, MKEMM (Multi-modal Knowledge-aware Event Memory Network), utilizes the multi-modal knowledge-aware network to obtain a shared representation of text, existing knowledge, and images (see Fig. 4 for the framework). An Event Memory Network (EMN) is used to obtain event-independent features as suggested in EANN [64] (see Sect. 9). MKEMM attempts to detect whether a claim is a rumor or not, where a claim comprises a sequence of correlated posts with timestamp associated with each post. Two major components of MKEMN are discussed below:

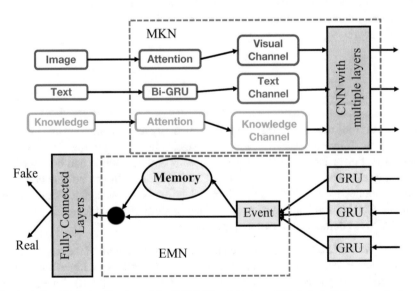

Fig. 4 A simplified visualization of the MKEMN architecture. *Filled circle* indicates concatenation operation

- *Multi-modal Knowledge-aware Network* (MKN): To capture four signals from a post $p = \{w_1, w_2, \cdots, w_n\}$ into the final embedding, four separate modules are designed—(i) *Text Encoder*, which takes a short text and uses a Bi-GRU to obtain a text embedding h_t; (ii) *Knowledge Encoder*, which first extracts entities from a post, then acquires concept information for each entity from existing knowledge graphs [65] and taxonomies [39], and finally obtains a concept knowledge vector k_t for each entity using an attention mechanism; (iii) *Visual Encoder*, which uses VGG-19 [57] to obtain the initial visual representation. A word-guided visual attention model is incorporated, which takes VGG-19 features and Bi-GRU embedding and projects regions that correspond to the highly relevant words to obtain a visual embedding v_t; and (iv) *Multi-modal knowledge-aware CNN*, which, instead of directly concatenating h_t, k_t, and v_t, uses two continuous transformation functions $\mathcal{H}_k(.)$ and $\mathcal{H}_v(.)$ to map k_t and v_t, respectively, to the word space keeping their semantic relation. Finally, a combined representation is obtained as $G = \begin{pmatrix} h_i \\ \mathcal{H}_k(k_i) \\ \mathcal{H}_v(v_i) \end{pmatrix}_{1 \leq i \leq n}^{3 \times n \times d}$. Afterward, multiple layers with different filters are applied to obtain the final representation of the post.

- *Event Memory Network* (EMN): To obtain event-independent features, EMN first generates an event representation x by passing the MKN embedding of posts related to the event through GRUs and feeding their outputs to a memory, which measures how dissimilar a query event is with respect to the previous events. The output of the memory network is concatenated with x to generate the new event representation X.

The final classification is performed by a deep neural network classifier $z = \mathcal{D}(X)$ using cross-entropy loss.

MKEMM achieved 0.870 and 0.814 F1-scores on the Twitter [35, 36] and PHEME [71] datasets, respectively, and outperformed six baselines including EANN.

6.4 SpotFake and SpotFake+

Singhal et al. [59] argued that existing (adversarial) models [64] are heavily dependent on the secondary tasks performed by the discriminator. An inappropriate choice of the secondary task may deteriorate the performance by up to 10%. The authors proposed SpotFake (Spotting Fake News), a multi-modal early fusion approach to combine texts and images.

- *Textual Feature Extractor*: SpotFake uses BERT [13] to obtain the embeddings of words, which are further concatenated to form the embedding of a sentence.
- *Visual Feature Extractor*: A pretrained VGG-19 model is adopted, and the output of the second last layer is passed through a FCL.

– *Multi-modal Fusion*: The outputs of the above two extractors are concatenated to obtain the final representation.

While comparing with nine baselines including att-RNN [24] (see Sect. 8), EANN [64], and MVAE [28] (see Sect. 10), SpotFake turned out to be outperforming others with 0.82 and 0.932 F1-scores on the Weibo-att and MediaEval datasets, respectively.

Singhal et al. [60] further extended SpotFake to a transfer learning framework and proposed SpotFake+.[17] It leverages a pretrained language transformer (XLNet [67]) and a pretrained ImageNet model (VGG-19) for feature extraction. The authors claimed that SpotFake+ is the first multi-modal approach that performs fake news detection on *full-length articles*. On the PolitiFact and Gossip Cop datasets, SpotFake+ achieved 0.846 and 0.854 F1-scores, respectively, outperforming four baselines including EANN, MVAE, and SpotFake.

6.5 MCE

Kang et al. [27] proposed MCE (Multi-modal Component Embedding) that focuses on the reliability of various multi-modal components and the relationship among them. A vector representation is learned for each modality whose magnitude and direction indicate "reliability" and "consistency." A news will have overall high magnitude if the sum of its component magnitudes is high and all of them are closely aligned (high consistency). MCE learns a latent space such that the magnitude of the real news would be higher than that of fake news. Text-CNN [29] and VGG-19 are used to extract textual and visual features, respectively. For event-related features, multilayer perceptron is used. The final representation of a news is the sum of the representation of its individual components.

MCE was reported to outperform three baselines with 0.9234 and 0.5915 F1-scores, respectively, on the MediaEval and BuzzFeed News datasets.

6.6 SAFE

Zhou et al. [70] also argued to measure the consistency between two modalities and hypothesized that fake news articles tend to contain uncorrelated/dissimilar text and image modalities. Their proposed model SAFE (Similarity-Aware FakE news detection method) attempts to combine the representations of two modalities along with their dissimilarities in an end-to-end framework, which is composed of three components.

---

[17]https://github.com/shiivangii/SpotFakePlus.

- *Multi-modal Feature Extraction*: Similar to MCE, Text-CNN is used for textual embedding F_t. However, for visual feature extraction, unlike other methods that directly apply pretrained VGG-19, SAFE first uses a pretrained image2sentence model [62] to obtain the initial embeddings that are further fed to a similar Text-CNN framework with an additional FCL to obtain the final visual embedding F_v.
- *Modal-Independent Fake News Prediction*: Two different representations are further concatenated to obtain the final representation, which is passed through a FCL with cross-entropy loss $\mathcal{L}_p$.
- *Cross-Modal Similarity Extraction*: This component independently assumes that texts and images are dissimilar in the case of fake news; thus, a loss can also be computed between the ground-truth and the similarity between two modalities. The similarity between F_t and F_v is computed using a modified cosine similarity measure as follows:

$$M_s(F_t, F_v) = \frac{F_t F_v + \|F_t\| \|F_v\|}{2\|F_t\| \|F_v\|}$$

The loss function calculated in this step assumes that news formed by dissimilar texts and images is more likely to be fake and thus is defined as follows:

$$\mathcal{L}_s = y \log(1 - M_s(F_t, F_v)) + (1 - y) \log M_s(F_t, F_v)$$

where $y = 1$ if the article is fake, 0 otherwise.
- *Model Integration and Joint Learning*: The model is jointly trained by combining both the losses: $\mathcal{L} = \alpha \mathcal{L}_p + \beta \mathcal{L}_s$, where α and β balance their corresponding components.

SAFE outperformed seven baselines including att-RNN and models obtained by dropping each modality in isolation from SAFE, by achieving 0.896 and 0.895 F1-scores on the PolitiFact and Gossip Cop datasets.

7 Late Fusion Approaches

7.1 AGARWAL

Agrawal et al. [1] detected fake multimedia tweets containing texts and images. They defined fake news as follows:

Definition 2 "A multimedia news is fake if the multimedia content (image/video) is unrelated to the texts."

The authors proposed a fusion technique (we call it AGARWAL) that concatenates the output of a ranking method with the other features of the tweet entities and

Content features		User Features	
length of tweet	num of words	num of friends	num of followers
contains question mark	contains exclamation mark	follower-friend ratio	num of times listed
num of question marks	num of exclamation marks	user has a URL	user is a verified user
contains happy emoticon	contains sad emoticon	num of tweets	
contains 1st order pronoun	contains 2nd order pronoun		
contains 3rd order pronoun	num of uppercase characters		
num of negative senti words	num of positive senti words		
num of mentions	num of hashtags		
num of URLs	num of retweets		

Fig. 5 Content and user features used by [1, 4] to characterize a tweet entity

feeds the concatenated features into a classifier. The other features of a tweet entity can be broadly categorized into three classes as follows:

- *Image-Based Features*: These features are often used to identify if an image is doctored [5, 17, 32]. The intuition is that a multimedia fake news is generally associated with doctored image(s). The used features are as follows:

  - Probability map of the aligned double JPEG compression
  - Probability map of the nonaligned double JPEG compression
  - Potential primary quantization steps for the first six DCT (discrete cosine transform) coefficients of the aligned double JPEG compression
  - Potential primary quantization steps for the first six DCT coefficients of the nonaligned double JPEG compression
  - Block artifact grid
  - Photo-response nonuniformity.

- *Twitter Content and User-Based Features*: These features (as shown in Fig. 5) are taken from Boididou et al. [4] to capture the social status of users who post the news and the lexicographic properties of tweet texts.
- *Tweet-Based Features*: Doc2vec [31] embedding method is trained on the Sentiment140 corpus [15] to obtain the vector representation of the text. The authors showed that document embedding outperforms n-gram-based features.

Various traditional classifiers (such as SVM, deep neural network, and logistic regression) were trained along with a ranking model. The ranking model was trained in such a way that it prefers genuine tweets more than fake tweets. The ranking model produces a score, which was further used as a feature along with the other features mentioned before. AGARWAL achieved 83.5% unweighted average recall in detecting fake multimedia tweets.

7.2 MVNN

Qi et al. [49] classified fake images into two categories: *tampered images* that have been modified digitally, and *misleading images* that are not modified, but content-wise they are misleading (outdated images used for current events, images taken in one country are used for another country, etc.). They defined fake news as follows:

Definition 3 "Fake news is a post that is intentionally and verifiably false. A fake-news image is an image that is attached to a fake news."

The authors proposed MVNN (Multi-domain Visual Neural Network) that combines frequency and pixel information for fake news detection. It is composed of three modules:

- *Frequency Domain Sub-network*: Discrete cosine transformer (DCT) is used to transfer images from pixel domain to frequency domain. A CNN (three convolutional blocks and a FCL) is used to process the output of DCT and return the final feature representation l_o.
- *Pixel Domain Sub-network*: This module is used to extract the visual features of the input image at the semantic level. A multi-branch CNN network is used to extract multiple levels of features, and a bidirectional GRU (Bi-GRU) network is utilized to model the sequential dependencies between features. The proposed CNN model is composed of four blocks, each having a 3×3 and a 1×1 convolution layer and a max-pooling layer. One CNN block feeds its input to the next CNN block. Furthermore, the outputs of all CNN blocks are fed to a Bi-GRU to obtain a strong dependency between features. The composite representation obtained from GRU is denoted by $L = \{l_1, l_2, l_3, l_4\}$, where l_i is the output of the ith GRU unit.
- *Fusion Sub-network*: All features extracted so far may not contribute equally. For instance, misleading images may not have gone through tampering; therefore, semantic features are more effective than pixel-level features. Fusion sub-network introduces an attention mechanism to weigh individual features.

Finally, the weighted feature vector is passed through a FCL (with cross-entropy loss) to make the final prediction.

Note that MVNN only considers image-related features for fake news detection.[18] It was compared with four baselines, and 0.832 F1-score was reported on the Weibo-att dataset. Furthermore, while the visual feature extraction module of att-RNN (Sect. 8), EANN (Sect. 9), and MVAE (Sect. 10) was replaced by MVNN, it improves the performance of the original methods. The highest accuracy was obtained with att-RNN+MVNN with 0.906 F1-score (see Fig. 3 for a comparative analysis).

---

[18] Although we avoid any method that solely uses image features for fake news detection, we intentionally add MVNN as it has widely been used as a baseline by other multi-modal fake news detection models. Moreover, it shows significant performance gain when being incorporated into the existing methods (see Table 3).

8   Hybrid Fusion Approach

Jin et al. [24] proposed att-RNN, a multi-modal deep fusion model to leverage
multiple modalities present in the tweets (see Fig. 6 for the schematic diagram of att-
RNN). It captures the intrinsic relations among three modalities—text, multimedia
content (image), and social context (metadata of the tweets). The model intrinsically
captures the coherence between these three modalities. The authors hypothesized
that images would have certain correlations with text or social context in genuine
tweets.

A tweet is represented as a tuple $I = \{T, S, V\}$, where T is the text of the tweet,
S is its social context (hashtag topic, mentions and retweets, emotion, sentimental
polarity, etc.), and V is the visual content. The model extracts features from each
of these modalities to obtain a combined representation. The model follows three
steps:

- *Step 1*: The text $T = \{T_1, T_2, \cdots, T_n\}$ and the social context S are fused using an
  RNN to obtain a joint representation as follows. A pretrained Word2Vec [38] is
  used to obtain the embedding R_{T_i} of each word T_i in the tweet. The social context
  vector R_S is passed through a FCL to match the dimension of R_{T_i} and to obtain
  $R_{S'} = W_{sf} R_S$, where W_{sf} is the weight matrix of a FCL. Next, for each time step
  (word), an LSTM cell takes $[R_{T_i}; R_{S'}]$ as an input, and the final representation
  R_{TS} is obtained by averaging the output neurons of all LSTM cells.
- *Step 2*: A visual representation R_V is obtained using deep CNN. The authors used
  the standard VGG-19 network in the initial layer and added back to back two
  512-dimensional FCLs to obtain R_V. In order to capture the correlation between
  the text/social context and image, a visual attention mechanism is incorporated.
  From every time step (word) in Step 1, the output hidden state h_i of LSTM is
  passed through two FCLs (the first FCL with ReLU and the second FCL with
  softmax function) to obtain the attention vector A_n (of the same dimension as
  that of R_V). The output of this step is an attention vector $R_{V'}$.

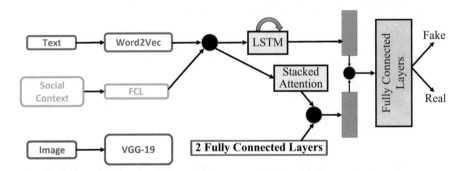

Fig. 6 Schematic diagram describing the architecture of att-RNN. *Filled circle* indicates concatenation operation

– *Step 3*: A combined representation for each tweet is obtained by concatenating R_{TS} and $R_{V'}$: $R_I = [R_{TS}; R_{V'}]$, which is fed to a softmax layer with cross-entropy loss.

The proposed method was evaluated on two datasets—Weibo-att and MediaEval; it achieved 0.764 and 0.689 F1-scores, respectively, for two datasets and outperformed seven baselines (including different variants of att-RNN).

9 Adversarial Model

Wang et al. [64] argued that most of the existing approaches tend to detect event-specific fake news; therefore, they fail miserably in detecting fake news on newly emerged and time-critical events (novel fake news). The proposed method EANN (Event Adversarial Neural Networks) attempts to overcome this problem by learning an event-independent feature representation of every tweet using an adversarial network (see Fig. 7). It consists of three components:

1. *Multi-modal Feature Extractor* (MEF): Text-CNN is used to encode tweet texts. A pretrained vector embedding is used to initialize each word. Multiple filers with various sizes are applied to extract textual features with different granularity. Following this, a FCL is used to ensure the same dimension of the text representation with that of the image representation (discussed below).
   For image-level feature extraction, the same architecture as proposed by [24] was adopted. These two features are then concatenated to form a multi-modal feature R_F.
2. *Fake News Detector* (FND): Given the multi-modal feature R_F, this module uses a FCL with softmax to predict if a post is real or fake. The cross-entropy loss is used to calculate the detection loss L_d.
3. *Event Discriminator* (ED): Given the multi-modal feature R_F, this module uses two FCLs to classify posts into one of the K events. Cross-entropy loss L_e is

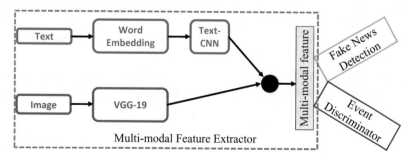

Fig. 7 Schematic diagram describing the architecture of EANN. *Filled circle* indicates concatenation operation

calculated to estimate the dissimilarities between the representations of different events—large loss indicates a similar distribution of the representations of events, which in turn ensures that the resultant representation is event-invariant.

Finally, the model integrator combines the two losses as follows: $L = L_d - \lambda L_e$, where λ balances between two losses. The combined loss ensures that MEF tries to fool ED to achieve event-invariant representations by maximizing $L_e(.)$, whereas ED tries to identify each event by minimizing $L_e(.)$.

On two datasets, namely, MediaEval and Weibo-att, ENVV outperforms six baselines including att-RNN with 0.719 and 0.829 F1-scores, respectively.

9.1 SAME

Cui et al. [12] argued that along with multiple modalities, the views of readers expressed on a particular post also play an important role to detect whether the post is fake or not. Users' viewpoints can be captured by the comments left for the post. The authors statistically validated that users tend to express more sentiment polarity on the comments related to fake news than real news. The proposed model, named SAME (Sentiment-Aware Multi-modal Embedding), consists of three components:

- *Feature Extractor*: To generate the embedding of images, texts, and user profiles, three different networks are designed—a pretrained VGG-19 is used to extract image feature, a pretrained Glove [48] embedding followed by a multilayer perceptron is used to extract text feature, and a two-layer multilayer perceptron is used to extract user profile (represented by a vector of discrete values such as topics) feature. These features are passed through the adversarial network (discussed below) before integrating using a FCL with three hidden units.
- *Adversarial Learning*: In order to bridge the gap between three modalities, an adversarial network is designed. It consists of two modality discriminators for image and profile features—one takes image and text features, and the other takes profile and text features, to discriminate whether the feature corresponds to the image or the profile. Here, the feature extractor acts as a generator.
- *Fake News Detector*: A FCL with cross-entropy loss is used to discriminate a news as fake or real.

SAME achieved 0.772 and 0.804 (macro) F1-scores while comparing with six baselines including EANN on the PolitiFact and Gossip Cop [55] datasets.

10 Autoencoder Model

Qi et al. [49] argued that existing methods [24, 64] do not have any explicit objective function to discover correlations across the modalities. The authors proposed MVAE (Multi-modal Variational Autoencoder) that consists of three modules (see Fig. 8):

- *Encoder*: Two sub-modules are used for encoding texts and images. The encoder architecture is similar to MEF in EANN [64]. Here, instead of using a CNN, the authors used stacked bidirectional LSTM units (Bi-LSTMs). Upon obtaining the embeddings of words from a pretrained word embedding model, the embedding vectors are passed through two Bi-LSTMs, followed by a FCL to get the textual embedding R_T.

  The visual encoder is the same as the image-level feature extractor in MEF of EANN, except in this case where two FCLs are used to pass the VGG-19 feature, which outputs a visual embedding R_V.

  The concatenated representation $[R_T; R_V]$ is passed through another FCL to obtain two vectors μ and σ, indicating the mean and variance, respectively, of the distribution of the shared representation. The final output of the encoder is a linear combination of μ and σ as follows: $R_m = \mu + \epsilon\sigma$, where ϵ is a random variable sampled from a Gaussian distribution.

- *Decoder*: The decoder module is just the reverse of the encoder. It also has two sub-modules—one for text and the other for image. These sub-modules try to reconstruct the original data from the sampled multi-modal representation. The text decoder takes R_m and passes it through a FCL followed by the stacked Bi-LSTMs to obtain the original text. Similarly, the image decoder passes R_m through two FCLs to reconstruct the image.

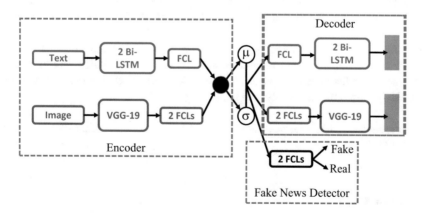

Fig. 8 Schematic diagram showing the flow in the MVAE model. *Filled circle* indicates concatenation operation

– *Fake News Detector*: The shared representation R_m is passed through two FCLs that minimize the cross-entropy loss for a binary classification.

The proposed VAE model and the fake news detector are trained jointly, and the combined loss is minimized in an end-to-end setting.

MVAE was evaluated on two datasets, Weibo-att and MediaEval, and compared with six baselines, including different variants of the original model, att-RNN and EANN. EANN outperforms all the baselines with 0.730 (MediaEval) and 0.837 (Weibo-att) F1-scores.

11   Summary of the Chapter

This chapter presented the current research on multi-modal fake news detection. We introduced various challenges that the existing methods deal with, which further open up opportunities for further research. We also summarized major datasets that are being used for multi-modal fake news detection. While summarizing the methods, we observed that

– Most of the methods adopted multi-modal fusion techniques, and feature-level fusion was incorporated at different positions of the architecture.
– MVNN as an image feature extractor turned out to be highly efficient, improving the performance of most of the methods significantly (Table 3).
– MAVE, although presents a completely different model paradigm, does not seem to be as effective as other fusion-based models.
– BERT-based embedding for text representation shows significant improvement in SpotFake.

We observed that there is still a scarcity of research on multi-modal approaches for large texts such as full-length news articles, blogs, etc. We also noticed that most of the methods have not been shown to be generalized across datasets of diverse domains. Model explainability is the other property that has not been addressed in any of the studies. Other modalities such as videos and audios should also be considered for fake news detection as these modalities are even more powerful and can easily communicate the story to the society.

Acknowledgment The author would like to acknowledge the support of Sarah Masud in writing the chapter.

References

1. Agrawal, T., Gupta, R., Narayanan, S.: Multimodal detection of fake social media use through a fusion of classification and pairwise ranking systems. In: 2017 25th European Signal Processing Conference (EUSIPCO), pp. 1045–1049. IEEE, New York (2017)

2. Angiani, G., Balba, G.J., Fornacciari, P., Lombardo, G., Mordonini, M., Tomaiuolo, M.: Image-based hoax detection. In: Proceedings of the 4th EAI International Conference on Smart Objects and Technologies for Social Good, pp. 159–164 (2018)
3. Arun Anoop, M.: Image forgery and its detection: a survey. In: 2015 International Conference on Innovations in Information, Embedded and Communication Systems (ICIIECS), pp. 1–9 (2015)
4. Boididou, C., Papadopoulos, S., Kompatsiaris, Y., Schifferes, S., Newman, N.: Challenges of computational verification in social multimedia. In: Proceedings of the 23rd International Conference on World Wide Web, pp. 743–748 (2014)
5. Boididou, C., Andreadou, K., Papadopoulos, S., Dang-Nguyen, D.T., Boato, G., Riegler, M., Kompatsiaris, Y., et al.: Verifying multimedia use at mediaeval 2015. MediaEval 3(3), 7 (2015a)
6. Boididou, C., Papadopoulos, S., Dang-Nguyen, D.T., Boato, G., Kompatsiaris, Y.: The certh-unitn participation@ verifying multimedia use 2015. In: MediaEval (2015b)
7. Boididou, C., Papadopoulos, S., Kompatsiaris, Y., Schifferes, S., Newman, N.: Challenges of computational verification in social multimedia (2015). http://www.multimediaeval.org/mediaeval2015/verifyingmultimediause/index.html
8. Bojanowski, P., Grave, E., Joulin, A., Mikolov, T.: Enriching word vectors with subword information. Trans. Assoc. Comput. Linguist. 5, 135–146 (2017)
9. Brank, J., Leban, G., Grobelnik, M.: Semantic annotation of documents based on wikipedia concepts. Informatica 42(1), 23–32 (2018)
10. Cao, J., Qi, P., Sheng, Q., Yang, T., Guo, J., Li, J.: Exploring the role of visual content in fake news detection (2020). Preprint. arXiv:200305096
11. Castillo, C., Mendoza, M., Poblete, B.: Information credibility on twitter. In: Proceedings of the 20th International Conference on World Wide Web, pp. 675–684 (2011)
12. Cui, L., Wang, S., Lee, D.: Same: sentiment-aware multi-modal embedding for detecting fake news. In: Proceedings of the 2019 IEEE/ACM International Conference on Advances in Social Networks Analysis and Mining, pp. 41–48 (2019)
13. Devlin, J., Chang, M.W., Lee, K., Toutanova, K.: BERT: pre-training of deep bidirectional transformers for language understanding. In: Proceedings of the 2019 Conference of the North American Chapter of the Association for Computational Linguistics: Human Language Technologies, vol. 1 (Long and Short Papers), pp. 4171–4186. Association for Computational Linguistics, Minneapolis, MN (2019). https://doi.org/10.18653/v1/N19-1423. https://www.aclweb.org/anthology/N19-1423
14. Glenski, M., Ayton, E., Mendoza, J., Volkova, S.: Multilingual multimodal digital deception detection and disinformation spread across social platforms (2019). Preprint. arXiv:190905838
15. Go, A., Bhayani, R., Huang, L.: Twitter sentiment classification using distant supervision. CS224N project report, Stanford 1(12) (2009)
16. Goebel, M., Flenner, A., Nataraj, L., Manjunath, B.: Deep learning methods for event verification and image repurposing detection. Electron. Imag. 2019(5), 530–531 (2019)
17. Goljan, M., Fridrich, J., Chen, M.: Defending against fingerprint-copy attack in sensor-based camera identification. IEEE Trans. Inf. Foren. Secur. 6(1), 227–236 (2010)
18. Gupta, A., Lamba, H., Kumaraguru, P., Joshi, A.: Faking sandy: characterizing and identifying fake images on twitter during hurricane sandy. In: Proceedings of the 22nd International Conference on World Wide Web, pp. 729–736 (2013)
19. Honnibal, M., Montani, I.: spacy 2: Natural language understanding with bloom embeddings. In: Convolutional Neural Networks and Incremental Parsing (2017)
20. Jaiswal, A., Sabir, E., AbdAlmageed, W., Natarajan, P.: Multimedia semantic integrity assessment using joint embedding of images and text. In: Proceedings of the 25th ACM International Conference on Multimedia, pp. 1465–1471 (2017)
21. Jaiswal, A., Wu, Y., AbdAlmageed, W., Masi, I., Natarajan, P.: AIRD: adversarial learning framework for image repurposing detection. In: Proceedings of the IEEE Conference on Computer Vision and Pattern Recognition, pp. 11330–11339 (2019)

22. Jin, Z., Cao, J., Zhang, Y., Zhang, Y.: MCG-ICT at mediaeval 2015: verifying multimedia use with a two-level classification model. In: MediaEval (2015)
23. Jin, Z., Cao, J., Zhang, Y., Zhou, J., Tian, Q.: Novel visual and statistical image features for microblogs news verification. IEEE Trans. Multimedia **19**(3), 598–608 (2016)
24. Jin, Z., Cao, J., Guo, H., Zhang, Y., Luo, J.: Multimodal fusion with recurrent neural networks for rumor detection on microblogs. In: Proceedings of the 25th ACM international conference on Multimedia, pp. 795–816 (2017)
25. Jin, Z., Cao, J., Guo, H., Zhang, Y., Luo, J.: Weibo dataset: Multimodal fusion with recurrent neural networks for rumor detection on microblogs (2017). http://mcg.ict.ac.cn/mm2017dataset.html
26. Jindal, S., Sood, R., Singh, R., Vatsa, M., Chakraborty, T.: Newsbag: a benchmark multimodal dataset for fake news detection. In: Proceedings of the Workshop on Artificial Intelligence Safety, co-located with 34th AAAI Conference on Artificial Intelligence, SafeAI@AAAI 2020, New York City, NY, February 7, 2020, CEUR Workshop Proceedings, AAAI, vol. 2560, pp. 138–145 (2020)
27. Kang, S., Hwang, J., Yu, H.: Multi-modal component embedding for fake news detection. In: 2020 14th International Conference on Ubiquitous Information Management and Communication (IMCOM), pp. 1–6. IEEE, New York (2020)
28. Khattar, D., Goud, J.S., Gupta, M., Varma, V.: MVAE: multimodal variational autoencoder for fake news detection. In: The World Wide Web Conference, pp. 2915–2921 (2019)
29. Kim, Y.: Convolutional neural networks for sentence classification (2014). Preprint. arXiv:14085882
30. Kwon, S., Cha, M., Jung, K., Chen, W., Wang, Y.: Prominent features of rumor propagation in online social media. In: 2013 IEEE 13th International Conference on Data Mining, pp. 1103–1108. IEEE, New York (2013)
31. Le, Q., Mikolov, T.: Distributed representations of sentences and documents. In: International Conference on Machine Learning, pp. 1188–1196 (2014)
32. Li, W., Yuan, Y., Yu, N.: Passive detection of doctored jpeg image via block artifact grid extraction. Signal Process. **89**(9), 1821–1829 (2009)
33. Li, Q., Zhang, Q., Si, L., Liu, Y.: Rumor detection on social media: datasets, methods and opportunities (2019). Preprint. arXiv:191107199
34. Liu, K., Li, Y., Xu, N., Natarajan, P.: Learn to combine modalities in multimodal deep learning (2018). Preprint. arXiv:180511730
35. Ma, J., Gao, W., Mitra, P., Kwon, S., Jansen, B.J., Wong, K.F., Cha, M.: Detecting rumors from microblogs with recurrent neural networks. In: Proceedings of the Twenty-Fifth International Joint Conference on Artificial Intelligence (IJCAI-16) (2016)
36. Ma, J., Gao, W., Mitra, P., Kwon, S., Jansen, B.J., Wong, K.-F., Cha, M.: Twitter dataset (2016). http://alt.qcri.org/wgao/data/rumdect.zip
37. Middleton, S.: Extracting attributed verification and debunking reports from social media: mediaeval-2015 trust and credibility analysis of image and video (2015)
38. Mikolov, T., Chen, K., Corrado, G., Dean, J.: Efficient estimation of word representations in vector space (2013). Preprint. arXiv:13013781
39. Miller, G.A.: Wordnet: a lexical database for English. Commun. ACM **38**(11), 39–41 (1995)
40. Müller-Budack, E., Theiner, J., Diering, S., Idahl, M., Ewerth, R.: Multimodal analytics for real-world news using measures of cross-modal entity consistency (2020a). Preprint. arXiv:200310421
41. Müller-Budack, E., Theiner, J., Diering, S., Idahl, M., Ewerth, R.: News400 dataset (2020b). https://github.com/TIBHannover/cross-modal_entity_consistency
42. Müller-Budack, E., Theiner, J., Diering, S., Idahl, M., Ewerth, R.: Tamperednews dataset (2020c). https://github.com/TIBHannover/cross-modal_entity_consistency
43. Nakamura, K., Levy, S., Wang, W.Y.: r/Fakeddit: a new multimodal benchmark dataset for fine-grained fake news detection (2019). Preprint. arXiv:191103854
44. Nakamura, K., Levy, S., Wang, W.Y.: Fakeddit dataset (2020). https://github.com/entitize/fakeddit

45. Oliva, A., Torralba, A.: Modeling the shape of the scene: a holistic representation of the spatial envelope. Int. J. Comput. Vis. **42**(3), 145–175 (2001)
46. ONION T: Trump admits 18 new states to increase competition for medical supplies (2020a). https://www.thepoke.co.uk/2020/04/08/queens-dress-perfect-green-screen-hilariously-exploited/
47. ONION T: Trump delays Easter to July 15 to keep promise on coronavirus (2020b). https://politics.theonion.com/trump-delays-easter-to-july-15-to-keep-promise-on-coron-1842566559
48. Pennington, J., Socher, R., Manning, C.D.: Glove: global vectors for word representation. In: Proceedings of the 2014 Conference on Empirical Methods in Natural Language Processing (EMNLP), pp. 1532–1543 (2014)
49. Qi, P., Cao, J., Yang, T., Guo, J., Li, J.: Exploiting multi-domain visual information for fake news detection (2019). Preprint. arXiv:190804472
50. Ramisa, A., Yan, F., Moreno-Noguer, F., Mikolajczyk, K.: Breakingnews: article annotation by image and text processing. IEEE Trans. Patt. Anal. Mach. Intell. **40**(5), 1072–1085 (2017)
51. Risdal, M.: Ti-news dataset (2016). https://www.kaggle.com/mrisdal/fake-news
52. Sabir, E., AbdAlmageed, W., Wu, Y., Natarajan, P.: Deep multimodal image-repurposing detection. In: Proceedings of the 26th ACM international conference on Multimedia, pp. 1337–1345 (2018)
53. Schroff, F., Kalenichenko, D., Philbin, J.: Facenet: a unified embedding for face recognition and clustering. In: Proceedings of the IEEE Conference on Computer Vision and Pattern Recognition, pp. 815–823 (2015)
54. Shu, K., Mahudeswaran, D., Wang, S., Lee, D., Liu, H.: Fakenewsnet dataset (2018a). https://github.com/KaiDMML/FakeNewsNet
55. Shu, K., Mahudeswaran, D., Wang, S., Lee, D., Liu, H.: Fakenewsnet: a data repository with news content, social context and dynamic information for studying fake news on social media (2018b). Preprint. arXiv:180901286
56. Shu, K., Mahudeswaran, D., Wang, S., Lee, D., Liu, H.: Politifact dataset (2018c). https://github.com/KaiDMML/FakeNewsNet
57. Simonyan, K., Zisserman, A.: Very deep convolutional networks for large-scale image recognition (2014). Preprint. arXiv:14091556
58. Singer-Vine, J.: Buzzfeednews dataset (2016). https://github.com/BuzzFeedNews/2016-10-facebook-fact-check
59. Singhal, S., Shah, R.R., Chakraborty, T., Kumaraguru, P., Satoh, S.: Spotfake: a multi-modal framework for fake news detection. In: 2019 IEEE Fifth International Conference on Multimedia Big Data (BigMM). IEEE, New York, pp. 39–47 (2019)
60. Singhal, S., Kabra, A., Sharma, M., Shah, R.R., Chakraborty, T., Kumaraguru, P.: Spotfake+: a multimodal framework for fake news detection via transfer learning. In: AAAI (Student Abstract), pp. 1–2 (2020)
61. The POKE: The queen's dress made the perfect green screen and people hilariously exploited it (2020). https://politics.theonion.com/trump-admits-18-new-states-to-increase-competition-for-1842708962
62. Vinyals, O., Toshev, A., Bengio, S., Erhan, D.: Show and tell: lessons learned from the 2015 MSCOCO image captioning challenge. IEEE Trans. Patt. Anal. Mach. Intell. **39**(4), 652–663 (2016)
63. Volkova, S., Ayton, E., Arendt, D.L., Huang, Z., Hutchinson, B.: Explaining multimodal deceptive news prediction models. In: Proceedings of the International AAAI Conference on Web and Social Media, vol. 13, pp. 659–662 (2019)
64. Wang, Y., Ma, F., Jin, Z., Yuan, Y., Xun, G., Jha, K., Su, L., Gao, J.: EANN: event adversarial neural networks for multi-modal fake news detection. In: Proceedings of the 24th ACM SIGKDD International Conference on Knowledge Discovery & Data Mining, pp. 849–857 (2018)

65. Wu, W., Li, H., Wang, H., Zhu, K.Q.: Probase: a probabilistic taxonomy for text understanding. In: Proceedings of the 2012 ACM SIGMOD International Conference on Management of Data, pp. 481–492 (2012)
66. Yang, Y., Zheng, L., Zhang, J., Cui, Q., Li, Z., Yu, P.S.: Ti-CNN: Convolutional neural networks for fake news detection (2018). Preprint. arXiv:180600749
67. Yang, Z., Dai, Z., Yang, Y., Carbonell, J., Salakhutdinov, R., Le, Q.V.: Xlnet: generalized autoregressive pretraining for language understanding (2019). http://arxiv.org/abs/1906.08237. Cite arxiv:1906.08237. Comment: Pretrained models and code are available at https://github.com/zihangdai/xlnet
68. Zhang, H., Fang, Q., Qian, S., Xu, C.: Multi-modal knowledge-aware event memory network for social media rumor detection. In: Proceedings of the 27th ACM International Conference on Multimedia, pp. 1942–1951 (2019)
69. Zhang, K., Zhang, Z., Li, Z., Qiao, Y.: Joint face detection and alignment using multitask cascaded convolutional networks. IEEE Signal Process. Lett. **23**(10), 1499–1503 (2016)
70. Zhou, X., Wu, J., Zafarani, R.: Safe: similarity-aware multi-modal fake news detection (2020). Preprint. arXiv:200304981
71. Zubiaga, A., Liakata, M., Procter, R.: Exploiting context for rumour detection in social media. In: Ciampaglia, G.L., Mashhadi, A., Yasseri, T. (eds.) Social Informatics. Springer International Publishing, Cham, pp. 109–123 (2017a)
72. Zubiaga, A., Liakata, M., Procter, R.: Pheme dataset (2017b). https://github.com/azubiaga/pheme-twitterconversation-collection

Deep Learning for Fake News Detection

Santhosh Kumar G

Abstract The widespread usage of fake news through social media outlets causes unpleasant societal outcomes. The research efforts to automatically detect and mitigate its use are essential because of their potential to influence the information ecosystem. A vast amount of work using deep learning techniques paved a way to understand the anatomy of fake news and its spread through social media. This chapter attempts to take stock of such efforts and look beyond the possibilities in this regard. The focus is given mainly to deep learning models and its use in fake news detection. A comprehensive survey of the current literature and datasets used, along with evaluation metrics, are highlighted. Finally, promising research directions toward fake news detection are mentioned.

Keywords Fake news · Deep learning · NLP · Social media

1 Introduction

Fake news is a type of *news that has no basis in fact but is presented as being factual* [1]. It may have deceptive, fraudulent, imposturous, manipulated, fabricated, or satirical content with the false intention to mislead people. Though there exist subtle differences among misinformation, deception, rumor, click-bait, hoax, and spam, this chapter considers all of them under one umbrella as *fake news*. The fake news problem is affecting many facets of human life and has become one of the most critical research field in artificial intelligence. In today's growing world of communication using social media, it has been quite easy to propagate fake news as well. The spread of fake news makes an adverse impression on human life. Hence it is essential to detect fake news to prevent its spread and thus eliminate the adversarial effects. Automated detection of fake news is a hard task to realize as it requires a detection model to comprehend gradations in natural language. Research on fake news detection from social media focuses mainly on approaches based on propagation, source analysis, and content analysis. Propagation-based research

© Springer Nature Switzerland AG 2021
Deepak P et al., *Data Science for Fake News*, The Information
Retrieval Series 42, https://doi.org/10.1007/978-3-030-62696-9_4

suggests behavioral changes in news items as it disseminates. The dissemination pattern indicated by a propagation map helps to signal a fake news item from a reliable one. Source analysis methods rely on information extracted from social media followed by stance detection and use source reliability check to mark the item as fake news. The content analysis mainly focuses on the linguistic and visual features present in the news item. It is possible to capture sensational headings or provoke images through content analysis to catch the article's fakeness.

Deep learning is effecting significant breakthroughs in certain challenging problems in natural language processing (NLP), visual understanding, speech processing, and many other well-known domains. It is evolving as a new branch of machine learning with cross-pollinating ideas from cognitive science, biology, statistics, computer science, and physics. Deep learning makes use of graphs coupled with appropriate transformations among neurons to develop multilayered learning models. The availability of computing power, big data, and storage have paved a way to design new learning models to make significant breakthroughs. While social networks have become a productive environment in disseminating news items, it also challenges the information ecosystem's trustworthiness. The volume, velocity, and veracity of news items have attracted automatic fake news detection techniques. This chapter primarily aims to inspect the deep learning methods, various architectures, and recent advancements in fake news detection and its mitigation.

1.1 Fake News Types

There are different formats of news types in today's world like text, multimedia, and embedded contents. The types of fake news can be broadly categorized as follows [2]:

- Visual Centered—In this context, the fake news is made-up images or video content, or even graphical representations of data. This is the most misused method of fake news due to the high usage of platforms like WhatsApp.
- Post Centered—The type of fake news which comes from social media like Facebook and Twitter can be considered as post centered. In this kind of post-centered fake news, all formats of news like text, multimedia, and hyperlinks have to be considered.
- Person/Group Centered—This is the category of fake news which is specifically aimed at a person or a group of persons (having similar interest or working in a network/organization or connected individuals in Facebook or LinkedIn).

The following section discusses early works on the fake news detection problem.

1.2   Early Works

Fake news detection is a classification problem that includes steps like preprocessing of the text, feature extraction, and feature selection, accompanied by model building, optimization, and evaluation. Many popular classifiers are successfully employed to develop a model for fake news detection. Rada et al. [3] applied NLP techniques for automatic classification of truth and falsehood expressed in written language. The linguistic cues like syntax, punctuation, and readability become features to build models to detect fakeness in the text. Classifiers like Naïve Bayes, Decision Trees, Support Vector Machines, and Random Forest work with various accuracies based on the datasets and the domain. Detecting suspicious memes of political abuse like astroturf in online social media using supervised learning and new ways of spreading rumors to a broader audience forms an early successful attempt. The features extracted from the topology of diffusion networks are used to find the spread of rumors. Sentiment analysis and crowdsourced annotations become the basis for classification [4, 5]. Twitter posts related to trending topics were analyzed and then classified to bring out the credibility of the information [1, 6]. The user's posting/re-posting attitude, content, and references are fed into a classifier to decide upon whether any newsworthy event is indicated by the tweets. Along with content and account-based features, the application used, and the location of the event, features are extracted from microblogs and used for rumor detection [7]. Identification of bursty temporal patterns of rumors on a social media stream has been studied [8]. A similar work based on modeling the changes in social context features as a time series for the detection of rumors [9].

Though there are works related to automated [1] and semiautomated systems [10, 11] for rumor detection, the real-time tracking of news shares from various sources and the determination of their accuracy are challenging. *wisdom of the crowds* from Twitter provides many cues to design an algorithm to debunk real-time rumors. The effectiveness of the method on Twitter streaming data is demonstrated [12]. Hoaxy [13] presents another system capable of collecting tweets containing links to specific websites from Twitter streaming API and determines the accuracy of the tweets by tracking the source.

By 2016, the stance detection technique became prominent for fake news detection. It seeks automatic detection of the relationship between two pieces of text [14] by classifying the stance of the news item toward a target into a set of labels. The labels generally indicate the agreement, disagreement, or neutrality of the facts given in the news item's body text with the headlines. The target could be an entity, opinion, an idea, or a claim, and it may or may not have an explicit mention in the news item. Stance detection gives importance to consistency rather than the veracity of a news item; hence, it forms a subtask of fake news detection. A survey on stance detection [15] gives a good insight into the state of the art of stance detection.

The field of fake news detection has attracted many researchers, and various methods are suggested. For a comprehensive review of the vast topics in this area, readers may refer to survey papers on fake news detection [16–22]. Kai Shu et al. [23] present an excellent volume of work on research opportunities and challenges in this field.

2 Deep Learning Methods

This section introduces popular architectures of deep learning, followed by the works that use the particular architecture for fake news detection.

Multilayer Perceptrons Multilayer perceptron (MLP) [24] is the simplest structure of a feed-forward artificial neural network consisting of a minimum of three layers: an input, a hidden, and an output layer. The input layer takes the data and passes to the next set of hidden layers to learn the representation in the data with the help of activation functions. It finally relates best to the predicted output variable. Each perceptron adjusts the weights based on the error between the expected output and the results produced when the input is processed. Hence, after several passes, the network learns a pattern.

Convolutional Neural Networks ConvNet/CNN is a multiplayer perceptron to recognize higher-dimensional shapes, usually visual imagery. CNN is a blend of several hidden convolutional layers, followed by a fully connected layer. Each of the convolutional layers has an activation function mostly, Sigmoid or RELU, consequently followed by pooling layers, fully connected layers, and normalization layers. These are known as hidden layers since the activation function and the final convolution mask their inputs and outputs. The main tasks of convolution layers are feature extraction, feature mapping, and subsampling. A summary of the main highlights of the CNN architecture for fake news detection is as follows:

- The extracted features like neural embeddings, bag of words, TF-IDF, n-grams, and external features (number of characters, number of 1 gram models) can be used as input to CNN.
- Feature engineering can be avoided by providing the sentence or document in the form of a matrix as input to the CNN architecture. The rows of the matrix are vectors (word embeddings like word2vec or GloVe) representing tokens.
- CNN is originally introduced to process images. The semantic relations in the adjacent pixels in an image can be captured, but the same rule is not true in NLP tasks. Hence, an architecture with just CNN layers may not help in detecting fake news.
- Adding a CNN layer after RNN can reduce computation time and resources required to train the model, enhancing the model's overall accuracy.
- Recently gated CNNs [25] based on stacked CNN concept and convolutional sequence to sequence learning [26] proved effective in NLP modeling.

Recurrent Neural Networks (RNNs) In this form of a neural network, a directed graph on a temporal sequence is formed between the nodes, inheriting temporal dynamic behavior. The hidden state captures the connections among the input sequence by leveraging the relationship between neighbors and the current node. RNNs have internal memory, which helps them to process the series of input. This ability to remember the state of the previous unit can help in natural language processing because it helps in understanding the language better.

Variants of RNNs Since the introduction of RNNs, various flavors of networks with multiple capabilities have appeared in the literature. Some of the essential structures are introduced below.

Bidirectional RNNs These RNNs are capable of processing the current state and its previous state. It requires looking at both sides of a sequence to find a missing word in a given context. Stacking of RNNs can achieve this, so bidirectional RNNs are mostly two RNNs stacked on top of each other in terms of time sequence. One network receives a series of input and the other receives in the reverse time order to capture the contextual relationship. The concatenated output at each time step allows the network to obtain the backward and forward information.

GRU/LSTM While RNNs are good at capturing recent information from a sequence of input words, when it comes to obtaining a piece of information that is far from the current context, a kind of long-term dependency, it works poorly. LSTM inherently tries to address the long-term dependency problem by carefully designing a handful of neural network structures that interact and connect with a chain structure. The cell state of LSTMs runs along the entire chain allowing minor direct interactions information to flow along with it unchanged. In addition to this, the LSTMs can include or exclude information to the cell state through Gated Recurrent Units (GRUs). Gates are engineered with reset variables to control the extent of the previous state, which is to be remembered. An update variable is also used to control the amount of new state retained from the old state. While the update gates capture long-term dependency in the sequence, the reset gates capture the short-term dependencies.

Bidirectional LSTM (Bi-LSTM) An LSTM is capable of retaining past information from a sequence. The natural extension of this for future information is Bi-LSTM. While unidirectional LSTM preserves the past, a Bi-LSTM with the help of two hidden states combined can see the past and future context of a word in a text sequence.

Attention Mechanism In the traditional encoder–decoder architecture, the decoder has no mechanism to pay attention selectively to the input tokens. The introduction of attention weights addresses this, and these weights can be learned from an additional feed-forward neural network within the architecture. The attention mechanism [27] improves encoder–decoder architecture. In this LSTM/RNN, units form the building blocks. The encoder LSTM placed at the last layer of the LSTM/RNN structure takes an input series and encodes as a fixed-length context vector to summarize the input. The decoder produces the words in a chain one after another. The attention mechanism combined with LSTM and CNN found useful in NLP classification problems. While the basic premise of attention mechanism is the semantics and internal relationship of words, the internal spatial relationship in words is captured by the self multi-head attention mechanism [28]. The main highlights of RNN and its variants for fake news detection are summarized in the following:

- While MLP examines the presence of words in a piece of text, RNNs can determine the authenticity of news items by checking the order of words.
- More linguistic features of the news items can be incorporated in the structure.
- A combination of LSTM with CNN improves the performance of the detection model.
- Attention mechanisms are useful to capture semantic level information.

Generative Adversarial Networks Generative modeling is an unsupervised method to learn the input's regularities and patterns. A Generative Adversarial Network (GAN) consists of a pair of interconnected networks that mutually compete for operations. The generative system generates candidates, while the discriminative network evaluates them. The generator's objective is to increase the discriminator's error rate, thereby being able to fool it. Training a GAN involves presenting the samples from a dataset until it attains an acceptable accuracy.

A variant of a GAN for natural language processing is Sequence GAN. This extends GANs with a mechanism to address the sequence generation problem. The discriminator signals a reward during each episode through the Monte Carlo method, enabling the generator to pick the action and subsequently learn the policy using comprehensive reward estimates.

The following summarizes the main highlights of GAN for fake news:

- GANs can generate content by learning the distribution of the data.
- The generated data is useful to construct a strong system to discriminate fake from real.

2.1   Fake News Detection Using CNN

The first multilayer perceptron machines proposed by LeCun in 1989 [29] became popular when computer vision researchers had a handle on higher computational power, the availability of data, and effective optimization techniques. Convolutional Neural Networks (CNNs) try a direct mapping of inputs to target labels to learn a hierarchical set of features, eliminating the need for handcrafted suboptimal features. In this hierarchical network, fully connected hidden layers are replaced with convolutional layers, enabling a locally connected layered network to capture the primitive cues and combine them to form higher-order features. The learning and generalization capability of these structures paved a way to design many unique

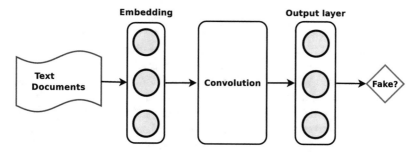

Fig. 1 Pipeline of CNN in a text classification task

deep learning architectures. Figure 1 shows a typical pipeline of a classification problem using CNN.

Text Content A general multilayer architecture for natural language processing (NLP) tasks suggested by Collobert in 2011 [30], along with dense vector representation of features, extended the use of deep learning to text processing research. TextCNN [31] introduced by Kim et al. has been used for sentence classification based on the CNN architecture. Though this work is not primarily on the detection of fake news, it has been adopted by many researchers for fake news detection.

Text Embedding There are many techniques developed for embedding natural language to feed those in deep learning input layers. Word2Vec [32] considers each word in the text as a context, and similar words appear nearer in the embedded lower-dimensional space. Similarly, Doc2Vec [33] tags the text, and tag vectors are produced so that the authors who use similar words will have corresponding vectors closer. fastText [34] considers an improved approach over the earlier method, where parts of words and characters are taken into account while building embeddings. In the GloVe [35] approach, embeddings are built assuming a combination of word vectors relates to their co-occurrence probability in the corpus. Unsupervised pre-training of word embeddings has become a valuable ingredient in deep learning for natural language processing.

The CNN architecture is incapable of obtaining connections among words in a long sentence. A self multi-head attention mechanism combined along with a CNN was found useful in discriminating fake news from real. The authors use the data gathered from *fakenews.mit.edu* and compared the experimental results with many baselines. A visualization of words that the classifier uses for the discrimination of news items is also attempted.

A semi-supervised approach using two-path CNNs is presented [37] to address the lack of labelled data. A two-path model containing three CNNs is jointly trained with labelled and unlabelled data. The model tested on the PHEME dataset for selected events yielded considerable results.

Multilevel CNN [38] incorporates local convolutional and global semantic features from the news articles to classify them. The proposed model consists of several convolutional layers with various kernel sizes. Hierarchical CNNs are useful in capturing sensitive words in the text, and the authors propose a method to calculate weights of these words to obtain better classification accuracy.

Kaliyar et al. [39] present FNDNet, a deep CNN-based architecture for fake news detection. Pretrained word embeddings generated from Glove are used as input to the convolution layers to learn the features. The output generated by the flatten layer is then given to two dense layers; subsequently, the last layer produces the output. Experiments carried out with the Kaggle fake news dataset report 98.36% accuracy.

The EMET model [40] presents a model to identify fake posts in social media by leveraging clues from online comments of the users. Tweets, along with comments and news items, are individually fed to multilingual encoder transformer followed by a series of convolution layers to max-pooling and fully connected layers to generate output vectors. Experiments yielded better results compared with the baselines [41–43].

Visual Content Complex and multiple ideas can be conveyed through visual content, as expressed via the English adage, "A picture is worth a thousand words." Despite the importance of visual content, its understanding is limited, especially in the domain of fake news. Many works in fake news detection utilizing text and images are discussed in a later section under multimode fake news detection. Qi et al. [41] propose an architecture called Multi-domain Visual Neural Network (MVNN) for fake news detection from visual content. This is achieved through leveraging frequency and pixel-level information to discriminate against the posts. While a CNN network trained with fake images captures patterns in the frequency domain, a CNN-RNN model is useful to capture visual clues from the pixel domain. A comprehensive survey emphasizing the role of visual content in fake news detection is presented [44]. Fake images seem to exhibit heavier recompression artifacts (block effects) in the frequency domain. Also, they often present periodicity due to manipulations. CNN is generally capable of capturing these artifacts. In the pixel domain, fake images exhibit visual impacts and emotional provocations. These semantic level artifacts could be captured through a CNN-RNN model.

2.2 Fake News Detection Using RNN and Its Variants

Memorizing the original sequential orders and grabbing the hidden semantic relationships are essential while analyzing a language. CNNs generally take fixed-size inputs and generate fixed-size outputs. Though not as powerful as CNNs, recurrent neural networks (RNNs) capture temporal information in sequential data such as text and speech. In a typical RNN, the units are connected in the form of a direct cycle forming an internal state of the network to allow them to acquire dynamic temporal signals.

Although RNNs are capable of handling arbitrary input/output lengths, they cannot learn long-distance temporal dependencies since their state is overwritten

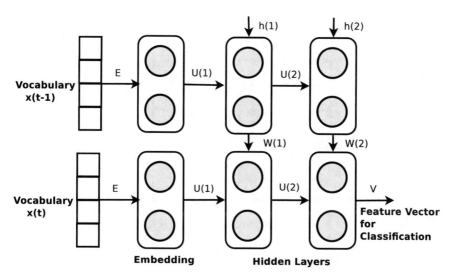

Fig. 2 Pipeline of RNN in a rumor detection task. Adapted from [48]

in each time step. RNNs are extended with internal memory structures for persistence, the two such units being Long Short-Term Memory (LSTM) [45] and Gated Recurrent Unit (GRU) [46]. The main task is to build models to capture temporal relations such as event-time, event-document, and event-event. The input to the system should generally be word embeddings, or cues to the classifier in the vectorized representation like token sequences, POS sequences, and both the token/POS sequences [47].

Ma et al. [48] present RNN as a base model to discriminate microblog events into rumors and non-rumors. The basic idea is to build a representation to capture variations of contextual information on relevant posts. Figure 2 shows the basic working model of rumor detection using RNN. Here $x(t-1), x(t)..., x(t+n)$ represent the input sequence for each time interval. The model updates the hidden states $h(1),\ h(2)...,$ to generate the output vector. $U, W,$ and V represent weight matrices of input-to-hidden, hidden-to-hidden, and hidden-to-output layers, respectively.

Considering the microblogs during a time interval as a single unit and a time series is constructed so as the recurrent units of RNN fit into the time intervals. In each interval, the top K terms of the Term Frequency-Inverse Document Frequency (TF-IDF) values of the vocabulary form the input. Experiments on different deep learning architectures with gated units, viz., tanh-RNN, 1-layer LSTM/GRU with embedding, and 2-layer GRU with embedding using datasets constructed from microblogging sites (Twitter, Sina Weibo), have shown good classification accuracies (82.7% on Twitter and 87.3% on Weibo). The performance study on early detection of rumors revealed that the RNN-based methods climb up accuracy

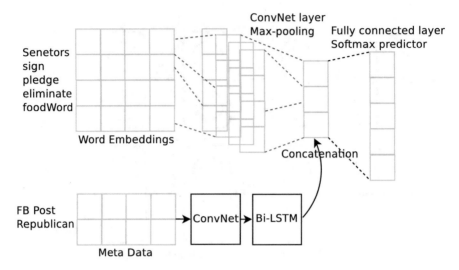

Fig. 3 A hybrid model for fake news detection. Adapted from [49]

in the early hours itself and quickly stabilize when compared with baseline methods [9].

W.Y. Wang [49] introduced LIAR dataset for fake news detection and made it publically available. The dataset has been built from news items ranging from 2007 to 2016 (https://politifact.com). Around 12.8K short statements in various contexts were curated, evaluated, and labelled to fall into six classes. A hybrid learning architecture, as shown in Fig. 3, is introduced to capture the dependency among the text and corresponding metadata.

CNN is used to capture the affinity among the text and metadata. The metadata embeddings are encoded with a randomly initialized matrix of embedding vectors. A max-pooling operation followed by the Bi-LSTM layer is applied to generate the vectors. The max-pooled text vectors are joined with that of metadata and given to the fully connected layer. A softmax activation function generates the final prediction. The model has been evaluated using the new LIAR dataset with many baselines. While the regularized logistic regression classifier (LR) and support vector machine classifier (SVM) showed significant accuracy, Bi-LSTM showed poor performance due to overfitting. CNN outperformed all the models. The overall claim of this work is that the combination of metadata with text is potentially significant for a compact fake news detection.

Ruchansky et al. [50] present a hybrid deep model to classify misinformation and identify groups of suspicious users by capturing three trivial aspects of fake news, viz., text, response, and source. A learning model encapsulating the above features tries to capture users' temporal engagements with articles and produce a label for each post and assign a suspicious score for corresponding users. The model shown in Fig. 4 consists of three modules, viz., Capture, Score, and Integrate. The Capture

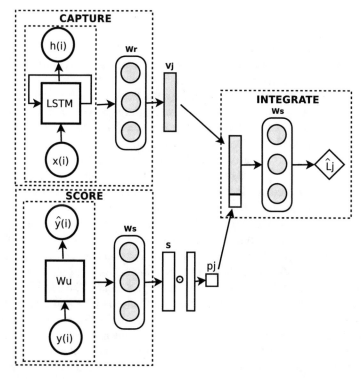

Fig. 4 CSI model [50]

module makes use of RNN to acquire temporal patterns of user activity on a given post.

The Capture module consists of LSTM for a single post, and the Score module acts over all users. An embedding layer converts the raw input features ready to be fed to LSTM given by $\tilde{x}_t = tanh(W_a x_t + b_a)$. W_a's are weights, and b_a represents the bias term. The low dimensional representation of temporal pattern of engagements v_j is computed as $v_j = tanh(W_r h_T + b_r)$. This vector is further used for article classification in the Integrate module. The scoring module tries to capture user behavior by computing each user's vector representation from a set of user features. The vector is calculated as $\tilde{y}_i = tanh(W_u y_i + b_u)$. Individual user scores are given by $s_i = Sigmoid(W_s^T \tilde{y}_i + b_s)$, and the set of s_i forms the vector s of user scores. The Integrate module combines v_j with user scores s_i to produce a prediction label $\hat{L}_j$. A mask m_j is applied to s to produce p_j representing the user's suspiciousness score, engaging with article a_j. The overall score p_j is concatenated with v_j to produce a vector c_j. The label is predicted as $\hat{L}_j = Sigmoid(W_c^T c_j + b_c)$. The joint training of Capture and Score modules helps the model to learn both the user behavior and article characteristics. This helps accurate prediction of fakeness.

Authors used Twitter and Weibo datasets for the experimentation and have shown the superiority of their model with many baseline models.

Shu et al. [51] present FakeNewsTracker architecture in the same line of the above work that uses linguistic and social engagements but with additional facilities for data collection, interactive visualization, and fake news detection.

Figure 5 depicts the framework for fake news detection. The social context part captures the richness in their social engagements through tweets and corresponding replies. Doc2Vec [33] embedding scheme represents the text content. The user-news engagement matrix is decomposed to get the latent user features. The output of the RNN gives the social context. The news content is an encoder–decoder framework that uses linguistic features to find clues to distinguish fake news from real news. LSTMs are used in both the encoder and the decoder. A social article fusion model blends the feature vectors produced by the autoencoder with the social context vector produced by the RNN. The single fused feature vector is then used for the classification through a softmax layer. The data from fact-checking websites like PolitiFact and BuzzFeed is used for checking social engagements. Most importantly, periodic collection of user interactions is collected for recent posts to get a clue on second-order temporal engagements. The training of feature learning and classification tasks are run simultaneously to make sure that the learned features are concerned with the detection task. When article content is combined with the social context, an improvement in classification metrics is seen. Classification tasks are run on well-known machine learning algorithms to test the quality of the features learned. It is concluded that the features learned from the social article model are suitable for fake news article classification.

Text summarization is applied to generate feature vectors capable of finding the central claims of articles [52]. The feature vectors thus obtained are fed to an RNN to classify misinformation.

DUAL [53] presents a hybrid attention model considering latent representations of the users' profile and reviewer's feedback along with the content of the news item. An attention-based bidirectional GRU learns the latent representation of the content, and a deep neural network learns the side information. A cross product of these latent vectors is obtained to extract the cross-domain information. The model is trained with standard datasets, and the results are compared with many baseline methods.

Ajao et al. [54] introduce a hybrid deep learning framework to identify and discriminate fake messages from Twitter posts. A 1D CNN layer is included after the word embedding layer. Also, max-pooling is applied to reduce the dimensionality and to avoid overfitting. Around 5.8K tweets focusing on five rumor stories were collected. The authors claim 82% accuracy on the PHEME dataset for a plain LSTM model and 74% for the hybrid LSTM-CNN model. The decrease in accuracy was ascribed to the absence of a large dataset for training and testing. The authors also claim that the model is domain-independent.

Thanos et al. [55] propose a deep learning architecture utilizing CNN and LSTM in combination with NLP techniques to classify real Twitter posts from the junk, which notifies an emergency like a forest fire. Figure 6 shows the overall architecture

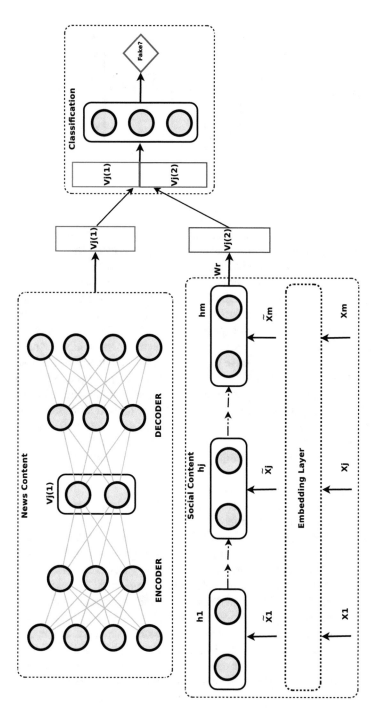

Fig. 5 FakeNewsTracker [51]

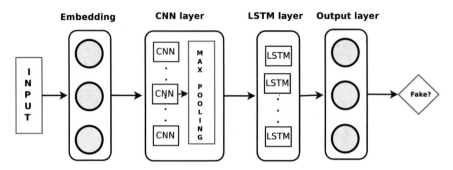

Fig. 6 CNN-LSTM model [55]

of the proposed model. The Twitter posts are given to a word embedding layer to form vectors equivalent to the sentences, subsequently given to a set of CNN layers and then to LSTM layers to capture the temporal dependencies. The classified posts are fed to an NLP module to process the place and time of the event.

Emotion content in a social media post can be useful in fake news detection. DEAN [56] framework utilizes the emotions of the publisher and the user in terms of emotional category, intensity, and expression. Bidirectional GRU forms the basis to model emotion features. Authors propose a new mechanism Gate_ N to control the emotion and semantics parts of the word to the latent representation. Experiments with Weibo and Twitter datasets have given good results compared to the baselines.

Adaptive Interaction Fusion Network (AIFN) has been proposed [57] to capture the semantics of news content by effective feature fusion. Encodings of words and emotions in the posts and comments are fused using the proposed semantic-level fusion self-attention networks. Experiments with RumourEval and PHEME datasets yielded better results than the state-of-the-art models.

FAKEDETECTOR [58] is a credibility label inference model defined to capture the unexplored aspects of fake articles and relations among the subject of the articles and their creators. A deep diffusive network model learns the credibility labels of news items, whereas Gated Diffusive Unit (GDU) models the correlations among the news articles, creators, and subjects.

A blend of CNN and LSTM architectures is proposed [59] for fake news detection. CNN is used to learn n-grams and the sequential correlations learned from LSTM. Experiments run on the Kaggle dataset yielded better results.

Huang et al. [60] propose an ensemble learning approach. The work considers embedding LSTM, depth LSTM (LSTM with two hidden layers), LIWC CNN and N-gram CNN, and extensive preprocessing, including tokenizing words, grammar, and text analysis, uni-gram and bi-gram extraction. The work emphasizes the importance of sentence depth in a text in discriminating fake text from true ones. The preprocessed vectors fed to the ensemble and appropriate algorithms optimize the classifiers' weights. Experiments with different datasets give better results in comparison with a few of the existing methods.

DeepFakE [61] is a deep learning approach considering news content and social context information for classifying fake news from true news items. The work differs from others because the data represents a tensor with the news item, user, and community information as dimensions. A third-order tensor $T_{ijk} = U_{ij} * C_{jk}$, where U_{ij} represents news-user engagement matrix and C_{jk} represents user-community interaction matrix. A coupled matrix-tensor factorization extracts the latent representation. The results of the deep neural network with four hidden layers trained with the FakeNewsNet dataset are compared with XGBoost method for different combinations of the input. For all combinations, DeepFakE outperformed in terms of classification metrics.

2.3 Multimodal Methods

Deep learning has already proven its effectiveness in feature representations from multiple aspects. Multimodal deep learning approaches for social networks learn features from the text, social contexts (in the form of texts), images, and videos that appear as content in microblogging sites.

Jin et al. propose [62] a deep learning-based fake news detection model, which constructs features from tweets, images, and social context and then fuses them by an attention mechanism called att-RNN. A tweet instance defines a tuple $I = \{T, S, V\}$ to represent textual content, social context, and visual content. Reliable representation of the tweet R_I is learned from the aggregation of T, S, and V. A joint representation R_{TS} is then formed by fusing R and S with an RNN. The visual feature R_V is learned through a deep CNN model. It is refined as R'_V at each time step through the attention of the RNNs output. Lastly, $R_T S$ and R'_V are concatenated to represent R_I.

The model utilizes curated datasets from Weibo and Twitter, which contained tweets with image attachments and social context. While an LSTM jointly learns the text and social context, a pretrained VGGNet learns the visual features. The experiments conducted on the datasets show that the att-RNN model effectively detects rumors based on multimedia content compared with traditional methods based only on text. However, the multimodal feature representations are constructed based on specific events in the dataset, making them inefficient in generalization and hence fails to identify new coming events.

Yang et al. [63] propose a unified model to analyze title, text, and image information contained in the news websites using convolutional neural networks. Apart from the specific features extracted from the text and visual content, two parallel CNNs extract latent features from text and images. Both features are projected into the same subspace to form a new set of representations, and then fused representation is used for classification. The dataset based on the US presidential election is used to build the model and testing. A comparison with many baseline methods proves the performance of the model.

An end-to-end architecture, Event Adversarial Neural Networks (EANN) for multimodal fake news detection, is presented [43]. The proposed model shown in

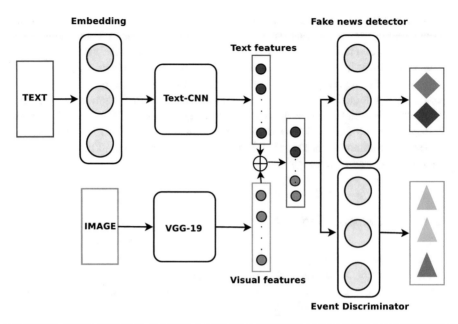

Fig. 7 EANN model [43]

Fig. 7 uses three significant components: a multimodal feature extractor, a fake news detector, and an event discriminator. The model is capable of learning multimodal and an event invariant representations. A TextCNN [31] fed with pretrained word embeddings generates features at multiple granularities with the help of different sized filters. The visual features are obtained from a pretrained VGG19. Both features are fused to form a multimodal feature representation. The fake news detector is a fully connected layer with softmax that takes the multimodal feature to classify the posts into fake or real. The model has to learn general feature representations rather than event-specific patterns from the dataset. When the uniqueness of each event is removed from the training set, an invariant feature can be obtained. This is achieved by measuring dissimilarities among the feature representations and removing them from further processing. The event discriminator consists of two fully connected layers with corresponding activation functions. This module attempts to correctly classify the posts into one of the K events based on the features learned. A min-max game defined between multimodal feature representation and the event discriminator helps to pick event-specific information with acceptable discrimination loss. Experiments with Twitter and Weibo datasets and comparison with baseline models such as VQA [64], NeuralTalk [65], and att-RNN [62] prove the effectiveness of the model. MVAE [42] presents another similar architecture inspired by EANN known as Multimodal Variational Autoencoder for Fake News Detection. The model consists of three components: an encoder, a decoder, and a fake news detection module. While a Bi-LSTM extracts text representation, image representation is built from VGG-19. The concatenated vector forms input to a decoder for the reconstruction. Experiments on Weibo and Twitter datasets report a 6% increase in accuracy and a 5% increase in F1 score compared to the baseline models.

SpotFake [66] is a multimodal framework for fake news detection. Text and image modalities are fused to build a vector for classification. The framework uses Bidirectional Encoder Representations from Transformers (BERT) [67] transformer blocks as the basis for textual feature extractor and a pretrained VGG-19 for visual features. The two features are fused using a concatenation technique to obtain the news representation vector for classification. Twitter MediaEval and Weibo datasets are used for training and testing. The model outperforms both EANN [43] and MVAE [42] by a margin of 6% accuracy on an average.

Zhou et al. [68] presented SAFE, a framework for multimodal fake news detection in the similar lines of the above work. SAFE leverages the role of the relationship (similarity) that exists between textual and visual information in news articles. The authors introduce a method to exploit multimodal and relational data to learn the representation jointly. The approach recognizes the falsity of the news items based on either text or images or the mismatch between them. Kim et al. extend TextCNN [31] with an introduction of a fully connected layer to extract the textual features from news articles. The visual information is derived with the help of a pretrained *imagetosentence* model [69], and the same architecture builds the representation. This approach provides insights into the relationship between text and images during computation. The relevance of text and visual content is measured using cosine similarity and cross-entropy-based loss function. Compared with baselines LIWC, VGG19, and att-RNN, the model gives better results over the datasets from PolitiFact and GossipCop.

3 Datasets and Evaluation Metrics

This section summarizes datasets available for practitioners and researchers for fake news detection tasks along with the evaluation metrics used. A summary of recent experiments conducted is available in Table 1.

3.1 *Datasets*

A short description of the datasets used for fake news detection is shown below.

FakeNewsNet A tool called FakeNewsTracker along with the dataset is available for collection, analysis, and visualization of fake news from social media.[1]

---

[1] https://github.com/KaiDMML/FakeNewsNet.

Table 1 Recent experiments on fake news detection

Study	Deep learning model	Dataset	Accuracy precision/recall F1 score
Agarwal et al. [59]	CNN+LSTM	Kaggle fake news	Precision 97.26%
Huang et al. [60]	Ensemble method	Satire	Accuracy 99.4%
Kaliyar et al. [39]	Deep CNN	Kaggle fake news	Accuracy 98.36%
Fang et al. [36]	CNN	MIT fake news	Recall 95.6%
Dong et al. [37]	Two-path CNN	PHEME	Macro recall 77.58%
Khattar et al. [42]	Variational autoencoder	Weibo & MediaEval	Accuracy 82.4%
Li et al. [38]	CNN	Twitter15 & Weibo	Average accuracy 91.67%
Li et al. [38]	CNN	LIAR & KaggleFN	Average accuracy 92.08%
Guo et al. [56]	GRU	Weibo	Accuracy 87.2%
Wu et al. [57]	Gated self-attention	RumorEval	F1 82.19%
Zhou et al. [68]	Multimodal CNN	PolitiFact	Accuracy 87.4%
Schwarz et al. [40]	Encoder transformer	Twitter	Accuracy 94.08%

Kaggle Fake News Dataset is provided as a csv file containing attributes id, title, author, text, and label. The origin of the data items is not specified.[2]

MIT Dataset A curated dataset of 16.4K (consisting of 9K real and 7.4K fake news) articles from *The NYT* and *The Guardian* in a 30-day interval during and after the US presidential election 2016.

PHEME A curated dataset consisting of rumors and non-rumors posted during breaking news in Twitter [70].[3]

Weibo The dataset curated from Weibo, a Chinese microblogging website. The crawled data consists of verified rumor posts from May 2012 to January 2016.

MediaEval Consists of around 9K rumor and 6K non-rumor tweets from different events. The data is a combination of text, images/video, and other social contexts.[4]

LIAR A benchmark dataset for fake news [49] consisting of a decade-long, 12.8K short statements from https://politifact.com, labelled manually into six classes.[5]

---

[2]https://www.kaggle.com/c/fake-news/data.

[3]https://www.zubiaga.org/datasets/.

[4]https://www.multimediaeval.org/datasets/.

[5]https://github.com/thiagorainmaker77/liar_dataset.

Twitter Curated microblog rumor dataset consisting of 5K claims collected from a large volume of posts [48].[6]

RumorEval RumorEval is a dataset released as part of SemEval 2017 shared task on rumor detection in social media.

PolitiFact PolitiFact represents political news collection of graded statements from fact-checkers. The data consists of around 10K statements.[7]

GossipCop GossipCop provides rated fact-checking stories collected from entertainment websites.[8]

Buzzfeed A curated dataset representing a sample of news published on Facebook from 9 news agencies over a week during the US presidential election 2016. The dataset consists of fake and real news with 91 observations and 12 features.

3.2 Evaluation Metrics

Most fake news detection approaches build a model for classification to predict whether a news article belongs to a real class or fake class.

Confusion Matrix Confusion Matrix is a square matrix that contains all classes through horizontal and vertical directions. The list of classes along the top is the predicted output, and the list on the left side is the target. The elements of the matrix at (i,j) show how many of the data points with class label C_i are misclassified into class C_j.

Accuracy Accuracy = $\frac{\#TP+\#TN}{\#TP+\#TN+\#FP+\#FN}$ is the number of correctly predicted data points out of all the data points, where TP, TN, FP, and FN represent True Positive, True Negative, False Positive, and False Negative, respectively. Accuracy gives the ratio of correct predictions to the total number of input samples.

Precision Precision = $\frac{\#TP}{\#TP+\#FP}$ represents the ratio of all identified fake news to those annotated as fake news.

Recall Recall = $\frac{\#TP}{\#TP+\#FN}$ represents the ratio of fake news to that of predicted results.

F1 Score F1 Score = $\frac{2*Precision*Recall}{Precision+Recall}$ gives overall prediction performance.

Specificity Specificity = $\frac{\#TN}{\#TN+\#FP}$ also called True Negative Rate relates to the classifier's ability to identify negative results.

---

[6]https://alt.qcri.org/~wgao/data/rumdect.zip.

[7]https://www.politifact.com/.

[8]https://www.gossipcop.com/.

ROC Curve ROC curve compares the performance of classifiers by looking at the trade-off in the False Positive Rate (FPR) and the True Positive Rate (TPR), where $TPR = \frac{\#TP}{\#TP+\#FN}$ and $FPR = \frac{\#FP}{\#FP+\#TN}$. From the ROC curve, the Area Under the Curve (AUC) shows the classifier's overall performance. $AUC = \frac{\sum(n_0+n_1+1-r_i)-n_0(n_0+1)/2}{n_0 n_1}$, where r_i is the rank of ith fake news piece and n_0 (n_1) is the number of fake (true) news pieces. The AUC score is more consistent and discriminating than accuracy and is mostly used while dealing with imbalanced classification problems. Interestingly, most datasets used for fake news detection are noted for their imbalanced distribution. Table 1 summarizes recent experiments on fake news detection.

3.3 Discussion

Fake news detection strategies generally use news content and social contexts. Linguistic features such as lexical and syntactic are generally derived from the news content. User profiles, tweets, retweets, and other interactions form the basis to learn social contexts. Most of the methods discussed in Sect. 2, including the classic machine learning techniques, use classification accuracy to claim the results. Although deep learning techniques further enhance performance with careful crafting of the network and use different types of embeddings, a few questions are worth addressing:

- Most of the models require carefully curated features as input. How to design a fully automated, end-to-end deep learning model?
- Most of the models claim the accuracy against a limited number of datasets, and the results are domain-dependent and biased. Are there models that are domain invariant? Also, how to identify fake news on newly emerged events?
- The networks use labelled claims from fact-checking sites for training, and there is no mechanism for credibility analysis. Is it possible to take evidence from external sources to support the claim?
- Although there are hybrid approaches to learn models from multimodal data, is it possible to extend the deep learning methods to heterogeneous data?

4   Trends in Fake News Detection Using Deep Learning

This section discusses some of the approaches to answer the above questions.

4.1 Geometric Deep Learning

Extending deep learning methods to non-Euclidean structural data (a.k.a. geometric data) empowers researchers to work on graph-structured data and process manifold data. Social networks are examples of graph-structured data in which nodes represent the user's information and edges represent the relationship between users. The manifold data usually represents geometric shapes, for instance, the landscape of a mountain. The critical properties of geometric data (1) are irregularly arranged and randomly distributed, (2) are heterogeneous, (3) and have an extraordinarily large scale. Social network data consists of the user profile, user activity over time, pattern of spreading, contacts, and content forming heterogeneous data suitable for the purpose. Gori et al. [71] first introduced the concept of neural networks on the graph. The early form of a Graph Convolutional Network (GCN) [72] has been optimized [73] to apply to practical purposes. GCN captures a graph's features using local permutation-invariant aggregation operation on the neighborhood of a node [74].

Figure 8 depicts the geometric deep learning architecture presented in [75]. Graph CNN used in this work consists of two convolutional layers with a head of graph attention and two fully connected layers for output prediction. Scaled Exponential Linear Units [76] is used as an activation function and max-pooling for dimensionality reduction. The nodes in the graph are defined from the embeddings corresponding to Twitter data with user profile details, user activity, social connections, and tweet text. The edges in the graph are defined based on the membership of the user in activities like news spreading and following. The authors report high accuracy and robustness of the geometric deep learning model presented.

Another work presents Graph Convolutional Networks-based Source Identification (GCNSI) [77] to spot multiple rumor sources in the absence of any prior knowledge of propagation. The proposed method assumes that nodes surrounded by infected nodes have the potential to become rumor sources, and nodes far from infected nodes are less likely to be infected. A stacked Graph CNN is used to grab features of multi-order neighboring nodes in the form of a multidimensional vector from the given graph. By defining an appropriate algorithm for input generation and

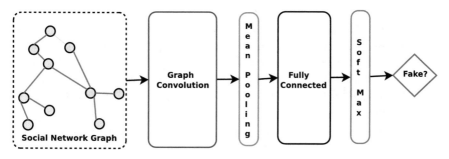

Fig. 8 Graph CNN model [75]

loss function, the network is trained until it converges. The results demonstrate that the GCNSI outperforms the baseline methods used for comparison.

A news-oriented heterogeneous graph network, News-HIN, models news articles, subjects, and creators as nodes and their relations as edges. The fake news detection problem becomes a node classification task on this graph [78]. A Hierarchical Graph Attention Network (HGAT) defined on node level and schema level attention uses aggregation mechanism on both levels to learn the latent vectors. Experiments with PolitiFact dataset yielded better results compared to the baseline methods used.

4.2   Explainable Fake News Detection

Neural network models generally criticized for their high opacity due to the weakness in interpreting the input to the output mapping process. Though the model parameters are accessible to the developer, and some observations can be drawn based on their position in the complex network, their interpretability is still limited. Du et al. [79] give an excellent review of current trends in explainable machine learning approaches. They classify the methods into two groups based on the time the model obtains the interpretability, viz., Intrinsic and Post-hoc explanation. In a deep network, the explainability is achieved at two levels, global or local, increasing the model transparency and predictability. The local interpretability tries to uncover the causal relationships between the input features and their prediction. The representations captured at the intermediate levels of a DNN provide global interpretability. Since the final latent vectors learned by a deep neural network are incomprehensible to humans, most of the explainability approaches are concerned with understanding the representations captured by the neurons at the intermediate layers.

The Language of Fake News The language of fake news is an early work [80] exploring the possibility of opening the "black box" and looking into what differences in terms of language help the classifier discriminate fake from real. They report bias in fake news as signatures of exaggeration and other forms of rhetoric to catch the reader's attention.

ClaimBuster ClaimBuster is work toward automated fact-checking focusing explainability of the claim [81]. The system uses NLP, machine learning, and query techniques on curated databases for claim spot. Though this work does not use deep learning techniques, it motivated many end-to-end fact-checking frameworks with evidence. The following are some methods developed in this direction based on deep learning.

DeClarE Debunking Claims with Interpretable Evidence [82] is end-to-end evidence-aware framework for debunking fake news and false claims. Given a natural language input claim, the model draws on evidence from external sources such as www to conclude its credibility. Also, the framework provides with annotated snippet with assertions and statistics automatically. A bidirectional LSTM represents input articles. An attention mechanism helps to focus on salient words in the article regarding the claim. An aggregation of the representations on the attention-focused article, claim source, and article source is combined to generate credibility. The model uses several fact-checking sources (Snopes, PolitiFact) and benchmarked datasets (NewsTrust, Sem-Eval-2017) for training and evaluation. The authors highlight the effectiveness of the method in terms of evidence-aware credibility calculations without using handcrafted features. The assessment of the model on baselines claims an accuracy of 67.32% on PolitiFact.

dEFEND Shu et al. [83] addressed the question of why a piece of news item is detected as fake. dEFEND framework proposed a computational framework to identify fake news from social media with proper explanation. This work presents a joint exploration of news content and user comments with the help of a co-attention mechanism. The sentences in news contents and corresponding user comments are encoded into appropriate feature vectors using RNN. The sentences are encoded with the help of bidirectional GRU to capture the contextual information in long sentences. The semantic affinity between sentences and comments, along with the weights, is learned. A fake news detection function is then defined and learned to maximize the prediction accuracy with explainable top-k ranked sentences.

4.3   Profiling Fake News Spreaders

Discriminating authors of social media who have shared some fake news piece from that of others who have not done this is an exciting idea to pursue. PAN[9] announced a shared task in this direction. Duan et al. [84] proposed supervised learning to profile fake news spreaders given the content of their tweet feed and ground truth information. For each user, from the tweet content, sentiment, hashtags, emojis, and political presence, a tweet-level representation vector is learned. An aggregated score then represents the profile-level vector. The dataset containing 300 Twitter profiles, each with 100 tweets, were fed to a pretrained BERT to get embeddings and subsequently given to GRU to produce the prediction probabilities.

4.4   Neural Fake News Detection

Deepfakes have amassed widespread attention for their use in many fields, including fake news generation and its spreading. Generative modeling techniques such as

---

[9]https://pan.webis.de/.

GANs [85] and Variational Autoencoders [86, 87] are capable of automated text image production and video news items. Though the developed methods are for authentic text generation and entertainment purposes, its use in social media for propaganda, defamation, and misinformation propagation poses a considerable challenge to the research community in fake news detection. Iqbal et al. [88] give a survey on text generation models using deep learning. Mirsky et al. [89] and Tolosana et al. [90] present comprehensive reviews on deepfake generation and detection techniques about the human face and body.

Toward the detection of neural fake news, only a few works appeared in the literature. Zellers et al. [91] present GROVER, a model for controllable text generation. From a given headline, the system is capable of writing stories more trustworthy than a human-written story. Interestingly, GROVER itself will act as the system for defense against such system-generated fake news.

Energy-based models (EBM) are used to model text sequences and applied to discriminate machine-generated text from human-generated ones [92]. The model learns from three corpora, viz., Toronto books corpus, CommonCrawl news dataset, and Wikitext, following which transformer-based network, convolutional networks, and GPT[10] are used to generate negative samples. The EBM framework considered Linear, Bi-LSTM, and Transformer architectures for energy function and reports classification accuracies of the generalized EBM framework.

Stylometric approaches [93] have been applied to distinguish fake articles from real ones. These methods pay attention to stylometric features such as readability indices, stop word n-grams, and statistics on the text. Though there are some promising works in this direction, Shuster et al. [94] have reported that stylometry is incapable of distinguishing models generated with a little bit of false information from authentic language models.

OC-FakeDect [95] presents deepfake classification based on one-class variational autoencoder. Variational encoders are capable of producing images through a stochastic generative process. An encoder–decoder structure learns from only one-class images (real images) with a low error rate for reconstruction. The fake images inherently distorted offer a higher error rate during the reconstruction and hence become detectable. The one-class technique is a promising direction toward building more generalized models.

4.5   Discussion

This section provided the most recent advancements in the field of fake news detection. Despite many promising attempts and results in the area, developing a fully automated system suitable for real scenarios still seems far away. The following summarizes the main points discussed in this section:

---

[10]https://openai.com/blog/better-language-models.

- The generalization capability of Deepfake detectors under unseen conditions is inadequate; it is also easy to manipulate the detectors to degrade their performance.
- Insights into the semantics of fake news are still an unexplored area of research.
- Most of the reported works are in the English language; methods to address fake news detection in low resource languages would be useful.
- Neural language generators are emerging as an essential tool in both generation and detection of fake news.

5 Conclusion

This chapter described the automatic detection of fake news using neural representation learning methods across diverse content and structures. Though deep learning techniques are promising, the research is trending toward the use of unsupervised methods, potentially relying on a limited amount of data seeking higher accuracy in the early stages of news propagation. The models discussed in the chapter considered text followed by multimodal features and their representations to build a classifier. The works on explainable models revealing why the model thinks that the news piece is flagged as fake, created on which topic, in what modality, etc. will accelerate the practical use of these methods. Machine-generated news items and their application to limit the spread of fake news are other challenges. There is a potential to explore more linguistic factors and the development of appropriate metrics taking bearings on interdisciplinary domains to measure fakeness. The propagation patterns of news items in the social network structures are useful in early detection. A combined approach would be very helpful; with regard to this, neural graph methods have shown much promise. Irrespective of all the progress, fake news phenomena remain daunting and call for considerable thought from the research community.

References

1. Castillo, C., Mendoza, M., Poblete, B.: Information credibility on twitter. In: Proceedings of the 20th International Conference on World Wide Web, pp. 675–684 (2011)
2. Rubin, V.L., Chen, Y., Conroy, N.J.: Deception detection for news: three types of fakes. In: Proceedings of the Association for Information Science and Technology, vol. 52(1), pp. 1–4 (2015)
3. Mihalcea, R., Strapparava, C.: The lie detector: Explorations in the automatic recognition of deceptive language. In: Proceedings of the ACL-IJCNLP 2009 Conference Short Papers, pp. 309–312 (2009)
4. Ratkiewicz, J., Conover, M., Meiss, M., Gonççalves, B., Patil, S., Flammini, A., Menczer, F.: Detecting and tracking the spread of astroturf memes in microblog streams. CoRR **abs/1011.3768** (2010)

5. Ratkiewicz, J., Conover, M.D., Meiss, M., Gonçalves, B., Flammini, A., Menczer, F.M.: Detecting and tracking political abuse in social media. In: Fifth International AAAI Conference on Weblogs and Social Media (2011)
6. Sikdar, S., Kang, B., ODonovan, J., Höllerer, T., Adah, S.: Understanding information credibility on twitter. In: 2013 International Conference on Social Computing, Alexandria, VA, pp. 19–24. IEEE (2013)
7. Yang, F., Liu, Y., Yu, X., Yang, M.: Automatic detection of rumor on Sina Weibo. In: Proceedings of the ACM SIGKDD Workshop on Mining Data Semantics, pp. 1–7. ACM (2012)
8. Kwon, S., Cha, M., Jung, K., Chen, W., Wang, Y.: Prominent features of rumor propagation in online social media. In: 13th International Conference on Data Mining, Dallas, TX, pp. 1103–1108. IEEE (2013)
9. Ma, J., Gao, W., Wei, Z., Lu, Y., Wong, K.-F.: Detect rumors using time series of social context information on microblogging websites. In: Proceedings of the 24th ACM International on Conference on Information and Knowledge Management (CIKM '15), pp. 1751–1754. Association for Computing Machinery, New York, NY, USA (2015)
10. Resnick, P., Carton, S., Park, S., Shen, Y., Zeffer, N.: Rumorlens: A system for analyzing the impact of rumors and corrections in social media. In: Proc. Computational Journalism Conference, vol. 5, p. 7 (2014)
11. Finn, S., Metaxas, P.T., Mustafaraj, E., O'Keefe, M., Tang, L., Tang, S., Zeng, L.: TRAILS: A system for monitoring the propagation of rumors on twitter. In: Computation and Journalism Symposium, NYC, NY (2014)
12. Liu, X., Nourbakhsh, A., Li, Q., Fang, R., Shah, S.: Real-time rumor debunking on twitter. In: Proceedings of the 24th ACM International on Conference on Information and Knowledge Management (CIKM '15), pp. 1867–1870. Association for Computing Machinery, New York, NY, USA (2015)
13. Shao, C., Ciampaglia, G.L., Flammini, A., Menczer, F.: Hoaxy: A platform for tracking online misinformation. In: Proceedings of the 25th International Conference Companion on World Wide Web, pp. 745–750 (2016)
14. Riedel, B., Augenstein, I., Spithourakis, G.P., Riedel, S.: A simple but tough-to-beat baseline for the Fake News Challenge stance detection task. Preprint (2017). arXiv:1707.03264
15. Kucuk, D., Can, F.: Stance detection: A survey. ACM Comput. Surv. CSUR 53(1), 1–37 (2020)
16. Zhou, X., Zafarani, R.: Fake news: A survey of research, detection methods, and opportunities. Preprint (2018). arXiv:1812.00315
17. Bondielli, A., Marcelloni, F.: A survey on fake news and rumour detection techniques. Information Sciences 497, 38–55 (2019)
18. Sharma, K., Qian, F., Jiang, H., Ruchansky, N., Zhang, M., Liu, Y.: Combating fake news: A survey on identification and mitigation techniques. ACM Trans. Intell. Syst. Technol. TIST 10(3), 1–42 (2019)
19. Meel, P., Vishwakarma, D.K.: Fake news, rumor, information pollution in social media and web: A contemporary survey of state-of-the-arts, challenges and opportunities. Expert Syst. Appl. 153 112986 Elsevier (2019)
20. Zhang, X., Ghorbani, A.A.: An overview of online fake news: Characterization, detection, and discussion. Inf. Process. Manag. 57(2), 102025 Elsevier (2020)
21. Collins, B., Hoang, D.T., Nguyen, N.T., Hwang, D.: Fake news types and detection models on social media a state-of-the-art survey. In: Asian Conference on Intelligent Information and Database Systems, pp. 562–573. Springer (2020)
22. Dwivedi, S.M., Wankhade, S.B.: Survey on fake news detection techniques. In: International Conference on Image Processing and Capsule Networks, pp. 342–348. Springer (2020)
23. Shu, K., Wang, S., Lee, D., Liu, H. (eds.): Disinformation, Misinformation, and Fake News in Social Media, LNSN. Springer, Cham (2020)
24. Rumelhart, D.E., Hinton, G.E., Williams, R.J.: Learning internal representations by error propagation, No. ICS-8506. California Univ. San Diego La Jolla Inst for Cognitive Science (1985)

25. Dauphin, Y.N., Fan, A., Auli, M., Grangier, D.: Language modeling with gated convolutional networks. In: International Conference on Machine Learning, pp. 933–941 (2017)
26. Gehring, J., Auli, M., Grangier, D., Yarats, D., Dauphin, Y.N.: Convolutional sequence to sequence learning. Preprint (2017). arXiv:1705.03122
27. Bahdanau, D., Cho, K., Bengio, Y.: Neural machine translation by jointly learning to align and translate. Preprint (2014). arXiv:1409.0473
28. Vaswani, A., Shazeer, N., Parmar, N., Uszkoreit, J., Jones, L., Gomez, A.N.: Attention is all you need. In: Advances in Neural Information Processing Systems, pp. 5998–6008 (2017)
29. LeCun, Y.: Generalization and network design strategies. Connectionism Perspect. **19**, 143–155 (1989)
30. Collobert, R., et al.: Natural language processing (almost) from scratch. J. Mach. Learn. Res. **12**, 2493–2537 (2011)
31. Kim, Y.: Convolutional neural networks for sentence classification. Preprint (2014). arXiv:1408.5882
32. Mikolov, T., Chen, K., Corrado, G., Dean, J.: Efficient estimation of word representations in vector space. In: ICLR Workshop (2013)
33. Le, Q., Mikolov, T.: Distributed representations of sentences and documents. In: International Conference on Machine Learning, pp. 1188–1196 (2014)
34. Bojanowski, P., Grave, E., Joulin, A., Mikolov, T.: Enriching word vectors with subword information. Trans. Assoc. Comput. Linguist. **5**, 135–146 (2017)
35. Pennington, J., Socher, R., Manning, C.D.: Glove: Global vectors for word representation. In: Proceedings of the 2014 Conference on Empirical Methods in Natural Language Processing (EMNLP), pp. 1532–1543 (2014)
36. Fang, Y., Gao, J., Huang, C., Peng, H., Wu, R.: Self multi-head attention-based convolutional neural networks for fake news detection. PLoS ONE **14**(9), e0222713 (2019)
37. Dong, X., Victor, U., Chowdhury, S., Qian, L.: Deep two-path semi-supervised learning for fake news detection. Preprint (2019). arXiv:1906.05659
38. Li, Q., Hu, Q., Lu, Y., et al.: Multi-level word features based on CNN for fake news detection in cultural communication. Pers. Ubiquit Comput. **24**, 259–272 (2020)
39. Kaliyar, R.K., Goswami, A., Narang, P., Sinha, S.: FNDNet—A deep convolutional neural network for fake news detection. Cogn. Syst. Res. **61**, 32–44 (2020)
40. Schwarz, S., Theóphilo, A., Rocha, A.: EMET: Embeddings from multilingual-encoder transformer for fake news detection. In: 2020 IEEE International Conference on Acoustics, Speech and Signal Processing (ICASSP), Barcelona, Spain, pp. 2777–2781 (2020)
41. Qi, P., Cao, J., Yang, T., Guo, J., Li, J.: Exploiting multi-domain visual information for fake news detection. In: 2019 IEEE International Conference on Data Mining (ICDM), Beijing, China, pp. 518–527 (2019)
42. Khattar, D., Goud, J.S., Gupta, M., Varma, V.: Mvae: Multimodal variational autoencoder for fake news detection. In: The World Wide Web Conference, pp. 2915–2921 (2019)
43. Wang, Y., Ma, F., Jin, Z., Yuan, Y., Xun, G., Jha, K., Su, L., Gao, J.: Eann: Event adversarial neural networks for multi-modal fake news detection. In: Proceedings of the 24th acm sigkdd International Conference on Knowledge Discovery & Data Mining, pp. 849–857 (2018)
44. Cao, J., Qi, P., Sheng, Q., Yang, T., Guo, J., Li, J.: Exploring the role of visual content in fake news detection. Preprint (2020). arXiv:2003.05096v1
45. Hochreiter, S., Schmidhuber, J.: Long short-term memory. Neural Comput. **9**(8), 1735–1780 (1997)
46. Cho, K., Van Merriënboer, B., Bahdanau, D., Bengio, Y.: On the properties of neural machine translation: Encoder-decoder approaches. Preprint (2014). arXiv:1409.1259
47. Dmitriy, D., Miller, T., Lin, C., Bethard, S., Savova, G.: Neural temporal relation extraction. In: Proceedings of the 15th Conference of the European Chapter of the Association for Computational Linguistics: Vol. 2, Short Papers, pp. 746–751 (2017)
48. Ma, J., Gao, W., Mitra, P., Kwon, S., Jansen, B.J., Wong, K.-F., Cha, M.: Detecting rumors from microblogs with recurrent neural networks. In: Proceedings of the 25th International Joint Conference on Artificial Intelligence (IJCAI 2016), pp. 3818–3824 (2016)

49. Wang, W.Y.: "liar, liar pants on fire": A new benchmark dataset for fake news detection. Preprint (2017). arXiv:1705.00648
50. Ruchansky, N., Seo, S., Liu, Y.: CSI: A hybrid deep model for fake news detection. In: Proceedings of the 2017 ACM on Conference on Information and Knowledge Management (CIKM '17), pp. 797–806. Association for Computing Machinery, New York, NY, USA (2017)
51. Shu, K., Mahudeswaran, D., Liu, H.: FakeNewsTracker: a tool for fake news collection, detection, and visualization. Comput. Math. Organ. Theory 25(1), 60–71 (2019)
52. Esmaeilzadeh, S., Peh, G.X., Xu, A.: Neural abstractive text summarization and fake news detection. Preprint (2019). arXiv:1904.00788
53. Dong, M., Yao, L., Wang, X., Benatallah, B., Sheng, Q.Z., Huang, H.: DUAL: A deep unified attention model with latent relation representations for fake news detection. In: Hacid, H., Cellary, W., Wang, H., Paik, H.Y., Zhou, R. (eds.) Web Information Systems Engineering – WISE 2018. Lecture Notes in Computer Science, vol. 11233. Springer, Cham (2018)
54. Ajao, O., Bhowmik, D., Zargari, S.: Fake news identification on twitter with hybrid CNN and RNN models. Preprint (2018). arXiv:1806.11316
55. Thanos, K.-G., Polydouri, A., Danelakis, A., Kyriazanos, D., Thomopoulos, S.C.A.: Combined deep learning and traditional NLP approaches for fire burst detection based on twitter posts. In: Text Mining-Analysis, Programming and Application. IntechOpen (2019)
56. Guo, C., Cao, J., Zhang, X., Shu, K., Liu, H.: DEAN: learning dual emotion for fake news detection on social media. Preprint (2019). arXiv:1903.01728
57. Wu, L., Rao, Y.: Adaptive interaction fusion networks for fake news detection. Preprint (2020). arXiv:2004.10009
58. Zhang, J., Dong, B., Philip, S.Yu.: Deep diffusive neural network based fake news detection from heterogeneous social networks. In: 2019 IEEE International Conference on Big Data (Big Data), pp. 1259–1266. IEEE (2019)
59. Agarwal, A., Mittal, M., Pathak, A., Goyal, L.M.: Fake news detection using a blend of neural networks: An application of deep learning. SN Comput. Sci. 1(143), 1–9 (2020)
60. Huang, Y.-F., Chen, P.-H.: Fake news detection using an ensemble learning model based on self-adaptive harmony search algorithms. Expert Syst. Appl. 159, 113584 (2020)
61. Kaliyar, R.K., Goswami, A., Narang, P.: DeepFakE: improving fake news detection using tensor decomposition-based deep neural network. J. Supercomput. (2020). https://doi.org/10.1007/s11227-020-03294-y
62. Jin, Z., Cao, J., Guo, H., Zhang, Y., Luo, J.: Multimodal fusion with recurrent neural networks for rumor detection on microblogs. In: Proceedings of the 25th ACM international conference on Multimedia (MM '17), pp. 795–816. Association for Computing Machinery, New York, NY, USA (2017)
63. Yang, Y., Zheng, L., Zhang, J., Cui, Q., Li, Z., Yu, P.S.: TI-CNN: Convolutional neural networks for fake news detection. Preprint (2018). arXiv:1806.00749
64. Antol, S., Agrawal, A., Lu, J., Mitchell, M., Batra, D., Lawrence Zitnick, C., Parikh, D.: VQA: Visual question answering. In: Proceedings of the IEEE International Conference on Computer Vision, pp. 2425–2433 (2015)
65. Vinyals, O., Toshev, A., Bengio, S., Erhan, D.: Show and tell: A neural image caption generator. In: Computer Vision and Pattern Recognition (CVPR), pp. 3156–3164. IEEE (2015)
66. Singhal, S., Shah, R.R., Chakraborty, T., Kumaraguru, P., Satoh, S.: SpotFake: A multi-modal framework for fake news detection. In: 2019 IEEE Fifth International Conference on Multimedia Big Data (BigMM), Singapore, pp. 39–47 (2019)
67. Devlin, J., Chang, M., Lee, K., Toutanova, K.: BERT: pre-training of deep bidirectional transformers for language understanding. CoRR abs/1810.04805 (2018)
68. Zhou, X., Wu, J., Zafarani, R.: SAFE: Similarity-aware multi-modal fake news detection. Preprint (2020). arXiv:2003.04981
69. Vinyals, O., Toshev, A., Bengio, S., Erhan, D.: Show and tell: Lessons learned from the 2015 MSCOCO image captioning challenge. IEEE Trans. Pattern Anal. Mach. Intell. 39(4), 652–663 (2016)

70. Kochkina, E., Liakata, M., Zubiaga, A.: PHEME dataset for Rumour Detection and Veracity Classification (2018). https://figshare.com. Dataset. https://doi.org/10.6084/m9.figshare.6392078.v1
71. Gori, M., Monfardini, G., Scarselli, F.: A new model for learning in graph domains. In: IJCNN, Vol. 2, pp. 729–734. IEEE (2005)
72. Duvenaud, D.K., Maclaurin, D., Aguilera-Iparraguirre, J., Gómez-Bombarelli, R., Hirzel, T., Aspuru-Guzik, A., Adams, R.P.: Convolutional networks on graphs for learning molecular fingerprints. In: NIPS, pp. 2224–2232 (2015)
73. Kipf, T.N., Welling, M.: Semi-supervised classification with graph convolutional networks. Preprint (2016). arXiv:1609.02907
74. Defferrard, M., Bresson, X., Vandergheynst, P.: Convolutional neural networks on graphs with fast localized spectral filtering. In: Proc. NIPS, pp. 3837–3845 (2019)
75. Monti, F., Frasca, F., Eynard, D., Mannion, D., Bronstein, M.M.: Fake news detection on social media using geometric deep learning. Preprint (2019). arXiv:1902.06673
76. Klambauer, G., Unterthiner, T., Mayr, A., Hochreiter, S.: Self-normalizing neural networks. In: Proc. NIPS, pp. 971–980 (2017)
77. Dong, M., Zheng, B., Hung, N.Q.V., Su, H., Li, G.: Multiple rumor source detection with graph convolutional networks. In: Proceedings of the 28th ACM International Conference on Information and Knowledge Management (CIKM '19), pp. 569–578. Association for Computing Machinery, New York, NY, USA (2019)
78. Ren, Y., Zhang, J.: HGAT: Hierarchical graph attention network for fake news detection. Preprint (2020). arXiv:2002.04397
79. Du, M., Liu, N., Hu, X.: Techniques for interpretable machine learning. Commun. ACM **63**(1), 68–77 (2020)
80. O'Brien, N., Latessa, S., Evangelopoulos, G., Boix, X.: The language of fake news: Opening the black-box of deep learning based detectors (2018). https://dspace.mit.edu/handle/1721.1/120056
81. Hassan, N., Arslan, F., Li, C., Tremayne, M.: Toward automated fact-checking: Detecting check-worthy factual claims by ClaimBuster. In: Proceedings of the 23rd ACM SIGKDD International Conference on Knowledge Discovery and Data Mining, pp. 1803–1812 (2017)
82. Popat, K., Mukherjee, S., Yates, A., Weikum, G.: Declare: Debunking fake news and false claims using evidence-aware deep learning. Preprint (2018). arXiv:1809.06416
83. Shu, K., Cui, L., Wang, S., Lee, D., Liu, H.: Defend: Explainable fake news detection. In: Proceedings of the 25th ACM SIGKDD International Conference on Knowledge Discovery & Data Mining, pp. 395–405 (2019)
84. Duan, X., Naghizade, E., Spina, D., Zhang, X.: RMIT at PAN-CLEF 2020: Profiling fake news spreaders on twitter. CLEF (2020)
85. Goodfellow, I., Pouget-Abadie, J., Mirza, M., Xu, B., Warde-Farley, D., Ozair, S., Courville, A., Bengio, Y.: Generative adversarial nets. In: Advances in Neural Information Processing Systems, pp. 2672–2680 (2014)
86. Kingma, D.P., Welling, M.: Auto-encoding variational Bayes. Preprint (2013). arXiv:1312.6114
87. Kingma, D.P., Welling, M.: An introduction to variational autoencoders. CoRR **abs/1906.02691** (2019)
88. Iqbal, T., Qureshi, S.: The survey: Text generation models in deep learning. J. King Saud University Comput. Inf. Sci. (2020). https://doi.org/10.1016/j.jksuci.2020.04.001
89. Mirsky, Y., Lee, W.: The creation and detection of deepfakes: A survey. Preprint (2020). arXiv:2004.11138
90. Tolosana, R., Vera-Rodriguez, R., Fierrez, J., Morales, A., Ortega-Garcia, J.: Deepfakes and beyond: A survey of face manipulation and fake detection. Preprint (2020). arXiv:2001.00179
91. Zellers, R., Holtzman, A., Rashkin, H., Bisk, Y., Farhadi, A., Roesner, F., Choi, Y.: Defending against neural fake news. In: Advances in Neural Information Processing Systems, pp. 9054–9065 (2019)

92. Bakhtin, A., Gross, S., Ott, M., Deng, Y., Ranzato, M., Szlam, A.: Real or fake? learning to discriminate machine from human generated text. Preprint (2019). arXiv:1906.03351
93. Potthast, M., Kiesel, J., Reinartz, K., Bevendorff, J., Stein, B.: A stylometric inquiry into hyperpartisan and fake news. CoRR **abs/1702.05638** (2017). https://arxiv.org/abs/1702.05638
94. Schuster, T., Schuster, R., Shah, D.J., Barzilay, R.: The limitations of stylometry for detecting machine-generated fake news. Computational Linguistics **46**(2), 499–510 (2020)
95. Khalid, H., Woo, S.S.: OC-FakeDect: Classifying deepfakes using one-class variational autoencoder. In: Proceedings of the IEEE/CVF Conference on Computer Vision and Pattern Recognition Workshops, pp. 656–657 (2020)

Dynamics of Fake News Diffusion

Tanmoy Chakraborty

Abstract Modeling information diffusion on social media has gained tremendous research attention in the last decade due to its impact in understanding the overall spread of news contents through network links such as followers, friends, etc. Those fake stories which gain quick visibility are deployed on social media in a strategic way in order to create maximum impact. In this context, the selection of initiators, the time of deployment, the estimation of the reach of the news, etc. play a decisive role to model the spread appropriately. In this chapter, we start by defining the problem of fake news diffusion and addressing the challenges involved. We then model information cascade in various ways such as a diffusion tree. We then present a series of traditional and recent approaches which attempt to model the spread of fake news on social media.

Keywords Fake news diffusion · Network structure · Information cascade

1 Introduction

Information diffusion/spreading/propagation has been an active interdisciplinary research problem across different fields—physics [12, 14], computer science [9, 11], mathematics [27], and social science [25]. It deals with how an information spreads online or offline due to the interactions of users or entities. For example, in the case of epidemic spreading, information is the "disease," and the diffusion model animates how many users are going to be affected by the disease after a certain time period. In the case of online fake news spreading, information is "fake message," and the diffusion model predicts how many online users are going to share the news or remain unaffected (are not going to share).

Misinformation spread due to a lack of scientific understanding and unawareness about the context is a major problem worldwide. This long-standing issue has wreaked havoc while we are fighting a deadly global pandemic such as COVID-19. While the linking of false conceptions to epidemics dates back to almost the

© Springer Nature Switzerland AG 2021

Deepak P et al., *Data Science for Fake News*, The Information
Retrieval Series 42, https://doi.org/10.1007/978-3-030-62696-9_5

dawn of history, the emergence of Web 2.0 has strengthened it to an overwhelming degree. Particularly, with the rise of social media, misinformation spreads at an unprecedented rate. For example, a misinformation in India stating "cow-urine can be used as a vaccine for COVID-19" spread like wildfire, resulting in people drinking it and getting sick.[1] Similar such false news in the UK stating that "the first volunteer in UK coronavirus vaccine trial, Elisa Granato, dies" released at the end of April 2020 went viral.[2]

Existing approaches studied the diffusion dynamics of fake news to identify both the "negative paths" (misinformation spreading paths) and the "positive paths" (spread combating). Misinformation spreading can be due to fraudulent external sources (news portals, an influential person making loose remarks, etc.), fraudulent groups of users acting in an organized manner, or, most of the time, both. Given the social network and a set of previously posted false claims, diffusion models explore the diffusion of those claims and assign a "fakeness score" to each user as well as external sources (if present) that identify their likelihood of spreading false claims. This score should also incorporate the influence of such users and sources in terms of misinformation spreading, measured by the degree of the spread they materialize. Such a score may govern some coercive containment strategy, i.e., suspending such users/sources and restricting their post-reach. Existing models also identify potential users who spontaneously post/share fact-checks [2]. Models also animate how users get attracted by online posts floating around social media [6].

This chapter summarizes existing studies dealing with fake news modeling.[3] It starts by outlining the major observations in fake news detection on Facebook (Sect. 2) and Twitter (Sect. 3)—two major social media platforms. Following this, it presents the role of bots in spreading online fake news in Sect. 4. It then presents how the modeling of information diffusion is conceptualized using the "tree" structure in Sect. 5. Section 6 describes how the sources of fake news are identified. Subsequently, Sect. 7 presents various mathematical models for fake news diffusion. Section 8 summarizes strategies to minimize the spread of fake news. Section 9 concludes the chapter with possible future directions.

---

[1] https://www.theweek.in/news/india/2020/04/02/gujarat-sees-increased-demand-for-cow-urine-amid-covid-19-scare.html.

[2] https://newsbreakng.com/first-volunteer-in-uk-coronavirus-vaccine-trial-elisa-granato-dies/.

[3] Note that there are many studies related to modeling misinformation. Fake news is a type of misinformation. Other types of misinformation include disinformation, spam, troll, clickbait, rumor, urban legend, etc. [20]. Here, we intentionally keep only the studies explicitly dealing with fake news propagation.

2   Fake News Diffusion on Facebook

Del Vicario et al. [5] studied the determinants behind the consumption of fake news by Facebook users. They collected 67 public pages divided between 32 about "conspiracy theories" (pages that disseminate alternative, controversial information) and 35 about "science news." They showed that the way social media users consume both types of news is almost the same—the propagation of both types of stories produces homogeneous and polarized communities which act as echo chambers. Users having the same profile and background and belonging to the same (polarized) community tend to consume similar types of news. Nguyen et al. [15] defined **echo chamber** as a group that, in addition to being insular (i.e., a community), exhibits strong single-mindedness when it comes to content.

One can model the information diffusion of a news as a directed tree where nodes represent users who shared the news, and an edge $\langle u, v \rangle$ indicates that user v shared the news posted by user u. A branch of the tree corresponds to a cascade. The length of a branch (measured by the number of nodes present in the branch) is the **size of the cascade**. The time difference between the submitting of the earliest (the root node) and the latest (the leaf node) posts on a branch indicates the **lifetime of a cascade**. Del Vicario et al. [5] showed that news assimilation differs according to the categories. Science stories have a high level of reach with large cascade size very quickly and have high longevity. However, this does not correlate with the high level of interactions among the users. Conversely, conspiracy rumors propagate slower than science stories, and there is a positive correlation between the longevity and the cascade size of the stories.

3   Fake News Diffusion on Twitter

Most of the studies related to fake news focus on the stories circulated around the 2016 US presidential elections. Jang et al. [10] collected 60 news stories (30 fake and 30 real), published between January 1, 2016, and April 30, 2017. Among 30 fake news selected for the analysis, 15 stories were written against Trump and another 15 stories intended to malign Clinton.

They observed that ordinary users (not bots) are responsible for generating these fake news. This is contradictory to the earlier finding which suggests that fake news websites themselves proactively spread their news [18]. Interestingly, half of the fake tweets contain non-credible news website links. Using a novel graph construction, called *evolution tree* or *phylogenetic tree* (discussed in Sect. 5), they corroborated the conclusion of Del Vicario et al. [5] that the velocity of the propagation of fake news is slower than that of real news on social media. While spreaders modify the content of fake news based on their opinion, real news is shared without much modification. This oftentimes challenges the automatic fake news detectors which are purely based on the content understanding and content similarity among news stories.

Glenski et al. [7] collected 11 million direct interactions (i.e., retweets and mentions) with the source accounts of 2 million unique users from 282 sources over 13 months between January 2016 and January 2017. They divided the news sources into five categories: Trusted News, Clickbaits, Conspiracy Theory, Propaganda News, and Disinformation News. They attempted to explain the following points:

– **Evenness of content spreading:** The initiators of each type of news sources are significantly less in number compared to the number of source news stories. News which contains "mention of users" (@mention) is more evenly propagated among users than direct retweets for trusted, conspiracy, and disinformation. For clickbait- and propaganda-related sources, direct tweets are more uniformly shared by users than tweets using @mentions. In contrast, disinformation is more unevenly disseminated than trusted news and propaganda stories. In general, the dissemination pattern across news sources is uneven and highly skewed—10% active users are responsible for the posting of the majority of new news stories and 40% of users are less active in the entire spreading process.
– **Types of content spreaders:** The number of source tweets is highly correlated with the number of users who retweet first, which indicates that each tweet is associated with one early spreader whose responsibility is to initiate the propagation of the news. There is also a significant overlap between the initial spreaders across different news types, except for conspiracy and disinformation sources. There are highest intersections with trusted sources for all types of deceptive sources. Users belonging to low-income zones (annual income below 35,000 USD) and users who are less educated (finished higher school studies) are primarily liable to share disinformation content more often than others. There is a high proportion of users who share clickbaits who also share propaganda news; however, the reverse is not true.
– **Velocity of content spreading:** The first 24 h after the posting of original tweets is crucial, regardless of the types of stories. The majority of the sharing (retweets and mentions) happen during this time. Suspicious sources tend to delay in retweeting than trusted sources. For trusted, conspiracy, and disinformation sources, the delay in retweeting the original tweets is shorter than the other types, whereas the same for clickbait and propaganda sources is significantly longer with propaganda sources having the longest wait after original postings. A much larger percentage of @mention tweets are shared within the first hour after the original posting occurs than the content retweeted directly from a source for all source types. More than 60% users who retweeted information from clickbait, conspiracy, propaganda, and disinformation sources also shared information from trusted sources.

Vosoughi et al. [24] presented a large-scale study on the diffusion of true, false, and mixed (partially true and partially false) news. The news stories verified by six independent fact-checking agencies (snopes.com, politifact.com, factcheck.org, truthorfiction.com, hoax-slayer.com, and urbanlegends.about.com) and posted on

Twitter from 2006 to 2017 were collected. It comprised ~126,000 rumor cascade spread by ~3 million people more than 4.5 million times. The cascade is characterized by four quantities:

- **Depth:** If we model the cascade as a tree where the root is the user of the original tweet and if user v retweets user u's tweet, there is a directed edge from u to v, the height of the longest branch indicates the depth of the cascade (see Sect. 5 for more discussion).
- **Size:** The number of users (if a user spreads a news multiple times, she will be counted multiple times) involved in the cascade over time indicates the size. Essentially, it is same as the number of nodes present in the cascade tree mentioned earlier.
- **Maximum breadth:** It is defined by the maximum number of users involved in the cascade at any breadth.
- **Structural virality (SV):** It is defined by the average distance between all pairs of nodes in a cascade [8].

$$SV = \frac{1}{n(n-1)} \sum_{i=1}^{n} \sum_{j=1}^{n} dist_{ij} \qquad (1)$$

where $dist_{ij}$ denotes the length of the shortest path between nodes i ad j, and n is the number of nodes in the cascade tree. A star-like structure has the lowest SV, whereas a line graph has the highest SV.

News stories are also divided into different categories based on their topics such as politics, entertainment, and natural disasters. Politics turns out to be the largest rumor category, followed by urban legends, business, terrorism, science, entertainment, and natural disasters.

While studying the diffusion dynamics of different types of news stories, Vosoughi et al. [24] observed that fake news stories diffuse significantly farther, faster, deeper, and more broadly than the truth in all categories of information. Whereas true stories rarely diffuse to more than 1000 people, the top 1% of false news cascades routinely diffuse to between 1000 and 100,000 users. False political news traveled deeper, more broadly, reached more people, and was more viral than any other category of false information. One may argue that the high reach of fake news stories is due to the structural properties of the propagators—they may have a large number of followers or followees; they may be highly active in terms of the number of tweets/retweets; or they may be spending significant time on social media. Surprisingly, Vosoughi et al. [24] observed that none of the above arguments is true. In fact, the propagators of fake news stories behave in just the opposite way. Users who spread false news tend to have significantly fewer followers, followed significantly fewer people, were significantly less active on Twitter, and had been on Twitter for significantly less time.

In order to understand the causes behind the fast and wide spread of fake news, Vosoughi et al. [24] explored the textual property of news stories. They observed that fake stories are novel and attractive, which may in turn engage new users. They also found that false information inspires replies expressing greater surprise corroborating the novelty hypothesis, and greater disgust, whereas the truth inspired replies that expressed greater sadness. In order to check if bots influence in the above inferences of the study, Vosoughi et al. [24] repeated the same experiments after removing bot traffic, i.e., removing all tweet cascades started by bots, including human retweets of original bot tweets. They concluded that none of the major conclusions changed—false news still spread farther, faster, deeper, and more broadly than real news. This in turn suggests that it is humans, not the bots, who are responsible for increasing the reach of false news. In the next section, we will discuss the role of bots in the spread of fake news.

4 Role of Bots in Spreading Fake News

Bots are accounts that communicate more or less autonomously on social media, often with the task of influencing the course of discussion and/or the opinions of readers. The algorithms controlling the activities of bots are written by human administrators who act behind the screen to spread a certain ideology or opinion. The bots and their administrators are often called "sockpuppets" and "puppetmasters," respectively [4, 21]. Shao et al. [19] conducted a large-scale study to show the role of social bots in spreading low-credible news. They collected 389,569 articles from 120 low-credibility sources and 15,053 articles from different fact-checking sources. They also collected 13,617,425 tweets linked to low-credibility sources and 1,133,674 linked to fact-checking sources. The Botometer system[4] was used to detect bots [23].

They noticed that relatively few accounts are responsible for spreading the misinformation largely. The majority of such accounts are likely to be bots. Bots are very active in the early stages once the news gets uploaded and publicize the news extensively until it becomes viral. To do so, they target influential accounts through replies and mentions and pretend to be genuine/normal/trustworthy user accounts by retweeting posts of other bots as much as a normal user does. Bots play an active role in amplifying the reach of misinformation to a point where it is statistically indistinguishable from that of fact-checking articles. These results directly contradict the findings of Del Vicario et al. [5] that bots alone do not entirely explain the success of false news. One of the reasons may be due to the methodological differences—Del Vicario et al. [5] did not consider the "resharing" activity by which a piece of news can be spread, which bots often choose to spread messages.

---

[4]botometer.iuni.iu.edu.

5   Trees for Modeling Information Diffusion

Generally, a "tree" is the simple data structure to model the spread of a social media post. A diffusion tree can be formed in different ways; some of them are explained below.

Sharing tree: Del Vicario et al. [5] proposed a sharing tree that is made up of the successive sharing of a news item. The *root* of the tree is the user who posted the message first on the social media. A directed link indicates the sharing or quoting the news story by a user. The *size* of the tree is the number of nodes or the number of shares (duplicate shares are also counted). The *height* of the sharing tree is the length of the longest branch of the tree.

Evolution tree: Jang et al. [10] borrowed the idea from the biological metaphor and proposed an evolution tree (*aka* phylogenetic tree) which models the evolutionary history and relationship among entities (tweets/posts in this case). The tree takes into account content-level similarity and timestamp information of tweets. The tree construction starts with the tweet whose posting time is earliest among all. All the other tweets are arranged in the chronological order of the posting time in a priority queue. At every iteration i, the first item t from the priority queue is taken, and content similarity between that tweet and each of the other $i - 1$ tweets (nodes) added so far in the tree is measured using the Q-gram string matching algorithm [22]. Tweet t is then attached to the node that has the maximum similarity to t.

6   Identifying the Sources of Fake News

Basu [2] utilized the concept of identifying codes to accurately spot the source(s) of misinformation propagation. Consider an undirected and unweighted graph $G(V, E)$ where:

1. **Node ($v \in V$):** user in a social network
2. **Edge ($e \in E$):** two users are connected by an edge $e \in E$ if they are friends.

Once the graph is created, multiple monitors are placed on each node of the graph. Here, a monitor indicates the trusted and verified agency (also known as fact-checking agency) who labels a news article as fake or real. Basu [2] considered two monitors: Politifact[5] and Media Bias/Fact Check (MBFC).[6] These monitors are placed in such a way that whenever a user becomes a source of misinformation, it can be easily identified, and all its friends or followers will get to know about the misinformation. Basu [2] considered the case where all the friends of the source user

---

[5]https://www.politifact.com/.

[6]https://mediabiasfactcheck.com/.

are involved in propagating the misinformation in the next time step. The problem can be framed as a Minimum Identifying Code Set (MICS) problem, where the task is to find the identifying code set of the smallest cardinality. The MICS problem can be understood easily by considering it as a variant of the classical graph coloring problem. Here, the goal is to inject as few colors as possible to the nodes in the graph such that every single node gets a color, and no two nodes have the same color. Mathematically, we can try to find the smallest subset $V' \subseteq V$ such that when colors are added to the nodes in the subset, it ensures that each node in the graph gets a unique color while propagating. For each node v_i, the neighborhood is denoted using $N(V_i)$, and an indicator variable is denoted using x_i as follows:

$$x_i = \begin{cases} 1 & ifacolorisinjectedatnodev_i \\ 0 & \text{otherwise} \end{cases}$$

Note that a node in the graph can receive a color in one of three ways: (1) the node is injected with a color, (2) the node receives a color from its neighborhood, and (3) the node is injected with a color and receives a color from its neighborhood.[7] Here, the following objective function needs to be minimized so that the fewest nodes are colored as mentioned below:

$$\text{Minimize} \sum_{v_i \in V} x_i$$

Basu [2] also mentioned about two constraints which ensure the following properties:

1. **Coloring constraint:** Every node will receive at least one color from its neighbor from the colors already injected. It is represented as follows:

$$\text{Minimize} \sum_{v_i \in N(v_j)} x_i \geq 1, \forall v_j \in V$$

2. **Unique coloring constraint:** In any pair of nodes in the graph, at least one of the nodes is injected with a color so that both will not receive same colors from its neighborhood. It is represented as follows:

$$\text{Minimize} \sum_{v_i \in [N(v_j) \oplus N(v_k)]} x_i \geq 1, \forall v_j \neq v_k, \in V$$

---

[7]Colors in the injected nodes are referred to as *atomic colors*, whereas colors generated by the combination of two or more colors are referred to as *composite colors*.

7 Modeling Fake News Diffusion

Traditional epidemic spreading models such as SIR (Susceptible-Infectious-Recovered), SIS (Susceptible-Infectious-Susceptible), and SIRS (Susceptible-Infectious-Recovered-Susceptible) [3] can be used to model information diffusion of news articles in online media. However, other types of models have also been proposed for modeling fake news stories. In this section, we first briefly define the SIR model, followed by three recent models.

7.1 Susceptible-Infected-Recovered (SIR) Model

The SIR model is a simple model used to explain the spread of disease within a population. At any given point in time, there is a group of infected population (I) carrying the contagion, and all the neighbors of the infected population are susceptible (S) to catch it. Meanwhile, every infected person has a chance of recovering (R) from the infection. Once an individual has recovered, she is immune to the contagion, i.e., she will neither catch it again nor will she be propagating it further. Based on varying rates of propagation of, and recovery from, the contagion, one can map the spread of the disease within a population (see Fig. 1).

The SIR model can be easily adapted from the real-world spread of disease to information diffusion in social media. Like the real world, the social network consists of an initial set of users (I) who create or come across a piece of new information (contagion) by interacting with it and expose their 1-hop neighbors (S) to the new information. Now some of the neighbors may find the information exciting/relevant and choose to spread it, while some may choose not to act on it. At any given point, a user may find the new information either not to her taste, or has by other means engaged with it beforehand via other initial users. Such users are immune to further exposure of the same news (R). As long as at least one user engages with the new information, the information continues to spread/diffuse.

It is important to note that the simple SIR model (online/offline) is static. It considers a snapshot of the network at a given time T_o and, for subsequent timestamps, does not consider any change in the network. Thus, no addition/removal

Fig. 1 The flow of contagion in an SIR model, with parameters infection and recovery rates. Only the paths indicated by the arrow are allowed

(birth/death) or addition/deletion of social ties (new friends/fall-outs) is considered. While for subsequent timestamps, the composition of the $S(t)$, $I(t)$, and $R(t)$ groups changes, and the overall population of the system remains constant, i.e., $S(t) + I(t) + R(t) = N$.

7.2 Dynamic Linear Threshold (DLT)

Litou et al. [13] proposed the Dynamic Linear Threshold (DLT) model. It provides a budgeted framework under which certain highly credible users can be enlisted to help spread counter arguments and stifle the spread of misinformation as early as possible. The core idea of DLT follows from the observation that social interactions (offline or online) are dynamic and multi faceted. The influence that one user has on another and the personal belief of an individual toward a subject vary over time. Thus, the objectives of DLT are two fold: Given a social network divided into non-overlapping groups of fake news spreaders, credible-news spreaders, and susceptible users:

– Derive a time-varying, multi faceted model that closely mimicks the fake news diffusion in OSN.
– Use the above model to build heuristics that can help contain the spread of fake news.

Network Construction

The network is represented as a directed and weighted graph $G(V, E)$, where direction represents the flow of information in the *to-from* format. Each node $u \in V$ has two node level scores. First is the pre computed static credibility score $c_u \in [0, 1]$ measuring the *trustworthiness* of the users. The other is *time-varying renouncement score* $r_u \in [0, 1]$, measuring the stubbornness/reluctance of a user toward a change of opinion when presented with new information. Along with the two node-centric scores c_u and r_u, we also have a pre computed edge weight $w_{uv} \in [0, 1]$, representing the level of influence of u on v. Influence can be measured in multiple ways (the level of interaction, how old the edge is, metadata, etc.). Users are divided into three categories:

– *Infected Users* (I): These users are more likely to adopt and propagate fake news. They are also called negatively influenced users.
– *Protected Users* (C): These users are more likely to adopt and propagate credible news. They are also called positively influenced users.
– *Inactive/Susceptible Users* (R): At any given time t, $R_t = V/(I_t - C_t)$.

Problem Definition

Given (1) a social network $G(V, E)$ along with the scores (c_u, r_u, w_{uv}), (2) an initial set I_o of misinformation originators, obtained via those whose c_u score is toward the lower end, and (3) a budget parameter k, the aim is to find a subset of credible users $S \in C$ ($|S| \leq k$) who by propagating counter arguments will help minimize the spread of misinformation. This use-case belongs to the category of *Influence blocking*.

Component I: Diffusion Dynamics

DLT is a variant of the Linear Threshold (LT) class of the diffusion model. Unlike the general case of LT that considers the edge weights (inter-user influence) and the user-engagement threshold (node property) to be static probabilities for all timestamps, DLT uses a time-window to update the inter-user influence and recomputes the user-level threshold to change the probabilities of adoption over time. The basic idea, however, remains the same. At any given timestamp t, a user in the network $u \in in(V)$ who is positively (or negatively) influenced, i.e., has become a supporter of fake (or credible) news, tried to exert her opinion into influencing her outgoing neighbors $v \in out(u)$ who are the receipts of the social network activities of u. A recipient user v, in turn, adopts/rejects an opinion based on the overall positive or negative influence she receives from all her incoming neighbors $u \in in(v)$. If the overall influence for v surpasses her threshold (here renouncement threshold is r_u), then v adopts the opinion and updates its group membership to either a misinformed user I or a credible user C. Once the user's belief is updated, the renouncement threshold is updated to reflect this change in opinion. In case the net effect of the incoming positive and negative influence is the same, the user does not change her opinion. At timestamp t, for user v, the overall influence $IF(v|t)$ can be represented in terms of the I, C, R set of users, via the following equation:

$$IF(v|t) = \sum_{u \in in(v)} B(u|t).p_{u,v}(t; \lambda).w_{uv} \tag{2}$$

where $in(v)$ denotes the set of incoming neighbors of v whose social activities are visible to v, and $B(u|t)$ represents the influence state of u at t. The state attained by u is in turn computed post its exposure at $t - 1$ as

$$B(u|t) = \begin{cases} 1 & u \in C \\ 0 & u \in I \\ 0 & u \in R \end{cases}$$

In Eq. (2), w_{uv} is the known inter-user influence which is subject to the time-varying component $p_{u,v} \in [0, 1]$. The probability that determines to what extent u influences v is computed from a Poisson distribution mapping

$$p_{u,v}(t; \lambda) = \frac{\lambda^t e^{-\lambda}}{t!} \tag{3}$$

where λ captures the variance in the time-interval of subsequent interactions between the users; $t_i \in T_{u,v} \in [1, n]$ is the ith interval among a total of n interactions. The value of $T_{u,v}$ and n is unique for each edge (u, v).

$$\lambda = \frac{1}{n} \sum_{i}^{n} (t_i - \mu) \tag{4}$$

The use of Poisson comes handy in (1) making each influence independent of the other. This follows from the observation that a user is not always compelled into following the opinions of her friends; while past interactions do play a role, each new opinion can be viewed as an independent incident of the influence. (2) As t increases, the factorial value of t in Eq. (3) causes $p_{u,v}$ to reduce. This again follows from two observations, that as more time elapses, the influence of u on v for the given opinion reduces, and a so-called interesting news becomes stale after a time.

Component II: Updating Personal Belief

Going back to Eq. (2), if the overall $IF(v|t)| \geq r_v$, i.e., if either the absolute positive or negative influence exceeds the threshold, then the user undergoes a change in opinion—$B(v|t)$ is updated to 1, if $IF(v|t) > 0$, -1 otherwise.

The influence of misinformed neighbors increases the chance of a user being negatively influenced while credible users have the opposite effect. This represents the multi faceted information dynamics of a user in the real world where opposing views coexist. Whenever $B(v)$ is updated, the renouncement score $r_v(t)$ is also updated as follows:

$$r_v(t) = 1 - (1 - r_o)^{y+1} \tag{5}$$

where r_o is the initial belief of the user (inherent bias), and y is the number of times a user undergoes a change in opinion. The renouncement score of user v increases every time she adopts a belief as one's acquired opinions are hard to change.

Component III: Misinformation Blocking

The aim of misinformation blocking is to present the spread of misinformation as early as possible and minimize count of the negatively influenced users. Thus, the most effective action will be to target the 1-hop neighbors of the misinformation originators, early referenced by I_o. These 1-hops are the first group of users to view the activities of I_o, and by hoping to positively influence them, we can altogether curb misinformation from spreading at all. Under the given budget k, we hope to find a set of users $S \in C$ whose influence on their neighbors will help minimize the misinformation propagation.

Now, based on Eq. (2), users v for $IF(v|t) < r_v$ are still susceptible to influence, and it is these sets of users that we hope to turn to the positive side. For $v \in out(I_o)$ we want to maximize $IF(v|t)$, while taking into account the number additional users I_k infected after k credible users have been added to S. The cost function we want to maximize is expressed as

$$g(S_k) = \sum_{v \in out(I_o)} IF(v|t) - (|I_k| - |I_o|) \tag{6}$$

$$= \sum_{v \in out(I_o)} \sum_{u \in in(v)} B(u|t).p_{u,v}(t;\lambda).w_{u,v} - (|I_k| - |I_o|) \tag{7}$$

$$= \sum_{v \in out(I_o)} \left(\sum_{u \in \{S \cup C\}} p_{u,v}.w_{u,v} - \sum_{u \in I_o} p_{u,v}.w_{u,v} \right) - (|I_k| - |I_o|) \tag{8}$$

Here, $(|I_k| - |I_o|)$ denotes the additional users influenced by credible set S_k. The first component of the equation with $u \in \{S \cup C\}$ captures the overall positive neighbors of v, and the second component $u \in I_o$ captures the overall negative neighbors of v. Thus, we aim to maximize $g(S_k)$ by either maximizing the net score of positive neighbors over negative ones or by minimizing the number of negative users reached by minimizing $(|I_k| - |I_o|)$. The more the number of positive users reached, the higher the chance of stopping the spread of misinformation.

Since *Influence Blocking* problem is NP complete, Litou et al. [13] proposed a greedy approach to obtaining the set S. At iteration n, a random node u_n is selected, and $g(S_k)$ is computed assuming $S_k = S \cup u_n$. Once n iterations complete, the best candidate u is added to $S = S \cup u$. This process repeats until k users are added to S, or there is no change in the number of impacted users.

7.3 Percolation Model

While current diffusion models only consider the edge weights and node threshold, we need a system that accounts for the homogeneity of users and groups, as well as the diversity of content. To this end, Del Vicario et al. [5] proposed a model accommodating the same.

Network Components

Consider a set of n users and m news items the users are exposed to. Each user $i \in n$ has an inherent opinion/polarization score $\omega_i \in [0, 1]$, while each news $j \in m$ item expresses an opinion $\theta_j \in [0, 1]$.

In addition, the model restricts the flow of information to homogeneous links only, considering varying fractions of homogeneous links in the network $\phi_{HL} \in [0, 1]$ defined as

$$\phi_{HL} = \frac{n_h}{M} \tag{9}$$

where M is the total links in the network, and $0 \leq n_h \leq M$ is the number of homogeneous links. Subsequently, $(1 - \phi_{HL})$ is the fraction of nodes with non-homogeneous links (mixed group setting).

Adoption of Information

The alignment of the user's opinion with the opinion expressed in the news is given by $|\omega_i - \theta_j|$. The smaller the difference, the higher the polarization alignment. For a user i with a threshold δ_i, she will share/engage with the news posts j iff:

$$|\omega_i - \theta_j| \leq \delta_i \tag{10}$$

Branching and Size of Cascade

Equation (10) can be rewritten as

$$-\delta_i \leq \omega_i \leq \delta_i \Rightarrow -\delta_i + \theta_j \leq \omega_i \leq \delta_i + \theta_j \tag{11}$$

With $0 \leq \omega_i \leq 1$, we get $0 \leq -\delta_i + \theta_j$ and $\delta_i + \theta_j \leq 1$.

Drawn from two independent, identical distributions (i.i.d), in general $f(\omega)$ and $f(\theta)$, capture the probability p that a user i will engage with news post j based on

$$p = min(1, \delta_i + \theta_i) - max(0, -\delta_i + \theta_i) \approx 2\delta_i \tag{12}$$

$$= f(\theta_i) \int_{max(0, -\delta_i + \theta_i)}^{min(1, \delta_i + \theta_i)} f(\omega_i) d\omega_i \tag{13}$$

So far, we have considered each user i to have its own threshold δ_i, which we require to parameterize n different values of δ, one per user. For the sake of simplicity, we consider each user to carry the same, globally defined threshold represented by δ.

In addition to δ, we also define the branching parameter z (neighborhood dimension), providing us with an average number of sharers (branching ratio) as

$$\mu = zp \approx 2z\delta \tag{14}$$

Subsequently, with a probability q of a neighbor with different polarization than the user under consideration, the branching factor becomes

$$\mu = z(1-q)p \approx (1-q)p\frac{\langle z^2 \rangle}{\langle z \rangle} \tag{15}$$

Equation (15) follows from the epidemic model with epidemic threshold $\lambda_c = \frac{\langle k^2 \rangle}{\langle k \rangle}$ with a system of degree distribution $p(k)$.

With a critical cascade size $S = (1-\mu)^{-1}$, a set of m initial sharer's distribution $f(m)$, and average $\langle m_f \rangle$, we get

$$S = \sum_m f(m)m(1-\mu)^{-1} = \frac{\langle m_f \rangle}{1 - 2z\delta} \tag{16}$$

Thus, the cascade size depends on the initial set of spreaders (some users are likely to garner more support from their followers/friends than others) and the branching factor where a high branching factor provides more opportunity for the spreading of information, and the user's threshold where a larger value of δ captures larger variation in opinion between the user and posts.

Parameter Estimation

As observed in Eq. (16), the parameters $z = 8$ and $\delta = 0.05$ are fixed beforehand. We are mainly left with the task of approximating the distribution for the first sharer. Del Vicario et al. [5] fit probability density functions of the first sharer (who directly contributes to the cascade size, as discussed above) on the whole sample of the scientific and conspiracy data with different distributions. Out of Inverse Gaussian (IG), Log-Normal, Poisson, and Uniform in (1100), IG seemed to be the best fit.

7.4 Spread-Stifle Model

Yang et al. [26] performed exhaustive pre- and post-modeling analysis on two large-scale Facebook networks and developed a modified SIR model to explain the observations. The model allows fake/false and credible/true news to spread simultaneously, and let a consumer of fake news redeem herself by switching to become a consumer of credible news instead. However, the model does not allow the supporters of credible news to switch their sides. In addition, Yang et al. [26] observed how the final density of the uninterested (non-spread) population, coupled with higher rates of credible information spreading, is majorly responsible for curbing the spread of fake news.

Two large Facebook networks consisting of ~500 k and ~720 k nodes/users are used for the experiments. The degree distribution of both these datasets reveals power-law-like distribution, with a majority of nodes having less than degree 10. These highly heterogeneous systems mean that the degree of an individual plays an important role in determining the spread of information. Unlike the situation where we have a large number of influential/hub nodes controlling the mass spread of information, and irrespective of the low rate of interaction, the sheer volume of engagement is still high. This is similar to how your friends on Facebook are the first (and at times the only) people who view/engage with your post, and any chance of your post reaching far and wide begins with your first hop neighbors. At any given point, a node will either consume or propagate information around its first hop neighbors only.

How the Spread-Stifle Model Differs from Others?

The Spread-Stifle model is a slight variant of the general SIR model with the following properties:

1. Unlike the general SIR model, the Spread-Stifle model accommodates a simultaneous spread of two contagious (true and false news).
2. Subsequently, there are four (instead of three) groups into which the population is divided. At any given point t, we have the ignorant/susceptible $I(t)$ users, the infected users who are spreaders of either false $S^F(t)$ or true $S^T(t)$ news, and the recovered/non-active/stifling $R(t)$ users who curb the spread of any form of information. The net population of the system is $I(t)+S^F(t)+S^T(t)+R(t) = N$, where N is the number of nodes/users in the system.
3. As discussed in the Facebook dataset, since the degree of a node plays an important role in the spread of information, the user groups described above are considered on the node-degree basis where nodes with the same degree are assumed to exhibit similar behavior. Further, instead of considering the population of the groups as a whole, we represent them as fractions to facilitate the computations. Thus, at any given time t, for degree k, the fraction of nodes of ignorant, spreader, and recovered population is represented as $I_k(t)$, $S_k^T(t)$, $S_k^F(t)$, $R_k(t)$, respectively. Also, $I_k(t)+S_k^T(t)+S_k^F(t)+R_k(t) = 1$.
4. If an ignorant user I comes in contact with a user who spreads real news, then based on the rate of adoption of real news $\lambda_T \in [0, 1]$, the ignorant user can adopt the real news, and herself become a member of the S^T group. Similarly, the ignorant user can be exposed to false news. With a rate of adoption of false news $\lambda_F \in [0, 1]$, it can update the membership to S^F. It is important to note that at any timestamp t, the ignorant user can be simultaneously exposed to both

true and false news via her true/false-spreader neighbors and can choose to act on either or none of them.

5. Unlike the percolation model in Sect. 7.3, which considers the cascades of true and false information independently, or the DLT model in Sect. 7.2 which allows for both fake and credible users to change opinions based on thresholds, the Spread-Stifle model considers a middle ground. *When a spreader of false news comes in to contact with a spreader of true news, the former always converts, and the vice versa is not allowed ($\lambda_{FT} = 1$, $\lambda_{TF} = 0$).* Such a behavior is similar to the situation where a user unintentionally propagates false information but upon confrontation renounces the false belief and adopts the truth. Setting $\lambda_{FT} = 1$, the Spread-Stifle model however naively assumes that every proponent of false news is uninformed and will change when provided with true information, which may not be true.

6. Let us consider the case where an active spreader of information (true or false) is no longer interested in the information and stops publicizing it, or the active spreader of information comes in contact with a person who is not interested in the information. In such case, no further propagation of information takes place, and the spreader updates to a recovered/stifler node. At all timestamps, the spreaders S_T and S_F convert to R based on the removal rate α_t and α_f, respectively. Here, the removal of each spreader is independent of the other. For the sake of simplicity, the removal rate for both type of spreaders is set to α.

Mean-Field Approach

The final property to consider, which forms the backbone of the Spread-Stifle model (and for the class of SIR models as well), is the macroscopic view of the diffusion dynamics. Instead of focusing on the susceptibility and immunity of every node separately, we consider the susceptibility and immunity of the group as a whole. Thus, instead of considering $N \times 3$ different parameters of $\lambda_t^i, \lambda_f^i, \alpha^i$ (where $i \in [1, N]$), we consider only three global parameters that affect the inter-groups movements as a whole. Consequently, instead of modeling and tracking the transitions of each individual node, we focus on the rate in change of population for each group. In the Spread-Stifle model, this is further broken on the basis of the degree of the nodes, yet considered at a group level. If our system is modeled as m timestamps $t_1, t_2, \ldots, t_m$, then, from a microscopic point of view, we would need to track $m \times N \times 4$ rates of change to fully capture the different state transitions for each node. Instead, using a macroscopic approach, we track $m \times 4$ rates of change; one for each group. This macroscopic view of diffusion that approximates the behavior of an individual to a group of users instead and focuses on modeling the rates of change of the groups is commonly known as the mean-field approach (an idea transferred to Network Science from Physics).

In the forthcoming sections, we will apply this approach to fit the Spread-Stifle model.

Reachability Probabilities

At a given timestamp t, $I_k(t)$, $S_k^T(t)$, $S_k^F(t)$, and $R_k(t)$ represent the fraction of users with degree k belonging to the four groups—ignorant, true-spreader, false-spreader, and removed, respectively. Then, the fraction of true-spreaders subject to k and t is

$$S_k^T(t) = \frac{\#\text{True} - \text{spreadernodeswithdegree}k\text{attime}t}{\#\text{Allnodetypeswithdegree}k\text{attime}t} \tag{17}$$

where the sum of all node types with degree k at time t can be denoted by N_k,

$$N_k = S_k^T(t) + S_k^F(t) + I_k(t) + R_k(t) \tag{18}$$

and the total population of the system N at given time t is obtained by

$$N = \sum_{k'} N_{k'} \tag{19}$$

Equation (17) can be extended for other groups as well. With a degree distribution $P(k)$, and the average degree of the network $\langle k \rangle = \sum_{k'} k' P(k')$, the probability of meeting a true-spreader with degree k at time t is

$$N_k \langle k \rangle \approx \text{Sumofdegreesoftype}k\text{attime}t$$
$$\approx \text{Alledgesalongdegree}k\text{attime}t \tag{20}$$

$$\#\text{True} - \text{spreaderswithdegree}k\text{at}t = S_k^T(t).N_k$$
$$kP(k)S_k^T(t)N_k = \text{Alledgesalongtrue} - \text{spreaderswithdegree}k\text{attime}t \tag{21}$$

Thus, the probability of reaching (i.e., coming in contact with) a true-spreader (of any degree) at time t is:

$$\sum_{k'} \frac{k'P(k')S_{k'}^T(t)}{\langle k \rangle} = \frac{\sum_{k'} k'P(k')S_{k'}^T(t)}{\langle k \rangle} \tag{22}$$

Equivalent is the probability of reaching (i.e., coming in contact with) a false-spreader (of any degree) at time t as

$$\frac{\sum_{k'} k'P(k')S_{k'}^F(t)}{\langle k \rangle} \tag{23}$$

Meanwhile, the probability of reaching a removed node is obtained via two scenarios:

$$\frac{\sum_{k'} k' P(k') S_{k'}^T(t)}{\langle k \rangle} + \frac{\sum_{k'} k' P(k') R_{k'}(t)}{\langle k \rangle} \text{ or } \frac{\sum_{k'} k' P(k') S_{k'}^F(t)}{\langle k \rangle} + \frac{\sum_{k'} k' P(k') R_{k'}(t)}{\langle k \rangle} \tag{24}$$

One reaches a removed node by either finding a removed node in the first place (reaching removed node via one of the edges of the active node) or by rendering an active node as removed. When an active true-spreader of a news item comes in contact with another active true-spreader of a news item, then one of the true-spreaders will recover as such an interaction is an indicator that the information is already widespread and may not warrant further propagation. The same applies to an interaction between two active false-spreaders. This is similar to the case where you, having already shared an interesting post, are less likely to share it again on the recommendation of a friend's timeline (who has also shared the same post).

Transition Probabilities

Based on the flow defined in Fig. 2 and the reachability equations derived in Sect. 7.4, we can obtain the transition probabilities along the various state transitions as:

- Moving from an ignorant state to true-spreader:

$$W_{I \to S^T}(t, k) = \lambda_T k N_k P(k) I_k(t) \frac{\sum_{k'} k' P(k') S_{k'}^T(t)}{\langle k \rangle} \tag{25}$$

Here, the first part of the equation $k N_k P(k) I_k(t)$ gives us how many edges (i.e., ignorant users along these edges with degree k) are present at time t. When one of these edges comes in contact with an active spreader (the probability of which is given by the second part of the equation), then with true-transfer rate λ_T, some ignorant users along the edges will convert to true-spreaders.
- Moving from an ignorant state to false-spreader:

$$W_{I \to S^F}(t, k) = \lambda_F k N_k P(k) I_k(t) \frac{\sum_{k'} k' P(k') S_{k'}^F(t)}{\langle k \rangle} \tag{26}$$

Similar to the previous case, here based on the false-transfer rate λ_F, some ignorant users transition to false-spreader.

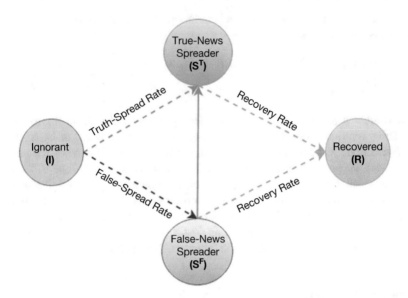

Fig. 2 The four-grouped Spread-Stifle model for simultaneous real/false information spread. The three parameters of the system are the real information spread rate, the false information spread rate, and the recovery rate. Only the paths indicated by the arrow are allowed. The dotted arrow indicates the probable paths, while the plain arrows indicate the absolute path. The color of the arrow is an indicator of the group it transitions to

- *Moving from a false-spreader to a true-spreader*: As discussed in one of the properties of the Spreader-Stifle model, whenever a false-spreader comes in contact with a true-spreader, the former *always* converts. Thus, with $\lambda_{FT} = 1$

$$W_{S^F \to S^T}(t, k) = \lambda_{FT} k N_k P(k) S_k^F(t) \frac{\sum_{k'} k' P(k') S_{k'}^T(t)}{\langle k \rangle} \qquad (27)$$

- *Moving from a true-spreader to removed/stifled*:

$$
\begin{aligned}
W_{S^T \to R}(t, k) &= \alpha k N_k P(k) S_k^T(t) \frac{\sum_{k'} k' P(k') S_{k'}^T(t)}{\langle k \rangle} \\
&\quad + \alpha k N_k P(k) S_k^T(t) \frac{\sum_{k'} k' P(k') R_{k'}(t)}{\langle k \rangle} \\
&= \alpha k N_k P(k) S_k^T(t) \frac{\sum_{k'} k' P(k') (S_{k'}^T(t) + R_{k'}(t))}{\langle k \rangle}
\end{aligned}
\qquad (28)
$$

True-spreaders coming in contact with other true-spreaders or removed nodes may lead to transitioning into removed state (with the rate of removal α).

- *Moving from a false-spreader to removed/stifled:*

$$W_{SF \to R}(t, k) = \alpha k N_k P(k) S_k^F(t) \frac{\sum_{k'} k' P(k') S_{k'}^F(t)}{\langle k \rangle}$$

$$+ \alpha k N_k P(k) S_k^F(t) \frac{\sum_{k'} k' P(k') R_{k'}(t)}{\langle k \rangle} \qquad (29)$$

$$= \alpha k N) k P(k) S_k^F(t) \frac{\sum_{k'} k' P(k') (S_{k'}^F(t) + R_{k'}(t))}{\langle k \rangle}$$

Similar to Eq. (28), false-spreaders coming in contact with other false-spreaders or removed nodes may lead to transitioning into removed group (with rate of removal α).
- *Moving from a false-spreader to removed/stifled.*

The above transitions probabilities depend on degree k, the overall transition probabilities between groups at time t is simply obtained via:

$$W_{I \to ST}(t) = \sum_{k'} W_{I \to ST}(t, k'),$$

$$W_{I \to SF}(t) = \sum_{k'} W_{I \to SF}(t, k'),$$

$$W_{SF \to ST}(t) = \sum_{k'} W_{SF \to ST}(t, k'), \qquad (30)$$

$$W_{ST \to R}(t) = \sum_{k'} W_{ST \to R}(t, k'),$$

$$\text{and } W_{SF \to R}(t) = \sum_{k'} W_{SF \to R}(t, k')$$

Mean-Field Rate of Change

Combining the information from the above transactions, we can obtain the mean-field rate of change for each of the four groups as follows:

- **Ignorant Group:** For the ignorant group, the net change in population is negative as any contact with the spreaders (true or false) causes some members of the ignorant group to move to the spreader's group. Using Eqs. (25) and (26), we get:

$$\frac{dI_k(t)}{dt} = -k I_k(t) \left(\lambda_T \frac{\sum_{k'} k' P(k') S_{k'}^T(t)}{\langle k \rangle} + \lambda_F \frac{\sum_{k'} k' P(k') S_{k'}^F(t)}{\langle k \rangle} \right) \qquad (31)$$

The negative sign in the above equation denotes the net loss in population.

- **True-Spreader Group:** For the true-spreader group, an increase in population is caused by the ignorant nodes and the false-spreader nodes moving into the true-spreader group. On the other hand, interactions among members and interactions with removed users lead to a part of true-spreaders being converted into removed users, causing a decrease in the population of the group. These movements, combined with Eqs. (25) and (27), give

$$
\frac{dS_k^T(t)}{dt} = (\lambda_T k I_k(t) + \lambda_F T k S_k^F(t)) \frac{\sum_{k'} k' P(k') S_{k'}^T(t)}{\langle k \rangle}
$$

$$
- \alpha k S_k^T(t) \frac{\sum_{k'} k' P(k')(S_{k'}^T(t) + R_{k'})}{\langle k \rangle}
$$

(32)

- **False-Spreader Group:** For this group, the only source of the increase in population is the movement of ignorant users to the false-group. In addition, due to the movements of users from false-group to the true-group or removed group, there is a decrease in population as well. Using Eqs. (26), (27), and (29), the rate of change for false-spreaders comes out to be

$$
\frac{dS_k^F}{dt} = \lambda_F k I_k(t) \frac{\sum_{k'} k' P(k') S_{k'}^F(t)}{\langle k \rangle}
$$

$$
- k S_k^F(t) \left(\frac{\lambda_{FT} \sum_{k'} k' P(k') S_{k'}^T(t)}{\langle k \rangle} + \alpha \frac{\sum_{k'} k' P(k')(S_{k'}^F(t) + R_{k'}(t))}{\langle k \rangle} \right)
$$

(33)

- **Removed Group:** There is always a net increase in the population of the removed group, as no node moves out. At each timestamp, a portion of the true/false-spreaders converts to stiflers. From Eqs. (28) and (29), the net gain in population of the removed group is given as

$$
\frac{dR_k(t)}{dt} = \alpha k \left(S_k^T(t) \frac{\sum_{k'} k' P(k')(S_{k'}^T(t) + R_{k'}(t))}{\langle k \rangle} + \right.
$$

$$
\left. S_k^F(t) \frac{\sum_{k'} k' P(k')(S_{k'}^F(t) + R_{k'}(t))}{\langle k \rangle} \right)
$$

(34)

In conclusion, the Spread-Stifle model can be easily extended to accommodate other possible transactions among the four groups, as well as incorporate metadata of the content other than its true-false nature. Based on the simulations run, Yang et al. [26] observed that in the case of multi-information spreading diffusion systems, degree and true-information rate seem to play an important role, impacting the overall density of the ignorant and the stifled population.

8 Strategies to Minimize the Spread of Fake News

To limit the spread of fake news, several approaches have been proposed. A simple way would be to employ content moderators and frequently scan the stories. However, it is expensive both in terms of manpower and computational complexity. The second option would be to choose some social media accounts which are highly suspicious and regularly scan their posts. The third option could be to employ honeypot accounts which are deemed to be vulnerable to fake news. When malicious users approach them to spread the fake news further, they will be detected automatically. A hybrid strategy would be to deploy both the honeypots and human employees to strategize the entire process.

Balmau et al. [1] proposed CREDULIX, a Bayesian approach to reduce the spread of fake news. From a high-level point of view, CREDULIX follows three steps:

1. First, the human fact-checking team creates the ground-truth by reviewing a few news items which were too viral in the past.
2. Second, each user's sharing behavior is probabilistically modeled based on the reactions on the already fact-checked items.
3. Third, CREDULIX predicts the probability of an unchecked news item to be fake or not based on the user behavior modeling. It further creates a threshold based on a cutoff probability threshold (p_0) which decides whether the news item should be shown in the timeline of the user or not, thus limiting the spread of misinformation.

Consider a user u in a social network. To model the user's behavior, CREDULIX computes two probabilities: $P_T(u)$, probability that u shares a news item if it is *true*, and $P_F(u)$, probability that u shares a news item if it is *false*. Note that these two probabilities are independent between users as the choice of sharing a news is solely determined by an individual. For a user u, CREDULIX creates a *User Credibility Record* (UCR), which is a tuple of the following values $(v_T(u), s_T(u), v_F(u), s_F(u))$, where $v_T(u)$ and $s_T(u)$ denote the number of fact-checked items viewed and shared by u, which are marked as *true*. Similarly, $v_F(u)$ and $s_F(u)$ denote the number of fact-checked items viewed and shared by u which are marked as *false*. Once the UCR is created, following functions are defined for u:

1. $\beta_1(u) = (s_T(u) + 1)/(v_T(u) + 2)$
2. $\beta_2(u) = (s_F(u) + 1)/(v_F(u) + 2)$
3. $\beta_3(u) = (v_T(u) - s_T(u) + 1)/(v_T(u) + 2)$
4. $\beta_4(u) = (v_F(u) - s_F(u) + 1)/(v_F(u) + 2)$

Based on Laplace's rule of succession, we have $P_T(u) = \beta_1(u)$ and $P_F(u) = \beta_2(u)$.

Finally, CREDULIX computes the likelihood of an unchecked news item. For an unchecked news item X, let V and S be two users who have viewed and shared the news item. The probability of X being fake is calculated as follows:

$$p(V, S) = g\pi_F(V, S)/(g\pi_F(V, S) + (1 - g)\pi_T(V, S)) \tag{35}$$

where

- $\pi_T(V, S) = \Pi_{u \in S}\beta_1(u)\Pi_{u \in V-S}\beta_3(u)$,
- $\pi_F(V, S) = \Pi_{u \in S}\beta_2(u)\Pi_{u \in V-S}\beta_4(u)$, and
- g is the estimated global fraction of fake news items in the social network, with $g \in (0, 1)$.

Note that g is randomly selected from the whole social network by fact-checking a set of news items.

To run CREDULIX in practice, Balmau et al. [1] proposed an improved version of using the *UCR scores* and *news item ratings*. In the improved version, *UCR* is updated (the values in the tuple mentioned previously) for u in the following way:

1. When the ground-truth of the news item is already available to the user (true or fake).
2. When u is exposed to fact-checking a news item.

The main idea behind incorporating these updates is that users who are not involved in encountering fact-checked items cannot contribute to CREDULIX. In addition to updating the *UCR*, one can also take into account the confidence of CREDULIX in determining an unchecked item as fake. We can compute *item ratings* ($\alpha(V, S)$) which is a similar metric like $p(V, S)$ as follows:

$$\alpha(V, S) = \pi_T(V, S)/\pi_F(V, S) \tag{36}$$

where V and S are the sets of users who viewed and shared the unchecked item X. We can also compute the cutoff rating threshold α_0 similar to the cutoff probability threshold p_0 as follows:

$$\alpha_0 = (1/p_0 - 1)/(1/g - 1) \tag{37}$$

The role of the rating threshold is that when $\alpha(V, S)$ for an unchecked news item X is less than the cutoff rating threshold, CREDULIX suppresses the unchecked news item X in the feed of the user.

9 Summary of the Chapter

This chapter presented state-of-the-art research on fake news diffusion. We first compared and contrasted the patterns of fake news diffusion across two major microblogging sites—Facebook and Twitter. The following observations are platform agnostic:

- Fake news propagates much faster and has wider spread compared to real news.
- Ordinary users (not the influential users or bots) play a major role in posting fake news.
- Bots play a significant role within the first few hours of the posting to spread the news rapidly. They target influential accounts through replies and mentions and pretend to be trustworthy users by retweets posts of other bots as much as a normal user does.
- Fake news is highly opinionated and moulded several times with the views of the spreaders, whereas real news is directly shared without much modification.

We also presented a dynamic linear threshold model, a percolation model, and a variant of the SIR model, called the Spread-Stifle model, to mimic the spread of fake news. We further showed how one can minimize the spread of fake news online.

However, many research questions are still unexplored; a few of them are mentioned below:

- It is well known that multi modal fake news (news containing text, audio, video, image) is more attractive than text-only fake news. There has been no study to understand whether adding multimedia content would help spread the message faster or not.
- It is not clear how the local topological structure affects the spread of fake news. For example, if a node is a part of a dense community vs. a part of a sparse community, which setting is more vulnerable for infecting the node quickly.
- There is no research which studies the relation between the psychological properties of users and the infection rate. For example, if a user has an agreeable personality, does she have a high chance to accept (get infected by) the fake news? It would be interesting to study how human personality (Big-5 model [16]) and values (Schwartz's values model [17]) affect the spread of fake news.
- None of the existing studies is able to explore the exact path through which a fake news will be propagated. It is very important for the security agency to know the exact route so that honeypots can be placed at appropriate places to minimize the spread.
- Early detection of fake news is important so that the damage can be prevented. Most of the misinformation detection and propagation models let the fake news remain active on social media for quite some time so that sufficient metadata can be collected and used in the models. However, the news may have caused sufficient damage by then.
- More research needs to be done on the prescription of efficient and cost-effective defensive mechanism. For example, if a political news has been fact-checked and marked as fake, what would be the spreading mechanism of the fact-checked news so that the damage can be reversely healed, i.e., those who have been affected by the fake news would come to know that the consumed news is fake.

Acknowledgment Tanmoy Chakraborty would acknowledge the support of Sarah Masud in writing the chapter.

References

1. Balmau, O., Guerraoui, R., Kermarrec, A.M., Maurer, A., Pavlovic, M., Zwaenepoel, W.: Limiting the Spread of Fake News on Social Media Platforms by Evaluating Users' Trustworthiness (2018). arXiv preprint: 180809922
2. Basu, K.: Identification of the source (s) of misinformation propagation utilizing identifying codes. In: Companion Proceedings of The 2019 World Wide Web Conference, pp 7–11 (2019)
3. Brauer, F.: An introduction to networks in epidemic modeling. In: Mathematical Epidemiology, pp 133–146. Springer, Berlin (2008)
4. Chetan, A., Joshi, B., Dutta, H.S., Chakraborty, T.: Corerank: ranking to detect users involved in blackmarket-based collusive retweeting activities. In: Proceedings of the Twelfth ACM International Conference on Web Search and Data Mining, pp. 330–338 (2019)
5. Del Vicario, M., Bessi, A., Zollo, F., Petroni, F., Scala, A., Caldarelli, G., Stanley, H.E., Quattrociocchi, W.: The spreading of misinformation online. Proc. Natl. Acad. Sci. **113**(3), 554–559 (2016)
6. Dutta, S., Das, D., Chakraborty, T.: Modeling engagement dynamics of online discussions using relativistic gravitational theory. In: Proceeding of the 2019 IEEE International Conference on Data Mining (ICDM), pp 180–189 (2019)
7. Glenski, M., Weninger, T., Volkova, S.: Propagation from deceptive news sources who shares, how much, how evenly, and how quickly?. IEEE Trans. Comput. Soc. Syst. **5**(4), 1071–1082 (2018)
8. Goel, S., Watts, D.J., Goldstein, D.G.: The structure of online diffusion networks. In: Proceedings of the 13th ACM Conference on Electronic Commerce, pp. 623–638 (2012)
9. Guille, A., Hacid, H., Favre, C., Zighed, D.A.: Information diffusion in online social networks: a survey. ACM Sigmod Rec. **42**(2), 17–28 (2013)
10. Jang, S.M., Geng, T., Li, J.Y.Q., Xia, R., Huang, C.T., Kim, H., Tang, J.: A computational approach for examining the roots and spreading patterns of fake news: evolution tree analysis. Comput. Hum. Behav. **84**, 103–113 (2018)
11. Jiang, C., Chen, Y., Liu, K.R.: Evolutionary dynamics of information diffusion over social networks. IEEE Trans. Signal Process. **62**(17), 4573–4586 (2014)
12. Kitsak, M., Gallos, L.K., Havlin, S., Liljeros, F., Muchnik, L., Stanley, H.E., Makse, H.A.: Identification of influential spreaders in complex networks. Nat. Phys. **6**(11), 888–893 (2010)
13. Litou, I., Kalogeraki, V., Katakis, I., Gunopulos, D.: Real-time and cost-effective limitation of misinformation propagation. In: Proceedings of the 2016 17th IEEE International Conference on Mobile Data Management (MDM), vol. 1, pp. 158–163. IEEE, New York (2016)
14. Miritello, G., Moro, E., Lara, R.: Dynamical strength of social ties in information spreading. Phys. Rev. E **83**(4), 045102 (2011)
15. Nguyen, H., Huyi, C., Warren, P.: The Propagation of Lies: Impeding the Spread of Misinformation by Identifying and Invading Echo Chambers in Networks Introduction and Objectives (2017)
16. Rothmann, S., Coetzer, E.P.: The big five personality dimensions and job performance. SA J. Ind. Psychol. **29**(1), 68–74 (2003)
17. Schwartz, S.H.: An overview of the Schwartz theory of basic values. Online Read. Psychol. Cult. **2**(1), 2307–0919 (2012)
18. Shao, C., Ciampaglia, G.L., Flammini, A., Menczer, F.: Hoaxy: a platform for tracking online misinformation. In: Proceedings of the 25th International Conference Companion on World Wide Web, pp. 745–750 (2016)
19. Shao, C., Ciampaglia, G.L., Varol, O., Yang, K.C., Flammini, A., Menczer, F.: The spread of low-credibility content by social bots. Nat. Commun. **9**(1), 1–9 (2018)
20. Shu, K., Sliva, A., Wang, S., Tang, J., Liu, H.: Fake news detection on social media: a data mining perspective. ACM SIGKDD Explor. Newsl. **19**(1), 22–36 (2017)
21. Solorio, T., Hasan, R., Mizan, M.: Sockpuppet detection in Wikipedia: a corpus of real-world deceptive writing for linking identities (2013). arXiv preprint: 13106772

22. Ukkonen, E.: Approximate string-matching and the q-gram distance. In: Sequences II, pp 300–312. Springer, Berlin (1993)

23. Varol, O., Ferrara, E., Davis, C.A., Menczer, F., Flammini, A.: Online human-bot interactions: detection, estimation, and characterization. In: Proceedings of the 11th International AAAI Conference on Web and Social Media (2017)

24. Vosoughi, S., Roy, D., Aral, S.: The spread of true and false news online. Science **359**(6380), 1146–1151 (2018)

25. Wang, D., Wen, Z., Tong, H., Lin, C.Y., Song, C., Barabási, A.L.: Information spreading in context. In: Proceedings of the 20th International Conference on World Wide Web, pp. 735–744 (2011)

26. Yang, D., Chow, T.W., Zhong, L., Tian, Z., Zhang, Q., Chen, G.: True and fake information spreading over the facebook. Phys. A Stat. Mech. Appl. **505**, 984–994 (2018)

27. Yuan, H., Chen, G.: Network virus-epidemic model with the point-to-group information propagation. Appl. Math. Comput. **206**(1), 357–367 (2008)

Neural Language Models for (Fake?) News Generation

Santhosh Kumar G

Abstract Recent progress in natural language processing (NLP), in conjunction with deep neural network models, has created a broad interest in natural language generation and many downstream applications of NLP. Deep learning models with multitudes of parameters have achieved remarkable progress in machine-generated news items indistinguishable from human experts' articles. Though the developed techniques are for authentic text generation and entertainment purposes, its potential use in social media for propaganda, defamation, and misinformation propagation poses a considerable challenge to the research community. This chapter attempts to present a study on various pre-trained neural models for natural language processing in general and their potential use in news generation. While showing these models' limitations, the chapter describes the future works in the NLP domain on language generation.

Keywords NLP · Neural language generation · Pre-trained language models · Fake news

1 Introduction

Natural language generation (NLG) produces a written or spoken narrative in a specific language like English from non-linguistic representations learned from structured or unstructured data. The automatic generation of NLG, thereby enabling seamless communication between a machine and humans, is a long continuing desire of computer scientists since Turing's famous question: Can machines think? The recent progress in natural language processing coupled with deep learning methods has given a fresh lease of enthusiasm to researchers working in this direction. The primary ingredients of natural language processing are models for natural language understanding (NLU) and NLG. The existing models and techniques have already proven their strength in tasks like translation, summarization, image/video annotation, and in conversational systems. However, endeavors connected with free text generation like story understanding, story generation, and other creative writing types at par with human abilities still pose a challenge. This chapter attempts to present a study on various neural models for natural language generation and their potential use in news generation.

2 Modeling Approaches

Natural languages are not like formal languages; they emerge, and changes are inevitable; moreover, they are prone to ambiguities. Therefore, devising a formal specification for a natural language is a bit challenging. Language models learn the syntactic order of a language from a given set of examples as probability values to predict a sequence of words. Language models operate in the context of characters, words, sentences, and even paragraphs, and among them, word-level models are the most common ones. A language model is an indispensable ingredient in many advanced natural language processing tasks such as text summarization, machine translation, language generation, speech recognition, parts-of-speech tagging, and parsing. Language models developed for NLP tasks have evolved from statistical language models to neural language models. Approaches such as n-gram and hidden Markov models have become the primary ingredient in developing statistical language models. In contrast, various deep learning neural networks build neural language models. Recent research efforts coupled with the constant improvement of deep learning architectures and current language models have outshined the past statistical language models in their efficiency and effectiveness.

The first attempts to use computers for other than number-crunching tasks were found successful after the public demonstration of the Georgetown-IBM for machine translation. The system succeeded in translating around 60 sentences from Russian to English with a dataset of 250 words and 6 grammar rules [1]. With the adoption of data-driven approaches and shallow artificial intelligence techniques, automatic learning features from the data came into prominence. In this, the machines were capable of automatically determining the patterns in the data, discarding the approaches of handcrafted features and rule bases followed in the earlier knowledge-based NLP systems. The IBM models [2] designed for machine translation based on statistical techniques were capable of extracting *alignments* between two language pairs and were able to learn parameters from the training data with the help of the expectation maximization (EM) algorithm. However, as the number of linguistic features increases, the model becomes complicated and challenging to compute. Though many discriminative methods have provided further improvements on many NLP tasks, the lack of models' generalization capability restricted the use of these models in practical applications. With the recent return of neural networks, deep learning-based models for NLP workloads trained with embedded representations have set unparalleled performances and reignited hopes of having computers understand the semantic underpinnings of natural language and their generation for various applications. Moreover, the neural methods offer an end-to-end trainable system.

2.1 Learning Paradigms for NLG

This section provides a brief account of machine learning paradigms of language generation that aim at improving the generalization capability of a model to various tasks with the help of other closely associated workloads. Though the learning methods equally apply to multiple problems, including language, vision, and speech, the discussion here will be on language generation.

Text-to-Text: This is to convert a piece of a source text into a target text.
Data-to-Text: This approach converts images, tables, records, graphs, and other ontologies into text.
Control free: This is for free-form text generation in creative domains like poetry writing.

Based on the length of the sentences generated, and the input modalities, the following classification of the NLG tasks is possible: *sentence-level, discourse-level*, and *cross-modal generation*. Some of the prominent tasks in a sentence-level generation are machine translation, text simplification, and production of paraphrase. In contrast, discourse-level generation considers abstract summarization, review generation, heading, and story generation. Image annotation and video captioning are tasks accomplished through cross-modal generation techniques. Figure 1 shows the types of text generation approaches.

2.2 Language Models

Language models and natural language generation have a close relationship. A typical language model is defined as

$$P(Y) = \prod_{t=1}^{T} P(y_t|y_{(<t)}) \tag{1}$$

In language modeling, the task is to predict the next token conditioned on the tokens generated until the current time steps. In the text generation tasks like story generation, this process is additionally conditioned, as shown in Eq. 2, with their input source. This source can be in any mode (text, tables, images, etc.). So, the text

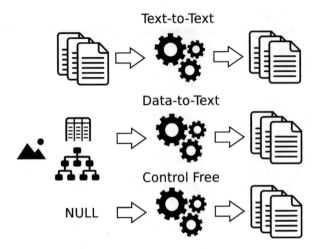

Fig. 1 Text generation approaches

generation task forms a natural extension to language modeling.

$$P(Y|X) = \prod_{t=1}^{T} P(y_t|X, y_{(<t)}) \tag{2}$$

Many modern neural natural language models fall into the category of the encoder–decoder with attention (The Transformer) [3] or autoregression model. It is interesting to note that the natural language generation only needs either a stack of encoders or decoders to perform the task.

2.3 Encoder–Decoder Attention

Typically, the generation part uses the paradigm of encoder–decoder attention mechanism as shown in Fig. 2. The modelers in the encoder and the decoder are usually composed of RNNs [4], CNNs [5], and such transformers. The conditioning on $y_{(<t)}$ time step is generally modeled with long short-term memory (LSTM) [6] or transformers using masking techniques to achieve the sequential context in the case of transforms.

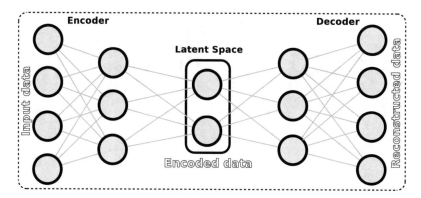

Fig. 2 Encoder–decoder model

2.4   Autoregression

The joint distribution of text Y over time steps T represented as factors is given by Eq. 1. The Bayesian network with no conditional independence assumptions follows *autoregressive* property. In an autoregressive generative model, the conditionals are written as $p_{\theta_i}(y_i|y_{(<i)})$, where θ_i's represent the fixed model parameters. To get more expressive power, the parameters can be made flexible with neural architectures. A typical autoregressive process with t time steps is shown in Fig. 3.

2.5   Seq2Seq Model

A sequence to sequence model maps a fixed-length input (x_n) to a fixed-length output (y_k), where n and k may differ. A stack of LSTM or Gated Recurrent Unit (GRU) [7] cells accepts each input element, processes it, and propagates it forward. An encoder vector represents the final representation learned from the input sequence. The decoder part again is a stack of recurrent units, trying to predict output at each time step based on the previous units' hidden states. The seq2seq

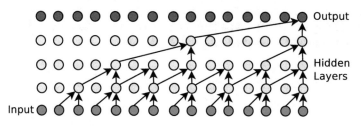

Fig. 3 Autoregressive generative model

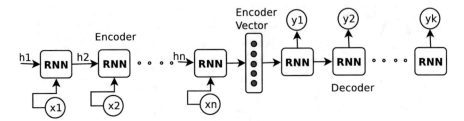

Fig. 4 Seq2seq model

models are used for pre-training language models for language generation. Figure 4 shows a typical encoder–decoder sequence to sequence model.

3 Learning Paradigms

Language models are built using multiple machine learning techniques. This section describes the main methods to construct them.

3.1 Supervised Learning Techniques

Bengio et al. [8] present one of the first neural probabilistic language models and the capacity of neural networks in text generation. In a typical supervised learning setting, the task is to train the model with maximum likelihood estimate (MLE) as objective with model parameters θ_i given by $-\sum_t^T logp(y_i|X, y_{<t}; \theta)$ for sequence generation. The length of the sentences generated is limited in this setting. $-\sum_t^T logp(y_i|X, y_{<t}; \theta)$ for sequence generation. The length of the sentences generated is limited in this setting. RNNs introduce a general approach for language modeling with Markov property enabling encoding of the variant length inputs into vectors to infer tokens in the next sequence. With the teacher forcing applied during the training, the difference in exposure to the ground truth in the training and inference stages leads to a problem called exposure bias. This makes teacher forcing (MLE) an infeasible method for the generation of long sentences. An alternative way of training RNNs [9] with a technique known as professor teaching improves the generated sequence length during the training phase.

3.2 Adversarial Learning Techniques

Adversarial learning [10] with competing objectives addresses the mismatch in conditions and thereby behavior during the training and inference stages when teacher forcing is applied. Adversarial domain adaptation brings the behavior of the training and sampling very close to each other. The two main components, the discriminator and the generator, enable this. The generator model generates data that is indistinguishable from the original data. At the same time, the discriminator estimates the probability that a sample came from the training data. Parameter sharing between the competing objectives achieves this, and it is shown that the professor forcing mechanism acts as a conventional regularizer. GANs are originally designed and applied for continuous data, and language space is discrete. It is shown that the gradient of the discriminator guides how to change the synthetic data or generated data and by what margin to make it more realistic. Generative adversarial networks (GANs) and many variants [11–14] can produce text that is stylistic and catchy, making these machine-generated texts a potential source for fake news generation.

3.3 Reinforcement Learning Techniques

Reinforcement based learning [15] coupled with policy gradient algorithms [16] address the mismatch in the objective function trained to optimize, and the metrics used to measure the quality of the generated text. However, the lack of perfect metric and the computational cost involved therein make these approaches inefficient.

3.4 Embedding Techniques

Neural network embeddings are useful for creating continuous vector representations learned from a discrete space and usually mapped from a higher dimensional space to a lower-dimensional space. There are many techniques developed for embedding natural language to feed those in a deep learning input layer. Word2Vec [17] considers each word in the text as a context, and similar words appear nearer in the embedded lower-dimensional space. Similarly, Doc2Vec [18] tags the text and produces tag vectors such that the authors who use similar words will have corresponding vectors closer. fastText [19] considers an improved approach over the earlier method, where parts of words and characters are taken into account while building embeddings. In the GloVe [20] approach, embeddings assume a combination of word vectors relating to their co-occurrence probability in the corpus. Unsupervised pre-training of word embeddings has become a valuable ingredient in deep learning for natural language generation and many other downstream tasks.

The success of pre-trained Convolutional Neural Networks (CNNs) and transfer learning in the computer vision domain has brought interest to test them in a discrete space like text. Many such models proved their representational ability, which forms the discussion in the following sections.

4 Pre-trained Language Models

Language models pre-trained on huge textbases, including a wide variety of sources, form the basis for many downstream NLP tasks. Though early statistical models have proven their effectiveness in language modeling, neural language models outperformed many of them. This success can be attributed to the availability of data and the inherent capacity of neural networks distributed representation to deal with the curse of dimensionality and, to some extent, the generalizability of the models. This section covers such models and their role in NLG.

4.1 Contextualized Word Vectors (CoVe)

CoVe [21] are word embeddings obtained from an encoder of a machine translation (MT) model. CoVe contextualizes the word vectors for the entire input sentence and captures the syntactic and semantic meaning. The addition of these context vectors (CoVe) has shown improved performance over the standard word embedding techniques that use unsupervised words and character vectors. For a given sequence of words represented by w, the sequence of CoVe vectors produced by the Machine Translation LSTM is $CoVe(w) = MT - LSTM(GloVe(w))$, where GloVe(w) represents the sequence of word vectors generated by the GloVe model. The difference between the GloVe and CoVe word embeddings is that GloVe learns word embeddings based on the ratios of global word co-occurrences and finds difficulty in capturing the sentence context. In contrast, CoVe captures contextual information as it is generated by processing text sequences. At first, a GloVe vector concatenates with its corresponding vector in CoVe in the task-specific model architecture. Later, it is fed into the downstream task model as additional features. This is represented as $\hat{w} = [GloVe(w); CoVe(w)]$. Since CoVe uses task-specific architecture, the performance is limited, and the pre-training process depends on the datasets available for supervised translation tasks. The information extracted by CoVe becomes equivalent to both the word-level information produced by GloVe and character-level information by n-gram embeddings.

4.2   Embeddings from Language Model (ELMo)

The ELMo [22] vectors are character-based deep representations learned based on the entire context combining all layers of a bidirectional language model pre-trained on a large dataset. While the forward model computes the probability of a token from the given history, the backward language model runs in the reverse order over a sequence and predicts the previous token given the future. The model $ELMo = \gamma \sum_{i=0}^{L} \frac{1}{Z} exp\lambda_i \mathbf{h}^k$, where L layers of the language model is formed from the linear combinations of λ_is and the scalar γ. The kth language model layer is given by $\mathbf{h}^k$, and Z is the normalization constant. The model has shown improvements in many NLP tasks compared to the existing works.

4.3   BERT

Bidirectional Encoder Representation of Transformer language (BERT) [23] model utilizes the encoder part of the encoder–decoder architecture to learn dependencies from both the left to right and the right to left. Google released two versions of BERT: BERT base having 12 transformer blocks and 110 million parameters and BERT large with 24 transformer blocks and 340 million parameters. Another version released by Google (multilingual BERT) accelerated the NLP domain research in most of the workloads. BERT uses the idea of masked language model to train the model for optimized loss. Rather than optimizing the next word's likelihood, BERT randomly masks the tokens with a unique token [MASK] at the designated proportion, and the model optimizes the cross-entropy loss to predict the correct words to replace the masked ones. It also uses a binary classification to decide whether a generated sentence naturally follows the previous one. Additionally, the pre-training stage uses [CLS] token for classification and [SEP] for sentence separation to integrate the sentences. The BERT model used preprocessed Wikipedia words and Book Corpus for training. Although BERT is a successful model for many natural language tasks, including sentence classification, question answering, and named entity recognition, the model needs many fine tunings for text generation. The model can generate text embeddings by concatenating the last few layers of the encoder.

4.4   RoBERTa

RoBERTa [24] released by facebook[1] is an improved version for training BERT models that outperform all of the post BERT methods. RoBERTa trains the model longer over more data by splitting it into bigger batches. The next sentence prediction layer of BERT is no longer needed for this model, and the ability to change the masking pattern during training dynamically is an added advantage. A large new dataset of CC-NEWS having a size similar to other private datasets is used to control training size effects. BERT models perform static masking by performing the masking task just once during the data preprocessing phase. The training data was duplicated ten times for masking the sequence in 10 different ways over 40 epochs in the training process. This training method is adopted to avoid using the same mask for each training instance in every epoch, and a training sequence is seen with the same mask four times during training. RoBERTa uses dynamic masking, where a new masking pattern is generated whenever a new sequence is fed into the model, and this is an essential step when the model is pre-trained for more number of steps with large datasets. RoBERTa improves the BERT model's performance by increasing the batch size and controlling the number of passes through the training data. The perplexity for the masked language modeling and end-task accuracy is improved by adopting large batches training. RoBERTa easily parallelizes large batches with distributed data parallel training and experiments with 8K sequences batches. RoBERTa presents alternatives for BERT's training strategies, which leads to better performance of various downstream tasks. This method indicates improved downstream tasks by using more data for pre-training and using a novel dataset, CCNEWS. Also, masked language model pre-training under the right design choices proved competitive with all the other published works.

4.5   Transformer-XL

Due to the use of fixed segment size text during an encoder's training process, the model is unable to predict the first few symbols. The context fragmentation problem thus prohibits to capture the long-term dependencies in the text. Dai et al. [25] address this with the introduction of a segment-level recurrent mechanism with relative positional encodings to reuse the previous hidden states during training and resolve the information bias.

---

[1] https://ai.facebook.com.

4.6 Larger Language Models

The potential of natural language generation in business applications is enormous. Companies like Google, Microsoft, Facebook, NVIDIA, and OpenAI are investing in building more complex language models to reap the benefits of language generation to their existing business applications. Following is a brief discussion on such existing models.

MegatronLM To advance the state-of-the-art NLP applications, larger and larger language models with vast amounts of parameters usually make it hard to train. In 2019, NVIDIA claimed the largest transformer-based model ever trained with 8.3 billion parameters [26]. The architecture uses both the encoder and decoder stacks toward text generation. They also use the idea of model parallelism along with data parallelism to split the parameters across 512 GPUs and execute the training efficiently.

Turing-NLG Turing natural language generation is from Microsoft with 17 billion parameters. Turing-NLG [27] has demonstrated free text generation, among other many NLP tasks. Microsoft has utilized the same NVIDIA MegatronLM framework but with fewer GPUs with the help of DeepSpeed[2] an open-source library that is compatible with PyTorch. Turing-NLG is a stack of 72 transformers with 4256 hidden states and 28 attention heads. The architecture also utilizes a novel memory optimization technique called Zero Redundancy Optimizer (ZeRO) [28]. The comparison with GPT-2[29] and MegatronLM on standard language tasks yielded better results.

Generative Pre-trained Transformer (GPT) GPT is a series of announcements of generative models from OpenAI.[3] The releases from GPT [30] to GPT-3 [31] have shown significant improvement in capturing linguistic features by processing long-term dependencies of a language model by pre-training on a broad category of datasets. GPT-3 is an autoregressive language model that uses a massive number of parameters compared to its previous versions. To explore the strength and weaknesses of the model, OpenAI has released APIs to benefit developers of NLP with a warning against abusive use of it. GPT-3 language model is a profound model with in-context learning capability with ten times more parameters than the past version, GPT-2. The model is pre-trained on about a large portion of a trillion words and accomplishes best in class execution on a few NLP benchmarks without calibrating. GPT-3 is a deep architecture with 175 billion parameters; it tailors and heightens the GPT-2 design; it additionally includes a balanced introduction, pre-standardization, and inconsistent tokenization. It considers an effective execution of different NLP undertakings and benchmarks in three unmistakable shots, such as zero-shot, one-shot, and fine-shot environments. GPT-3 can execute most of the

---

[2]https://github.com/microsoft/DeepSpeed.
[3]https://openai.com.

natural language jobs, even without requiring tweaking for a particular assignment. Authors claim that the model is powerful enough to generate samples of news articles in which individual evaluators have difficulty differentiating from articles written by humans.

4.7 XLNet

BERT follows an autoencoding-based pre-training technique to generate text from the corrupted text (some of the input words are masked). Due to this, the model is inherently incapable of modeling long-term dependencies in natural languages since the BERT model implies that [MASK] tokens are independent of each other, given the unmasked tokens [25]. Yang et al. proposed XLNet [32], an autoregressive approach toward language modeling. It tries to address the shortcomings of the autoencoder-based BERT model. Autoregressive models are well suited to generative NLP tasks since they generate the context in the forward direction. XLNet proposes a new idea called permutation language modeling: by performing permutations on the given sequence and trying to predict the $(t+1)$th token with the help of previous t tokens. It uses an attention mask to permute factorization order. XLNet can only use either forward or backward context, not both at the same time. Also, to predict the token at time step t, the model should see only the token position at time step t and not the content of the token at t. Similarly, to predict the token at time step t, the model should encode all tokens before step t as content. XLNet uses two self-attention models, content stream attention, and query stream attention to manage this. The proposed model beats all the versions of BERT and RoBERTa in many NLP tasks.

4.8 GROVER

Zellers et al. [33] present GROVER as a controllable text generation model. From a headline, the system is capable of writing stories that are more trustworthy than a human-written story. Authors represent each news article as a joint probability distribution on various metadata such as the article's domain, author names, published date, article heading, and the body text associated with a news article. The sampling is done concerning a set of defined fields with start and end tokens. During article generation, each context is generated from a partial context. The whole idea is to describe the probability of the forthcoming token dependent on the metadata. The model is trained to optimize cross-entropy, while tokens in each context are generated in the required order. GROVER uses the same architecture that of GPT-2. The Common Crawl dumps of news articles from 5000 domains indexed by Google News are used for training the model. The disinformation stories generated by GROVER tested with human evaluators brought exciting results. The GROVER is capable of generating stories with trustworthiness, style, and sensibility. The success brought out the importance of research in neural fake news generation and its mitigation. Interestingly, GROVER itself will act as the system for defense against such system-generated fake news.

4.9 CTRL

Keskar et al. [34] propose a Conditional TRansformer Language (CTRL) model for controllable text generation. The model incorporates many control codes for a domain, task-specific behavior, content, and style of the document to control the generated text explicitly. The language model shown in Eq. 1 is modified to incorporate the control code c as

$$P(Y|c) = \prod_{t=1}^{T} P(y_t|y_{(<t)}, c) \tag{3}$$

The model learns the parameterized $P_\theta(y_t|y_{(<t)}, c)$ by training on a sequence of input with control codes appended to it. Since the model uses explicit control codes, the relationship between the generated model and the training data can be reasoned. The authors also discuss the ethics of releasing the code and models for the scientific community for responsible use. The model trained on huge data collected from varieties of sources yielded better results on many NLP tasks.

Table 1 summarizes recent releases of pre-trained language models for text generation.

Table 1 Recently released pre-trained language models

Model name	Pre-training method	Released by	Parameters
ELMo [22]	Bidirectional LSTM	Allen inst. for AI	94 million
GPT [30]	Transformer encoder	OpenAI	110 million
MT-DNN [35]	Multi-task learning	Microsoft	330 million
BERT-Large [23]	Bidirectional encoder	GoogleAI	340 million
XLNet [32]	Autoregressive	CMU	340 million
RoBERTa [24]	Dynamic masking	Facebook	355 million
GROVER [33]	Transformer	University of Washington	1.5 billion
GPT-2 [29]	Transformer decoder	OpenAI	1.5 billion
CTRL [34]	Transformer	Salesforce	1.6 billion
Megatron [26]	Encoder–decoder	NVIDIA	8.3 billion
Turing-NLG [27]	Transformer	Microsoft	17 billion
GPT-3 [31]	Autoregressive	OpenAI	175 billion

4.10   Seq2Seq Pre-training Models

UniLM Dong et al. [36] propose a language model trained on three types of tasks: uni- and bidirectional and sequence to sequence prediction. UniLM is a unified approach applicable to NLU as well as to NLG. It is essentially a multi-layer transformer optimized jointly to meet many NLP objectives. A new seq2seq language model, which is bidirectional encoding, is followed by a unidirectional decoder. UniLMv2[37] utilizes both autoencoding and partially autoregressive methods along with a pseudo-masked language model to efficiently pre-train the model. While an autoregressive model emits tokens one at a time, the proposed model generates one or multiple tokens. It is shown that UniLM beats base models of BERT, RoBERTa, XLNet, and BART upon standard benchmark tests.

MASS Masked seq2seq pre-training model for language generation [38] attempts to improve upon BERT by using a different strategy for masking the input sentence and fully utilize the encoder–decoder architecture. Complementary masking encourages joint training so that the encoder masks the sequence of length k, and the decoder predicts the same sequence of length k, and every other token is masked for the decoder. The tokens masked in the decoder are the tokens that are not masked in the encoder. The encoder supports the decoder by extracting the useful information from the masked sentences, including the model's NLU capability. Since the sequence of length, k is decoded consecutively, and the NLG capability is improved. When $k = 1$, the model behaves like BERT, which is biased toward the encoder, and when $k = sentencelength$, the model is biased toward the decoder, which is the behavior of the GPT model.

BART BART [39] generalizes BERT and GPT language models. The input document encodes with a bidirectional model and computes the likelihood of the input document with an autoregressive decoder's help. The primary approach is to denoise the corrupted text with random noise and learn to reconstruct the original text. The suggested transformations for document corruption are token masking, token deletion, token infilling, sentence permutation, and document rotation. BART performs well in sequence generation tasks such as summarization and abstractive question answering.

4.11   Discussion

The following points summarize the findings obtained from the review of pre-trained neural language models.

– Most of the language models use a stack of encoders or decoders to generate text. Transformers use attention mechanisms to enhance model efficiency.
– Though the autoencoder scheme is useful in text generation, the trend is toward autoregressive techniques.

- Most of the models have many layers and vast numbers of parameters, and the text generation process is resource-hungry (CPU/GPU) and inefficient concerning memory and latency.
- Deep neural networks suffer from *overthinking*, a property defined considering the network as a whole by Wang et al. [40], and as a property closely associated with the internal state of the network by Kaya et al. [41]. Overthinking is computationally wasteful and can affect the quality of the final output. Works on shallow–deep learning by modifying the CNN layers and early exit schemes [42] are useful in this direction.
- DNNs are highly susceptible to carefully crafted adversarial attacks. Though it is difficult to attack the models that operate in the discrete domain such as text compared to those that work in the continuous domain (image, speech, video), the lack of metrics to assess the machine-generated text's originality is a big challenge.

5   (Fake?) News Generation and Future Prospects

The surge of interest in recent times to develop neural language generation models with the claim of unprecedented success in news generation raises concerns about the potential use of these models for fake news generation. The leading media have already warned their readers about deep fakes and misleading stories circulating online. Despite the success claimed by GPT-3, GROVER, Turing-NLG, and the similar architectures, all are far from the goal of reaching artificial general intelligence. The following are some of the reasoning behind this thought, along with future directions in this regard.

- Bisk et al. [43] postulated the idea of World Scope (WS) to examine the progress of NLP research and attempted to formalize corpora's limitations used to train the huge language models in terms of the language and experience. Primarily, the language models trained on datasets curated from the Internet lack human experience captured through many other forms, including embodiment, perception, and social connections. Despite the big datasets or computationally efficient platforms, these aspects are hard to catch. Currently, NLP applications benefitted through the models' ability to capture symbolic co-occurrences by learning distributions from the massive datasets. The next level of research in NLP to capture contextual meaning, visual context, and the emergence of language through social connections is highly required.
- The most extensively used metrics for fake news discovery in the NLP domain include Precision, Recall, F1, and Accuracy. These metrics, unfortunately, let only the evaluation of the performance of the classifier from different perspectives. It would be interesting to check the accuracy levels claimed by Turing-NLG, Megatron, and GPT-2 on LAMBADA [44] dataset that captures human intuition by predicting accuracy in the next word prediction. All the

models' prediction accuracy reported around 68%, which is far from the perfect score. Although the GPT-3 with 175 billion parameters reports mean human accuracy in detecting the machine-generated article somewhere around 52%, it is dramatically less than that of a smaller model with 125 million parameters. This fact indicates that the model requires more training examples to capture the necessary features. Moreover, the same model failed at the arithmetic calculation task of five-digit addition and subtraction given to it despite many examples existing in the training set. Additionally, most models still generate nonfactual and unintelligible articles.

– All the developed language models lack evidence on learned semantic representations that better capture abstract concepts. The world model available to the machine chiefly manifested by the Internet data is mostly in text format. Lake et al. [45] argue that building semantic representations derived from psychology that capture the perception of the world and ability to change it based on the input, generation of words based on the internal desire, goals, and appropriate response to the instructions would overcome the inability of text-only representations. Since these psychological aspects are inherently multimodal, the development of machine learning techniques that ground words to the context and higher-order semantics available through other modes is essential. Trott et al. [46] bring out the importance of psycholinguistic and cognitive semantics in language modeling and highlight the importance of construal—the process of human perception, understanding, and interpretation of the world around them.

– The openness of research in the field of AI, algorithmic regulations, trust, and confidence building is highly required to minimize the potential misuse of the fruits of this remarkable field.

6 Conclusion

An attempt to review the existing neural language generation schemes revealed many lines of attempts to develop the pre-trained language models. Deep learning systems supersede traditional statistical approaches. One line of technique uses the encoder–decoder framework with an attention mechanism. A stack of either encoder or decoder is useful in text generation, but some models use both the encoder and decoder blocks. The sequence to sequence models with masking techniques yield good results. Autoregressive schemes are alternative to GAN models and usually perform better when compared with the other models. There is an increasing trend in announcing more expensive neural generation models from the AI community. These models have to improve on the representation schemes to comprehend the semantics of the language. Finally, openness in research and the availability of artifacts thus generated in this field are of utmost importance to drive the world toward a better future.

References

1. Hutchins, J.: The first public demonstration of machine translation: the Georgetown-IBM system, 7th January 1954 (2005)
2. Brown, P.F., Cocke, J., Della Pietra, S.A., Della Pietra, V.J., Jelinek, F., Lafferty, J., Mercer, R.L., Roossin, P.S.: A statistical approach to machine translation. Computational Linguistics **16**(2), 79–85 (1990)
3. Vaswani, A., Shazeer, N., Parmar, N., Uszkoreit, J., Jones, L., Gomez, A.N., Kaiser, L., Polosukhin, I.: Attention is all you need. In: Advances in Neural Information Processing Systems, pp. 5998–6008 (2017)
4. Mikolov, T., Karafit, M., Burget, L., Cernocky, J., Khudanpur, S.: Recurrent neural network based language model. In: Eleventh Annual Conference of the International Speech Communication Association, pp. 1045–1048 (2010)
5. Collobert, R., Weston, J.: A unified architecture for natural language processing: deep neural networks with multitask learning. In: Proceedings of the 25th International Conference on Machine Learning, ICML '08, pp. 160–167 (2008)
6. Hochreiter, S., Schmidhuber, J.: Long short-term memory. Neural Computation **9**(8), 1735–1780 (1997)
7. Cho, K., van Merrienboer, B., Gulcehre, C., Bahdanau, D., Bougares, F., Schwenk, H., Bengio, Y.: Learning phrase representations using RNN encoder-decoder for statistical machine translation (2014). arXiv:1406.1078
8. Bengio, Y., Ducharme, R., Vincent, P., Jauvin, C.: A neural probabilistic language model. J. Mach. Learn. Res. 1137–1155 (2003)
9. Lamb, A.M., Goyal, A.G., Zhang, Y., Zhang, S., Courville, A.C., Bengio, Y.: Professor forcing: A new algorithm for training recurrent networks. In: Advances in Neural Information Processing Systems, pp. 4601–4609 (2016)
10. Goodfellow, I., Pouget-Abadie, J., Mirza, M., Xu, B., Warde-Farley, D., Ozair, S., Courville, A., Bengio, Y.: Generative adversarial nets. In: Advances in Neural Information Processing Systems, pp. 2672–2680 (2014)
11. Lantao, Yu., Weinan, Z., Jun, W., Yong, Yu.: Seqgan: Sequence generative adversarial nets with policy gradient. In: Thirty-First AAAI Conference on Artificial Intelligence (2017)
12. Che, T., Li, Y., Zhang, R., Devon Hjelm, R., Li, W., Song, Y., Bengio, Y.: Maximum-likelihood augmented discrete generative adversarial networks. Preprint (2017). arXiv:1702.07983
13. Guo, J., Lu, S., Cai, H., Zhang, W., Yu, Y., Wang, J.: Long text generation via adversarial training with leaked information. In: Thirty-Second AAAI Conference on Artificial Intelligence (2018)
14. Fedus, W., Goodfellow, I., Dai, A.M.: MaskGAN: Better text generation via filling. Preprint (2018). arXiv:1801.07736
15. Sutton, R.S., Barto, A.G.: Reinforcement Learning: An Introduction, vol. 1. MIT Press, Cambridge (1998)
16. Sutton, R.S., et al.: Policy gradient methods for reinforcement learning with function approximation. In: NIPS (1999)
17. Mikolov, T., Chen, K., Corrado, G., Dean, J.: Efficient estimation of word representations in vector space. In: ICLR Workshop (2013)
18. Le, Q., Mikolov, T.: Distributed representations of sentences and documents. In: International Conference on Machine Learning, pp. 1188–1196 (2014)
19. Bojanowski, P., Grave, E., Joulin, A., Mikolov, T.: Enriching word vectors with subword information. Trans. Assoc. Comput. Linguist. **5**, 135–146 (2017)
20. Pennington, J., Socher, R., Manning, C.D.: Glove: Global vectors for word representation. In: Proceedings of the Conference on Empirical Methods in Natural Language Processing (EMNLP), pp. 1532–1543 (2014)
21. McCann, B., Bradbury, J., Xiong, C., Socher, R.: Learned in translation: Contextualized word vectors. In: Advances in Neural Information Processing Systems, pp. 6294–6305 (2017)

22. Peters, M.E., Neumann, M., Iyyer, M., Gardner, M., Clark, C., Lee, K., Zettlemoyer, L.: Deep contextualized word representations. In: Proc. of ACL (2018)
23. Devlin, J., Chang, M.-W., Lee, K., Toutanova, K.: Bert: Pretraining of deep bidirectional transformers for language understanding. Preprint (2018). arXiv:1810.04805
24. Liu, Y., Ott, M., Goyal, N., Du, J., Joshi, M., Chen, D., Levy, O., Lewis, M., Zettlemoyer, L., Stoyanov, V.: Roberta: A robustly optimized BERT pretraining approach. Preprint (2019). arXiv:1907.11692
25. Dai, Z., Yang, Z., Yang, Y., Cohen, W.W., Carbonell, J., Le, Q.V., Salakhutdinov, R.: Transformer-xl: Attentive language models beyond a fixed-length context. Preprint (2019). arXiv:1901.02860
26. Shoeybi, M., Patwary, M., Puri, R., LeGresley, P., Casper, J., Catanzaro, B.: Megatron-lm: Training multi-billion parameter language models using GPU model parallelism. Preprint (2019). arXiv:1909.08053
27. Project Turing: https://msturing.org/. Last accessed 20 August 2020
28. Rajbhandari, S., Rasley, J., Ruwase, O., He, Y.: Zero: Memory optimization towards training a trillion parameter models. Preprint (2019). arXiv:1910.02054
29. Radford, A., Wu, J., Child, R., Luan, D., Amodei, D., Sutskever, I.: Language models are unsupervised multitask learners. OpenAI Blog **1**(8), 9 (2019)
30. Radford, A., Narasimhan, K., Salimans, T., Sutskever, I.: Improving language understanding by generative pre-training. https://s3-us-west-2.amazonaws.com/openaiassets/research-covers/languageunsupervised/language understanding paper. pdf (2018)
31. Brown, T.B., Mann, B., Ryder, N., Subbiah, M., Kaplan, J., Dhariwal, P., Neelakantan, A., Shyam, P., Sastry, G., Askell, A., Agarwal, S.: Language models are few-shot learners. Preprint (2020). arXiv:2005.14165
32. Yang, Z., Dai, Z., Yang, Y., Carbonell, J., Salakhutdinov, R.R., Le, Q.V.: Xlnet: Generalized autoregressive pretraining for language understanding. In: Advances in Neural Information Processing Systems, pp. 5753–5763 (2019)
33. Zellers, R., Holtzman, A., Rashkin, H., Bisk, Y., Farhadi, A., Roesner, F., Choi, Y.: Defending against neural fake news. In: Advances in Neural Information Processing Systems, pp. 9054–9065 (2019)
34. Keskar, N.S., McCann, B., Varshney, L.R., Xiong, C., Socher, R.: Ctrl: A conditional transformer language model for controllable generation. Preprint (2019). arXiv:1909.05858
35. Liu, X., He, P., Chen, W., Gao, J.: Multi-task deep neural networks for natural language understanding. Preprint (2019). arXiv:1901.11504
36. Dong, L., Yang, N., Wang, W., Wei, F., Liu, X., Wang, Y.,Gao, J., Zhou, M., Hon, H.-W.: Unified language model pre-training for natural language understanding and generation. In: 33rd Conference on Neural Information Processing Systems (NeurIPS 2019) (2019)
37. Bao, H., Dong, L., Wei, F., Wang, W., Yang, N., Liu, X., Wang, Y., Piao, S., Gao, J., Zhou, M., Hon, H.W.: Unilmv2: Pseudo-masked language models for unified language model pre-training. Preprint (2020). arXiv:2002.12804
38. Song, K., Tan, X., Qin, T., Lu, J., Liu, T.Y.: Mass: Masked sequence to sequence pre-training for language generation. Preprint (2019). arXiv:1905.02450
39. Lewis, M., Liu, Y., Goyal, N., Ghazvininejad, M., Mohamed, A., Levy, O., Stoyanov, V., Zettlemoyer, L.: Bart: Denoising sequence-to-sequence pre-training for natural language generation, translation, and comprehension. Preprint (2019). arXiv:1910.13461
40. Wang, X., Luo, Y., Crankshaw, D., Tumanov, A., Yu, F., Gonzalez, J.E.: IDK cascades: Fast deep learning by learning not to overthink. Preprint (2017). arXiv:1706.00885
41. Kaya, Y., Hong, S., Dumitras, T.: Shallow-deep networks: Understanding and mitigating network overthinking. In: Proceedings of the 36th International Conference on Machine Learning, PMLR, vol. 97, pp. 3301–3310 (2019)
42. Zhou, W., Xu, C., Ge, T., McAuley, J., Xu, K., Wei, F.: BERT loses patience: Fast and robust inference with early exit. Preprint (2020). arXiv:2006.04152

43. Bisk, Y., Holtzman, A., Thomason, J., Andreas, J., Bengio, Y., Chai, J., Lapata, M., Lazaridou, A., May, J., Nisnevich, A., Pinto, N.: Experience grounds language. Preprint (2020). arXiv:2004.10151
44. Paperno, D., Kruszewski, G., Lazaridou, A., Pham, N.-Q., Bernardi, R., Pezzelle, S., Baroni, M., Boleda, G., Fernández, R.: The LAMBADA dataset: Word prediction requiring a broad discourse context. In: Proceedings of the 54th Annual Meeting of the Association for Computational Linguistics (Volume 1: Long Papers), pp. 1525–1534 (2016)
45. Lake, B.M., Murphy, G.L.: Word meaning in minds and machines. Preprint (2020). arXiv:2008.01766
46. Trott, S., Torrent, T.T., Chang, N., Schneider, N.: (Re)construing meaning in NLP. Preprint (2020). arXiv:2005.09099

Fact Checking on Knowledge Graphs

Weichen Luo and Cheng Long

Abstract Fact checking, which verifies whether a given statement is true, could play a vital role in fake news detection. For example, for a given piece of news, a potential solution could involve a series of steps, including extracting statements from the news via text parsing, checking the validity of the extracted statements (i.e., fact checking), and classifying the news as fake if some statements have been confirmed to be false and performing further fake news detection processes otherwise. Considering that knowledge graphs are a popular way of representing knowledge, which could be used for verifying or counter-verifying statements, several solutions have been proposed that make use of knowledge graphs for fact checking. In this chapter, recent studies on fact checking with the help of knowledge graphs are reviewed, and three representative solutions, namely, Knowledge Linker, PredPath, and Knowledge Stream, are introduced with some details. Specifically, Knowledge Linker utilizes the semantic proximity metrics for mining knowledge graphs, PredPath employs the link prediction method and introduces a newly defined metric, and Knowledge Stream models the fact-checking problem as an optimization problem and uses flow theory for solving the problem.

Keywords Fact checking · Knowledge graph · Knowledge linker · Predicate path · Knowledge stream

1 Introduction

Rumors, misinformation, and fake news fill the Internet and social media these days, mostly due to the inability to identify fake news in large amounts of data quickly and accurately. These rumors and fake news will not only have negative impacts on public opinion but also affect people's judgment if they cannot be identified and corrected in a timely manner [13, 18, 23]. In order not to be misled, it is important

© Springer Nature Switzerland AG 2021
Deepak P et al., *Data Science for Fake News*, The Information
Retrieval Series 42, https://doi.org/10.1007/978-3-030-62696-9_7

to separate true news from a large scale of information mixed with fake news.

Dozens of methods or models have been proposed to detect and prevent the spread of rumors or fake news [47]. Most approaches are based on the contextual indicators of fake news for detecting the veracity of information, such as the abundance of inquiry tweets, the credibility of the information source, and the temporal patterns of news spread. A closely related issue is the evaluation of those statements that are presented in news media. This issue is called *fact checking*.

In order to be able to utilize as much information or fact data as possible, knowledge graphs (KGs) [31] are introduced to structure the existing knowledge and facts. Several models have been proposed for the fact-checking problem, which are based on knowledge graphs, including Knowledge Linker (KL) [11], PredPath [37], Knowledge Stream (KS) [39], PRA [21], Katz [20], TransE [9], Adamic & Adar [2], and Jaccard coefficient [24]. Most of these models rely on the traversal of the knowledge graph. For example, PRA [21] utilizes random walk, Knowledge Linker (KL) [11] employs the shortest path method, and PredPath [37] uses path enumeration.

In this chapter, we first review some preliminary knowledge of knowledge graphs and then introduce the three most recent and representative methods that use knowledge graphs for fact checking.

2   Preliminaries

While the idea of the "knowledge graph" can be traced to 1972 [34], the modern definition of the knowledge graph was first put forward by Google [42] in 2012. There are further developments of knowledge graphs by other companies, such as Facebook [30], LinkedIn [15], and Microsoft [40].

2.1   Knowledge Graph

A knowledge graph represents a data graph, which accumulates and transmits knowledge gathered from a real-world database [8]. The nodes of knowledge graphs denote entities, and each edge denotes the relationship between two entities. Most knowledge graphs are extracted from external knowledge bases containing numerous true statements. These statements can be divided into simple statements, such as "Sacramento is the capital of California," and qualitative statements, such as "capitals are cities." Simple statements can serve as edges in knowledge graphs.

There are two types of knowledge graphs: open knowledge graph and enterprise knowledge graph. Open knowledge graph refers to one that is published online and is freely accessible. Some open knowledge graphs may accumulate data directly from Wikipedia (such as DBpedia [22] and YAGO2 [16]), while others use crowd-sourcing methods to gather knowledge from volunteers collaboratively (such as Freebase [7] and Wikidata [46]). There are also some open knowledge graphs on specific topics, such as government [35], news [33], tourism [25], and geography [43]. The enterprise knowledge graph is mostly for internal use and/or commercial purposes. Based on applications, enterprise knowledge graphs can be classified into commerce (such as Uber[1] and eBay [2]), finance (such as Bloomberg[3] and Accenture[4]), social network (such as LinkedIn[5] and Facebook [30]), etc.

2.2   RDF

To allow the computer to better understand the information contained in statements, resource description framework (RDF) triples in the form of ⟨subject, predicate, object⟩ have been proposed [27]. Predicate illustrates the binary relationship between subject and object. For example, the statement "Sacramento is the capital of California" could be represented by an RDF triple ⟨Sacramento, CapitalOf, California⟩. RDF can build a labeled directed graph, where nodes denote entities (i.e., subject and object) and directed edges denote predicates. Different edge labels denote various predicates.

A formal definition of a knowledge graph constructed with RDF triples is as follows.

Definition 1 (Knowledge Graph) A knowledge graph is a directed graph $G = (V, E, \mathcal{R}, \mathcal{O}, g, h)$, where V denotes a node set, E denotes an edge set, $\mathcal{R}$ denotes the relation set, and $\mathcal{O}$ denotes the ontology set. $g: E \rightarrow \mathcal{R}$ is the labeling function, which maps edges to predicates, and $h: V \rightarrow \mathcal{O}$ is the function, which maps nodes to ontologies.

---

[1] https://eng.uber.com/uber-eats-query-understanding/.

[2] https://www.ebayinc.com/stories/news/cracking-the-code-on-conversationalcommerce/.

[3] https://speakerdeck.com/emeij/understanding-news-using-thebloomberg-knowledge-graph.

[4] https://www.accenture.com/us-en/insights/digital/data-to-knowledge.

[5] https://engineering.linkedin.com/blog/2016/10/building-the-linkedin-knowledge-graph.

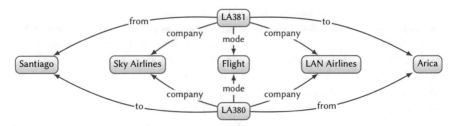

Fig. 1 A directed edge-labeled graph of companies that offer flights between Santiago and Arica [17]

Figure 1 shows an example of a knowledge graph constructed with triples.

3 Models

Quite a few models have been proposed to mine the knowledge graph for fact check-ing, including Knowledge Linker (KL) [11], PredPath [37], Knowledge Stream (KS) [39], PRA [21], Katz [20], TransE [9], Adamic & Adar [2], and Jaccard coefficient [24]. In this chapter, three models, namely, KL, PredPath, and KS, are introduced. KL utilizes the semantic proximity metrics for mining knowledge graphs. While it uncovers relationships among some nodes, it neglects the predicate between each pair of nodes. Sometimes, the results are difficult to interpret. PredPath employs the link prediction method and introduces a newly defined metric. KS models the fact-checking problem as an optimization problem and uses flow theory for solving the problem.

4 Knowledge Linker

The model Knowledge Linker (KL) [11] is based on the simple idea that fact checking on a knowledge graph aims to check whether the statement serves as an edge of the knowledge graph or if there exists a path to connect the target's subject and its object in the knowledge graph.

When checking a statement, it is seldom the case that a corresponding edge exists in the knowledge graph. Therefore, it is important to deduce the relation between the subject node and the object node by effectively mining the connectivity of the knowledge graph. KL adopts the epistemic closure theory [26]. The epistemic closure theory refers to a set of entities closed under logical implication, which means that a given statement could be deduced to be true through the entailment from what is already known. Generally, it can be regarded as a specific example of link prediction in knowledge graphs [29].

Semantic Proximity After establishing the link prediction method for fact checking in knowledge graphs, the next question is how to define the "path length" of different paths that connect the subject node and the object node. A path containing a lot of generic entities may sometimes provide weak information or even the wrong information. To illustrate, consider the following example.

The paths connecting entities "Sacramento" and "California" can be as follows:

· {Sacramento} $\xrightarrow{CityOf}$ {the United States} $\xleftarrow{StateOf}$ {California}
· {Sacramento} $\xleftarrow{Headquarter}$ {California State Police Department} $\xrightarrow{Jurisdiction}$ {California}

The entity "the United States" is a generic one, which means it can be related to many entities, thus providing little information. For any city in California or even in the United States, the first path could connect two entities, such as "Los Angeles" and "California" or "Chicago" and "California." Subsequently, the paths made in this way are of little value for checking the statement "Chicago is a city of California."

In the second path, however, two entities are connected to the middle entity, "California State Police Department." In addition, the entity "California State Police Department" has much fewer entities associated with it than the entity "the United States." Therefore, the second path depicts the special correlation information between these two entities. In fact, the statement "Chicago is a city of California" would be confirmed as a false statement with the second path.

From the example above, the length of a path can be defined by the generality of the nodes that comprise it. When a node is related to many nodes, such as "the United States," it has a higher generality score. There are three possible ways to illustrate whether two nodes are related:

1. If they are connected with the specific edge in the knowledge graph
2. If there exists a path connecting the two nodes in the knowledge graph
3. If the shortest path connecting the two nodes in the knowledge graph has a shorter length than the preset threshold

For the first way, the relation established contains less information and is inconsistent with the epistemic closure principle. The second and third ways both use intermediate nodes to establish relations. In addition, the third way takes into account the fact that the relevance decreases as the number of intermediate nodes increases, which seems to be more rational than the first way, but it is too computationally intensive to be practical and the threshold may be difficult to preset. Therefore, the second way is adopted in [41].

Since KL is based on the epistemic closure theory, when KL considers the relations, it does not care much about the predicate, and it models the knowledge graph as an undirected graph $G = (V, E)$, where V and E are the same as in Definition 1.

Definition 2 (Transitive Closure) $G = (V, E)$ is an undirected knowledge graph. Two nodes $a, b \in V$ are regarded as adjacent if there exists an edge $e = (a, b) \in E$. Two nodes $a, b \in V$ are regarded as connected if there is a sequence of nodes $(a = v_1, v_2, \ldots, v_n = b)$ connecting a and b ($n \geq 2$). $G^* = (V, E^*)$ is the transitive closure of G. The node sets of G and G^* are the same. Two nodes in G^* are determined to be adjacent iff the two nodes are connected in G.

A statement in the form of RDF triple $c = < s, p, o >$, where s denotes a subject, o denotes an object, and p denotes a predicate, is extracted from the transitive closure G^* of an undirected knowledge graph G. A path connecting subject s and object o is denoted as $P_{s,o} = (s = v_1, v_2, \ldots, v_n = o)$. The length of the path $P_{s,o}$ is defined as follows:

$$\mathcal{L}\left(P_{s,o}\right) = \mathcal{L}\left(v_1 \ldots v_n\right) = \left[1 + \sum_{i=2}^{n-1} \log k\left(v_i\right)\right]^{-1},$$

where $k(v_i)$ represents the entity v_i's degree, which means the number of appearances of the statement in the knowledge graph. With the help of the degree, the generality of an entity in the knowledge graph is defined. If c truly exists as an edge connecting entity s and entity o in the knowledge graph, then the corresponding value surely should be assigned the maximum value, i.e., $\mathcal{L}(P_{s,o}) = 1$. The semantic proximity $\mathcal{L}$ would be assigned the value 1 *iff* $n = 2$ since there are no nodes between the subject and the object.

When considering an alternative principle *the widest bottleneck* of the optimization problem, the length of the path $P_{s,o}$ could be measured with a new method:

$$\mathcal{L}'\left(P_{s,o}\right) = \mathcal{L}'\left(v_1 \ldots v_n\right) = \begin{cases} 1 & n = 2 \\ \left[1 + \max_{i=2}^{n-1}\{\log k\left(v_i\right)\}\right]^{-1} & n > 2, \end{cases}$$

where the function $k(v_i)$ has the same definition as above.

Since there could be several paths between the subject and the object, the truth value of a statement $c = < s, p, o >$ could be measured by finding the shortest path between the subject s and the object o [5, 28]. Formally, it is defined as follows:

$$\tau(c) = \max \mathcal{L}\left(P_{s,o}\right) (or \max \mathcal{L}'\left(P_{s,o}\right)).$$

Figure 2 shows an example of a path on the knowledge graph for a statement which has a low truth value.

Case Study Results *Wikipedia Knowledge Graph* (WKG) is built upon three datasets, namely, the DBpedia ontology dataset, the properties dataset, and the types dataset. The triples in the DBpedia ontology dataset all have the predicate "SubClassOf." The triples in the properties dataset are extracted from the Wikipedia

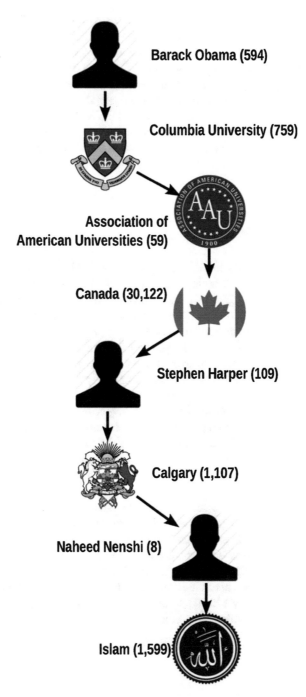

Barack Obama (594)

Columbia University (759)

Association of
American Universities (59)

Canada (30,122)

Stephen Harper (109)

Calgary (1,107)

Naheed Nenshi (8)

Islam (1,599)

infoboxes.[6] The triples in the types dataset are all in the form of ⟨subject, is-a, Class⟩, and *Class* is derived from the DBpedia[7] ontology.

The experiment is to compute the truth values of different statements such as "*a* belongs to *b*," where *a* is a US Congress member and *b* is an ideology. A matrix $\mathcal{M} = \{v_{i,j}\}_{n \times m}$ is defined, where n rows represent n members of Congress and m columns represent m ideology nodes in the WKG.

The matrix is computed by the definition of $\mathcal{L}(.)$ with the help of a force-directed layout [19]. The paths connecting blue or red nodes with gray nodes shown in Fig. 3 are all ranked in the top 1% of the truth value. The results shown in Fig. 3 are very much consistent with the results derived from blogs [3] and Twitter [12].

5 PredPath

As discussed in Sect. 3.1, the fact-checking problem based on the knowledge graph can be translated into a link prediction problem. The model Predicate Path (PredPath) [37] (KL) takes the connectivity (i.e., the degree of correlation between the nodes in a knowledge graph) and type information (i.e., the ontologies of each node) into consideration. Specifically, KL mines the knowledge graph based upon not only the connectivity and type information but also the interactions of predicates. The model aims to extract a set of discriminative paths that could illustrate the correlation between two entities uniquely in the knowledge graph.

Note that there exist some association mining methods [1, 14] and link prediction methods [6] on the knowledge graph, but when applied in fact checking, these methods would have drawbacks where in the derived results are general and lack specificity. For example, consider the predicate *CapitalOf* between two entities. Both the link prediction methods and the association mining methods would return the result that the predicate *LargestCityOf* is most related to the predicate *CapitalOf*. To some extent, the predicate *LargestCityOf* can be an alternative to the predicate *CapitalOf*. For example, given the statement "Columbus is the capital of Ohio," Columbus is truly the largest city and capital of Ohio. However, because the statement "Los Angeles is the largest city of California" is true does not mean that the statement "Los Angeles is the capital of California" is true. In fact, California's capital is Sacramento. The PredPath model could derive a discriminative path for statements where cities are capitals of states. If the intermediate nodes in a path have the city's headquarters and own jurisdiction in the located state, we can say that it is equal to the predicate *CapitalOf*.

$$\{Columbus\} \xrightarrow{LargestCityOf} \{Ohio\} \Longrightarrow \{Columbus\} \xrightarrow{CapitalOf} \{Ohio\}$$

---

[6]https://en.wikipedia.beta.wmflabs.org/wiki/Infobox.
[7]https://wiki.dbpedia.org/.

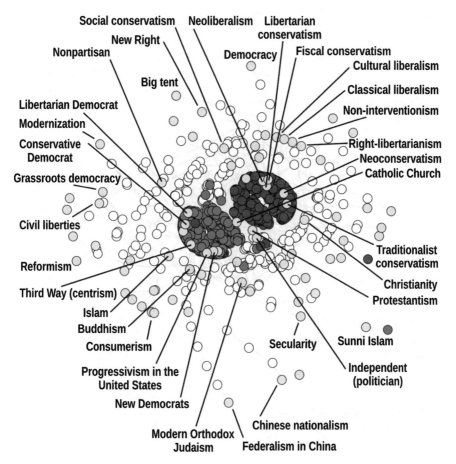

Fig. 3 Ideological map of Congress members [11]. The blue nodes represent the members of the Democratic Party and the red nodes represent the members of the Republican Party. The gray nodes denote the ideologies and the white nodes denote the intermediate nodes. The nodes' positions are calculated by a force-directed layout [10]. Only the most significant paths whose truth values rank in the top 1% are shown

. {Los Angeles} $\xrightarrow{LargestCityOf}$ {California} $\nRightarrow$ {Los Angeles} $\xrightarrow{CapitalOf}$ {California}

. {Sacramento} $\xleftarrow{Headquarter}$ {California State Police Department} or {California State Department of Transportation} etc. $\xrightarrow{Jurisdiction}$ {California} $\Rightarrow$ {Sacramento} $\xrightarrow{CapitalOf}$ {California}

To determine a statement's truthfulness, PredPath mines the connectivity characteristics of a knowledge graph by employing the principles of network closure, similarity search, and link prediction. There exist approaches which are based on meta paths and use the similarity property of paths in a knowledge graph. These approaches show brilliant results for solving problems such as clustering, recommendation, and classification [21, 38, 44, 45]. However, they all require users to know the domain of the problem in advance and the relevant meta paths before conducting the analysis [44]. PredPath can obtain a set of discriminative paths, which describe the relationship between two entities in a path of a knowledge graph uniquely.

Definitions Based on the previous definition of knowledge graph G, an entity in the knowledge graph can be mapped to multiple ontologies. When knowledge bases such as DBpedia[8] are used to build the knowledge graph, entities such as *Columbus*, *Los Angeles*, and *Sacramento* form the node set V; predicates such as *CapitalOf* and *LargestCityOf* form the predicate set $\mathcal{R}$; type labels such as *state* and *city* form the ontology set $\mathcal{O}$; and the edge set E represents the set of links each between two nodes in the knowledge graph G.

With the type information in the knowledge graph, an intensive set of connections named meta path to depict how the type labels can connect entities is defined as follows.

Definition 3 (Meta Path) A meta path $\mathcal{M}_n$ in the knowledge graph G is denoted as a typed, directed sequence of entities and edges: $o_1 \xrightarrow{p_1} o_2 \xrightarrow{p_2} \ldots \xrightarrow{p_{n-1}} o_n$, where o_i represents the ontology of entity e_i, p_i denotes the predicate that links entity e_i to e_{i+1}, and n represents the generalized length of the meta path.

To reduce storage and computational complexity, the intermediate type nodes in a meta path are ignored, but the predicates and endpoints are reserved. An anchored path vividly illustrates the structure of the path, which comprises the start node, the end node, and the predicates linking them.

Definition 4 (Anchored Path) The anchored path $\mathcal{A}_n$ of a meta path $\mathcal{M}_n$ is denoted as a directed path with typed sequences of edges and only the typed endpoints: $\mathcal{A}_n = o_1 \xrightarrow{p_1} \xrightarrow{p_2} \ldots \xrightarrow{p_{n-1}} o_n$.

Discriminative predicate paths are targets of the PredPath model, which are defined as follows.

Definition 5 (Discriminative Predicate Path) We use $D_n(o_u, o_v)$ to represent the set of discriminative predicate paths. It consists of all those anchored paths which could express the given statement $o_u \xrightarrow{p} o_v$ alternatively, and the paths' maximum generalized length is k.

Consider an example for illustration. One meta path that links the two entities "Sacramento" and "California" could be $\mathcal{M}$: $\{city\} \xleftarrow{Headquarter^{-1}} \{state$

---

[8] See footnote 7.

agency} $\xrightarrow{Jurisdiction}$ {state}. The anchored path $\mathcal{A}$ anchored by {city} and {state} for this meta path is $\langle$ Headquarter^{-1}, Jurisdiction$\rangle$. The corresponding discriminative predicate path set comprises many other anchored predicate paths connecting {city} and {state}.

The meta paths tend to own more type label information and can thus be more prone to involve labeling error. Therefore, anchored paths are used since they are more tolerant of labeling error.

PredPath solves the fact-checking problem by performing a supervised link prediction task, i.e., it determines a statement triplet $c = <s, p, o>$ to be true or not by first computing the discriminative path set $D_k(o_u, o_v)$, where the subject s's and the object o's ontologies are o_u and o_v, respectively, and then checking whether the edge $s \xrightarrow{p} o$ can be implied in the knowledge graph G. If $p \in \mathcal{R}$, then the positive path set H^+ and the negative path set H^- can be generated as the node pair sets, where $H^+ = \{(u, v)|u \xrightarrow{p} v \in G\}$, $H^- = \{(u, v)|u \xrightarrow{p} v \notin G\}$, $o_u = o_s$, and $o_v = o_o$. When $p \notin \mathcal{R}$, H^+ and H^- ought to be provided by humans.

PredPath considers both the generality and the context dependency of paths for discovering the most discriminative paths. Generality means whether the entities connected by the predicate p are of the same or similar type. The context dependency represents the similarity of different paths, which link the entities of the same or similar type.

Path Extraction Most existing meta path-based models need hand annotation [38] or exhaustive enumeration [21] when extracting the paths from knowledge graphs. In contrast, PredPath can extract the paths automatically, employing a constrained graph traversal algorithm. Though the amount of data in a knowledge graph can be massive, only a small part of the data is truly useful for the given task. Among the extracted meta paths, there are only a few discriminative paths for a certain predicate. When it checks the fact "Sacramento is the capital of California," it only considers those meta paths which start from the ontology city and end at the ontology state. A constrained graph traversal algorithm extracts anchored paths by traversing the graph from the subject entity to the object entity with the length less than k instead of traversing all the possible paths.

The anchored path sets $\mathbf{A}^+_{(o_u,o_v)}$ and $\mathbf{A}^-_{(o_u,o_v)}$ are extracted separately by using the depth-first traversal algorithm. This algorithm is implemented with the help of a closure function $\mathbb{C}$:

$$\mathbb{C}_p(v) = \{v'| (v, p, v') \in G\} \cup \{v'| (v', p, v) \in G\},$$

where v denotes an entity in a knowledge graph and p represents the predicate related to the closure. The function $\mathbb{C}$ returns all the entities that can be reached by v from predicate p or p^{-1}. With the definition of closure function, a transition function $\mathbb{T}(v_i)$ could be defined, which returns all the next nodes v_{i+1} of entity v_i.

$\mathbb{T}(v)$ returns all the entities that can be reached from $\mathbb{C}_p(v)$ without those that have already been visited:

$$\mathbb{T}(v_i) = \left\{ \cup_{p \in \mathcal{R}} \mathbb{C}_p(v_i) \setminus \cup_{j=1}^{i} \{v_j\} \right\}.$$

With the definitions of functions $\mathbb{C}_p(v)$ and $\mathbb{T}(v_i)$, the path set $\mathbb{P}$ could be derived with all the paths whose lengths are less than n: $\mathbb{P} = \cup_{i=1}^{n} \mathbb{P}^n$, where

$$\mathbb{P}^n = \{s, \mathbb{T}(v_1), \mathbb{T}(v_2), \dots, \mathbb{T}(v_{n-2}), o |$$

$$(s, o) \in \mathbf{T}, v_1 = s, v_i \in \mathbb{T}(v_{i-1}), o \in \mathbb{T}(v_{n-1})\}.$$

The next issue is how to measure the importance/helpfulness of a path. This problem is tackled with a regression model.

Path Selection Given the predicate path sets P^+ and P^-, the aim is to select the most discriminative predicate path set $\mathbf{D}$. The training matrix is defined as X, where the i-th row of matrix $X_{n \times m}$ denotes an instance anchored by u and v such that $o_u = o_s$ and $o_v = o_o$. Every member $X_{i,j}$ of the matrix X represents the number of anchored paths P_j anchored by u and v.

The goal for the path selection lies in deriving a new matrix $X'_{n \times m'}$, where the columns for the new matrix X' only contain the most discriminative paths' power:

$$\mathbf{X}' = f(\mathbf{X}, \mathbf{w}, \delta) = \mathbf{X}_{1:n, \{j | j \in 1:m, w_j \geq \delta\}},$$

where $\mathbf{w}$ is a feature importance vector with m dimensions and δ represents a threshold that controls importance.

The element $w_j \in \mathbf{w}$ is the important vector of an anchored predicate path $P_j \in \mathbf{P}$, which is defined by the information gain of $X_{:,j}$ and y:

$$I(\mathbf{X}_{:j} : \mathbf{y}) = \sum_{x_{i,j} \in \mathbf{X}_{i,j}} \sum_{y_i \in \mathbf{y}} p(x_{i,j}) p(y_i) \log \left(\frac{p(x_{i,j}, y_i)}{p(x_{i,j}) p(y_i)} \right),$$

where y is the label vector for the feature vector $X_{:,j}$, $x_{i,j}$ denotes the value of element $X_{i,j}$ [32], and the threshold δ is set empirically.

With the definition of matrix X', the validation of the statement of a fact can be solved by a logistic regression model [36].

Fact Interpretation Not all paths are intuitive enough to describe important information. For example, the statement of fact ⟨Sacramento, CapitalOf, California⟩ may generate some meaningless predicate paths such as ⟨location^{-1},location⟩ and ⟨deathPlace^{-1},deathPlace⟩, which represent the statements that "a capital's location is in the state" and that "City is place of death for a person who died in the state." In this example, the paths generated do not provide much information related to

"CapitalOf." Therefore, it is necessary to select those vital discriminative predicate paths that only depict the predicate in question.

This could be done by sorting out the predicate paths with the importance vector **w**: $P_i \prec P_j$ if and only if $w_i \geq w_j$. After ranking the predicate paths, we can remove those unimportant and off-topic predicate paths:

$$\mathbf{D}^* = \left\{ P \left| P \in \mathbf{D} \backslash \left\{ P_j | P_j \in \mathbf{P}^-, \sum_{i=0}^{i=n} \mathbf{X}_{i,j} \geq \theta \right\} \right. \right\},$$

where θ represents the threshold, which is chosen empirically between 10 and 20. The function introduced above is able to select a discriminative path set $\mathbf{D}^*$, which contains the paths that can specifically define the predicate provided. The discriminative predicate paths in the top 5 for the predicate "CapitalOf" are listed in Table 1.

Comparison Between Meta Path and Predicate Path Different from those existing studies which use meta paths on heterogeneous networks, PredPath uses the anchored predicate paths. As a knowledge graph can be much more complicated, an entity in the knowledge graph can own multiple labels. For example, Boston's type label is {city, settlement, populated place}, and Sacramento's type label is {settlement, populated place}, though they are, respectively, the capital of Massachusetts and California. Because the type labels do not match exactly, the

Table 1 Top discriminative paths for "CapitalOf"

Rank	Meta Path $\mathcal{M}$
1	{city, settlement} $\xrightarrow{location^{-1}}$ {state agency} $\xrightarrow{location}$ {state}
2	{city, settlement} $\xrightarrow{deathPlace^{-1}}$ {person} $\xrightarrow{deathPlace}$ {state}
3	{city, settlement} $\xrightarrow{headquarter^{-1}}$ {state agency} $\xrightarrow{jurisdiction}$ {state}
4	{city, settlement} $\xrightarrow{location^{-1}}$ {state agency} $\xrightarrow{jurisdiction}$ {state}
5	{settlement} $\xrightarrow{location^{-1}}$ {state agency} $\xrightarrow{jurisdiction}$ {state}
	Anchored Path D
1	$\langle headquarter^{-1}, jurisdiction \rangle$
2	$\langle location^{-1}, jurisdiction \rangle$
3	$\langle headquarter^{-1}, regionServed \rangle$
4	$\langle garrison^{-1}, country \rangle$
5	$\langle deathPlace^{-1}, deathPlace \rangle$
	Discriminative Anchored Path D^8
1	$\langle headquarter^{-1}, jurisdiction \rangle$
2	$\langle location^{-1}, jurisdiction \rangle$
3	$\langle garrison^{-1}, country \rangle$
4	$\langle headquarter^{-1}, parentOrganisation \rangle$
5	$\langle location^{-1}, parentOrganisation \rangle$

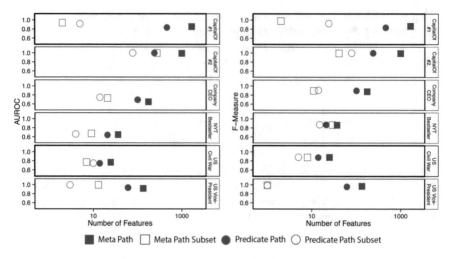

Fig. 4 Comparison experiment between Meta Path and Predicate Path [37]

PredPath model will treat the two paths differently, which could result in high overlap.

As shown in Fig. 4, the performance of four models for fact checking on DBpedia,[9] namely, Meta Path, Meta Path Subset, Predicate Path, and Predicate Path Subset, is shown. Meta Path Subset and Predicate Path Subset both represent the paths selected by the function mentioned above.

According to these results, the predicate path performs almost as accurately or even better than the meta path, though it has fewer features with entities removed. The subset selected by the importance selection function performs better than the original set.

6 Knowledge Stream

Knowledge Stream (KS) [39] is based on a novel and unsupervised network flow framework for fact checking. The model measures the trustworthiness of a statement in the form of a RDF triple. For the problem of fact checking on knowledge graphs, many approaches involve some traversal on knowledge graphs. Knowledge Linker utilizes the shortest path algorithm, and PredPath utilizes the path enumeration algorithm. KS is based on the fact that the information carried by multiple paths can provide more semantic context information than a single path as the non-disjoint paths may send additional flow on the knowledge graph. The KS model can

---

[9]See footnote 7.

automatically extract the meaningful patterns and contextual facts with a broader structure.

As shown in Fig. 5, the paths drawn in different colors form a stream of knowledge for the RDF triple ⟨David and Goliath, WrittenBy, Malcolm Gladwell⟩. To visually represent the flow of information on the knowledge graph, each path has been assigned a different width based on the amount of evidence it can offer for the RDF triple. KS would assign larger flows to those paths that provide more and discriminative information.

KS can be vividly interpreted as a network flow model. Given an RDF triple (s, p, o), it could be regarded as the knowledge flow starting from the subject entity s through the network and ending up at the object entity o. The remaining issue is to quantify the capacity and cost for each edge in the network. The capacity quantifies the amount of knowledge or information carried related to statement (s, p, o). The cost can be regarded as a constraint for the knowledge to pass a certain edge, which ensures that the paths extracted by KS are short. KS aims to extract the set of paths which can provide maximum flow of knowledge and minimize the cost.

The capacity of each edge $e' \in E$ in a knowledge graph can be intrinsic. With the definition above, the edge e' will be mapped to a certain predicate p'. For the statement (s, p, o) to be checked, the capacity of each edge in the knowledge graph is quantified as the relevance or similarity between the target predicate p and p' in the knowledge graph. The more relevant or similar p is p', and a higher capacity can be assigned to edge e'. It then measures the capacity of each path by the minimum capacity of all the edges on the path, i.e., the bottleneck [4]. The bottleneck can be interpreted as the least relevant or similar triple to the target statement in the path. Since there could be many paths connecting subject entity s and object entity o, the sum of their bottlenecks corresponds to the upper bound of knowledge flow through these paths. For KS, the path length is defined by not only the number of entities in the path but also the degrees of the connections from entities to other entities in the graph [11].

Relational Similarity Different from the models Knowledge Linker and PredPath, Knowledge Stream treats the knowledge graph as an undirected graph, only considering whether two entities are connected. The line graph $L(G) = (V', E')$ is defined on the undirected graph $G = (V, E)$. The node set V' of $L(G)$ is defined as $V' = E$, in which two new nodes are adjacent if and only if the corresponding edges in set E are connected by the same node in G, i.e., $E' = \{(e_1, e_2)|e_1, e_2 \in E, e_1 \cap e_2 \neq \phi\}$. With the definition of line graph, the edge-labeled graph G could be transformed into a node-labeled graph $L(G)$. In addition, a contracted line graph $L^*(G)$ could be defined, which is an edge-weighted graph that substitutes two nodes with a new node if the new nodes' set of neighbors corresponds to the union of the sets of the two nodes' neighbors. For illustration, an example is shown in Fig. 6.

An adjacency matrix $C \in \mathbb{N}^{R \times R}$ is defined for the contracted line graph $L^*(G)$, where $R = |\mathcal{R}|$. Matrix C is defined as the co-occurrence matrix of $\mathcal{R}$. The similarity between two relationships is measured by computing the cosine value between two corresponding rows of vectors in C. Similar to information retrieval,

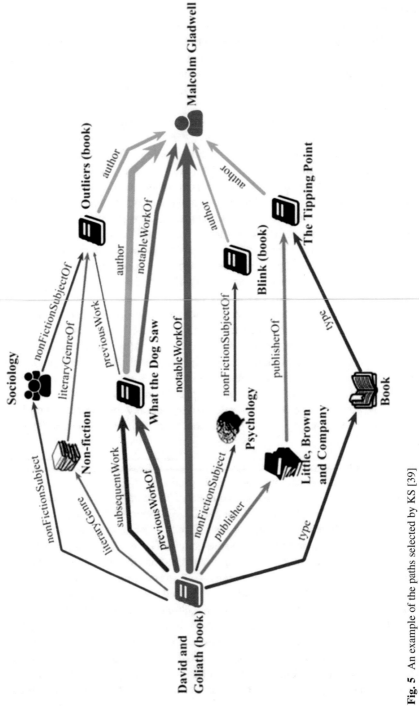

Fig. 5 An example of the paths selected by KS [39]

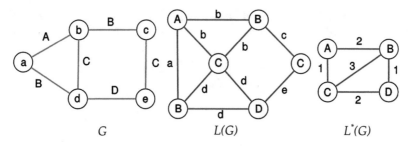

Fig. 6 Example of line graph and contracted line graph [39]

IF and IDF terms could be defined based on matrix C:

$$\text{TF}\left(r_i, r_j\right) = \log\left(1 + C_{ij}\right),$$

$$\text{IDF}\left(r_j, \mathcal{R}\right) = \log\frac{R}{\left|\{r_i \mid C_{ij} > 0\}\right|},$$

$$C'\left(r_i, r_j, \mathcal{R}\right) = \text{TF}\left(r_i, r_j\right) \cdot \text{IDF}\left(r_j, \mathcal{R}\right),$$

where $C_{i,j}$ denotes the count of co-occurrences between $r_i, r_j \in \mathcal{R}$, as discussed before. Then, the relational similarity $u(r_i, r_j)$ is computed as the cosine value between i-th and j-th rows of C'.

Fact Checking as a Network Flow Problem As discussed before, the fact-checking problem can be viewed as a problem of finding an optimal approach to transfering the knowledge across the knowledge graph under certain constraints, which could be modeled as a minimum cost maximum flow problem.

The next question lies in how to specifically utilize the knowledge stream for fact checking. As the long chain path may lead to a general or obvious result, we need to define the specificity of a path $P_{s,p,o}$. The specificity $S(P_{s,p,o})$ is defined proportionally to the inverse of the sum of degrees:

$$S\left(P_{s,p,o}\right) = \frac{1}{1 + \sum_{i=2}^{n-1} \log k\left(v_i\right)}.$$

Combined with the definitions above, the net flow $\mathcal{W}(P_{s,p,o})$ of a path $P_{s,p,o}$ can be set as the product of its bottleneck and specificity:

$$W\left(P_{s,p,o}\right) = \beta\left(P_{s,p,o}\right) \cdot S\left(P_{s,p,o}\right).$$

To check whether the statement triple (s, p, o) is true or not, KS derives the truth score $\tau^{KS}(s, p, o)$ of the triple by summing all the paths' flow in the net together:

$$\tau^{KS}(s, p, o) = \sum_{P_{s,p,o} \in \mathcal{P}_{s,p,o}} \mathcal{W}\left(P_{s,p,o}\right)$$

$$= \sum_{P_{s,p,o} \in \mathcal{P}_{s,p,o}} \beta\left(P_{s,p,o}\right) \cdot \mathcal{S}\left(P_{s,p,o}\right).$$

7   Conclusion and Future Work

In this chapter, three methods, namely, Knowledge Linker (KL), Predicate Path (PredPath), and Knowledge Stream (KS), which utilize knowledge graphs for fact checking, are introduced. There are quite a few future research directions. One possible future direction is to introduce deep learning models such as GNN into the fact-checking problem on knowledge graphs given that deep learning has been used successfully for many complex problems such as those in computer vision, natural language processing, and control. In addition, it seems necessary to bring the temporal dimension into consideration for fact checking. Take the capital of the Roman Empire, for example; the statement "Roman Empire's capital is Rome" is correct only before 323 CE, because the capital was later changed to Constantinople.

References

1. Abedjan, Z., Naumann, F.: Synonym analysis for predicate expansion. In: Extended Semantic Web Conference, pp. 140–154. Springer, New York, (2013)
2. Adamic, L.A., Adar, E.: Friends and neighbors on the web. Soc. Netw. **25**(3), 211–230 (2003)
3. Adamic, L.A., Glance, N.: The political blogosphere and the 2004 US election: divided they blog. In: Proceedings of the 3rd International Workshop on Link Discovery, pp. 36–43 (2005)
4. Ahuja, R.K., Magnanti, T.L., Orlin, J.B., Weihe, K.: Network flows: theory, algorithms and applications. ZOR-Methods Models Oper. Res. **41**(3), 252–254 (1995)
5. Aiello, L.M., Barrat, A., Schifanella, R., Cattuto, C., Markines, B., Menczer, F.: Friendship prediction and homophily in social media. ACM Trans. Web **6**(2), 1–33 (2012)
6. Barabási, A.L., Albert, R.: Emergence of scaling in random networks. Science **286**(5439), 509–512 (1999)
7. Bollacker, K., Tufts, P., Pierce, T., Cook, R.: A platform for scalable, collaborative, structured information integration. In: International Workshop on Information Integration on the Web (IIWeb'07), pp. 22–27 (2007)
8. Bonatti, P.A., Decker, S., Polleres, A., Presutti, V.: Knowledge graphs: new directions for knowledge representation on the semantic web (dagstuhl seminar 18371). Schloss Dagstuhl-Leibniz-Zentrum fuer Informatik (2019)
9. Bordes, A., Usunier, N., Garcia-Duran, A., Weston, J., Yakhnenko, O.: Translating embeddings for modeling multi-relational data. In: Advances in Neural Information Processing Systems, pp. 2787–2795 (2013)

10. Cheong, S.H., Si, Y.W.: Force-directed algorithms for schematic drawings and placement: a survey. Inf. Visual. **19**(1), 65–91 (2020)
11. Ciampaglia, G.L., Shiralkar, P., Rocha, L.M., Bollen, J., Menczer, F., Flammini, A.: Computational fact checking from knowledge networks. PloS ONE **10**(6), e0128193 (2015)
12. Conover, M.D., Ratkiewicz, J., Francisco, M., Gonçalves, B., Menczer, F., Flammini, A.: Political polarization on twitter. In: Fifth International AAAI Conference on Weblogs and Social Media (2011)
13. Flanagin, A.J., Metzger, M.J.: Perceptions of internet information credibility. Journal. Mass Commun. Quart. **77**(3), 515–540 (2000)
14. Galárraga, L.A., Teflioudi, C., Hose, K., Suchanek, F.: AMIE: association rule mining under incomplete evidence in ontological knowledge bases. In: Proceedings of the 22nd International Conference on World Wide Web, pp. 413–422 (2013)
15. He, Q., Chen, B., Argawal, D.: Building the LinkedIn knowledge graph. In: LinkedIn (2016)
16. Hoffart, J., Suchanek, F.M., Berberich, K., Lewis-Kelham, E., De Melo, G., Weikum, G.: Yago2: exploring and querying world knowledge in time, space, context, and many languages. In: Proceedings of the 20th International Conference Companion on World Wide Web, pp. 229–232 (2011)
17. Hogan, A., Blomqvist, E., Cochez, M., d'Amato, C., de Melo, G., Gutierrez, C., Gayo, J.E.L., Kirrane, S., Neumaier, S., Polleres, A., et al.: Knowledge graphs (2020). Preprint. arXiv:2003.02320
18. Howell, L., et al.: Digital wildfires in a hyperconnected world. WEF Rep. **3**(2013), 15–94 (2013)
19. Kamada, T., Kawai, S., et al.: An algorithm for drawing general undirected graphs. Inf. Process. Lett. **31**(1), 7–15 (1989)
20. Katz, L.: A new status index derived from sociometric analysis. Psychometrika **18**(1), 39–43 (1953)
21. Lao, N., Cohen, W.W.: Relational retrieval using a combination of path-constrained random walks. Mach. Learn. **81**(1), 53–67 (2010)
22. Lehmann, J., Isele, R., Jakob, M., Jentzsch, A., Kontokostas, D., Mendes, P.N., Hellmann, S., Morsey, M., Van Kleef, P., Auer, S., et al.: DBpedia–a large-scale, multilingual knowledge base extracted from Wikipedia. Seman. Web **6**(2), 167–195 (2015)
23. Lewandowsky, S., Ecker, U.K., Seifert, C.M., Schwarz, N., Cook, J.: Misinformation and its correction: continued influence and successful debiasing. Psychol. Sci. Publ. Int. **13**(3), 106–131 (2012)
24. Liben-Nowell, D., Kleinberg, J.: The link-prediction problem for social networks. J. Am. Soc. Inf. Sci. Technol. **58**(7), 1019–1031 (2007)
25. Lu, C., Laublet, P., Stankovic, M.: Travel attractions recommendation with knowledge graphs. In: European Knowledge Acquisition Workshop, pp. 416–431. Springer, New York (2016)
26. Luper, S.: The epistemic closure principle (2008)
27. Manola, F., Miller, E., McBride, B., et al.: RDF Primer. W3C Recommend. **10**(1–107), 6 (2004)
28. Markines, B., Menczer, F.: A scalable, collaborative similarity measure for social annotation systems. In: Proceedings of the 20th ACM Conference on Hypertext and Hypermedia, pp. 347–348 (2009)
29. Nickel, M., Murphy, K., Tresp, V., Gabrilovich, E.: A review of relational machine learning for knowledge graphs. Proceedings of the IEEE **104**(1), 11–33 (2015)
30. Noy, N., Gao, Y., Jain, A., Narayanan, A., Patterson, A., Taylor, J.: Industry-scale knowledge graphs: lessons and challenges. Queue **17**(2), 48–75 (2019)
31. Paulheim, H.: Knowledge graph refinement: a survey of approaches and evaluation methods. Seman. Web **8**(3), 489–508 (2017)
32. Quinlan, J.R.: C4.5: Programs for Machine Learning. Morgan Kaufmann Publishers Inc., Burlington (1992)
33. Raimond, Y., Ferne, T., Smethurst, M., Adams, G.: The BBC world service archive prototype. J. Web Semant. **27**, 2–9 (2014)

34. Schneider, E.W.: Course modularization applied: the interface system and its implications for sequence control and data analysis (1973)
35. Shadbolt, N., O'Hara, K.: Linked data in government. IEEE Intern. Comput. **17**(4), 72–77 (2013)
36. Shewhart, W.A., Wilks, S.S.: Applied Logistic Regression, 2nd edn. Wiley, New York (2005)
37. Shi, B., Weninger, T.: Discriminative predicate path mining for fact checking in knowledge graphs. Knowl.-Based Syst. **104**, 123–133 (2016)
38. Shi, C., Kong, X., Huang, Y., Philip, S.Y., Wu, B.: Hetesim: a general framework for relevance measure in heterogeneous networks. IEEE Trans. Knowl. Data Eng. **26**(10), 2479–2492 (2014)
39. Shiralkar, P., Flammini, A., Menczer, F., Ciampaglia, G.L.: Finding streams in knowledge graphs to support fact checking. In: 2017 IEEE International Conference on Data Mining (ICDM), pp. 859–864. IEEE, New York (2017)
40. Shrivastava, S.: Bring rich knowledge of people, places, things and local businesses to your apps. Bing blogs (2017)
41. Simas, T., Rocha, L.M.: Distance closures on complex networks. Netw. Sci. **3**(2), 227–268 (2015)
42. Singhal, A.: Introducing the knowledge graph: things, not strings. Official google blog **16** (2012)
43. Stadler, C., Lehmann, J., Höffner, K., Auer, S.: Linkedgeodata: a core for a web of spatial open data. Seman. Web **3**(4), 333–354 (2012)
44. Sun, Y., Han, J., Yan, X., Yu, P.S., Wu, T.: PathSim: meta path-based top-k similarity search in heterogeneous information networks. Proc. VLDB Endowm. **4**(11), 992–1003 (2011)
45. Sun, Y., Norick, B., Han, J., Yan, X., Yu, P.S., Yu, X.: Pathselclus: Integrating meta-path selection with user-guided object clustering in heterogeneous information networks. ACM Trans. Knowl. Discov. Data **7**(3), 1–23 (2013)
46. Vrandečić, D., Krötzsch, M.: Wikidata: a free collaborative knowledgebase. Commun. ACM **57**(10), 78–85 (2014)
47. Zubiaga, A., Aker, A., Bontcheva, K., Liakata, M., Procter, R.: Detection and resolution of rumours in social media: a survey. ACM Comput. Surv. **51**(2), 1–36 (2018)

Graph Mining Meets Fake News Detection

Kaiqiang Yu and Cheng Long

Abstract Nowadays, diversified services on social media diffuse news at higher rates and larger volumes, which poses unique challenges in terms of efficiency, scalability, and accuracy in fake news detection. To solve these issues, graph mining, as a promising direction of data mining, has successfully attracted the attention of recent studies. In this chapter, we present a comprehensive study on recent graph-based fake news detection approaches and show how graph mining performs the task. We first introduce different kinds of information related to fake news and then divide the existing graph-based approaches into two scenarios, where various graphs and graph patterns are introduced to model the information on social media and characterize features of fake news, respectively.

Keywords Fake news detection · Anomaly detection · Suspicious behavior · Graph mining · Cohesive subgraph

With rapidly increasing information, online social media, such as Twitter and Facebook, have become the key tool for people to seek and share knowledge. As producing and diffusing content online are much easier and faster, groups of anomalous users have strong incentives to publish fake news which take various forms, e.g., disinformation, misinformation, and junk news [28]. For example, users of Twitter can buy thousands of zombie followers to exaggerate their impact. Fake reviews on Amazon would mislead consumers about products. To tackle these issues, various studies on fake news detection, also known as anomaly detection, are proposed to detect the anomalies, e.g., fake news and abnormal users, on social media.

As a useful tool to model real-world entities and relationships, graph mining has generated a lot of interest in recent studies. A data mining task can be efficiently processed by solving the transformed graph mining problem, as there are many graph mining algorithms. For example, the frequent closed itemset mining problem

© Springer Nature Switzerland AG 2021
Deepak P et al., *Data Science for Fake News*, The Information
Retrieval Series 42, https://doi.org/10.1007/978-3-030-62696-9_8

[40], which aims to find all frequent closed itemsets from a transaction database, can be equivalently regarded as the maximal biclique enumeration problem, which proposes to list all maximal bicliques in a bipartite graph. Here, we use the bipartite graph to model the transaction database and employ bicliques to characterize frequent closed itemsets. To solve fake news detection problems in such a way, we need to (1) model different kinds of information on social media as a graph, (2) find measurements or properties of a subgraph to characterize the target anomalies, and (3) propose algorithms for the formulated graph mining problem.

To better show the effects of graph analytics on fake news detection problems, we confine our discussion to the first two steps, i.e., graph models and measurements or properties of a subgraph. We briefly present the related graph mining algorithms when necessary, as it is slightly beyond our scope. In this chapter, we discuss the major challenges faced by fake news detection in Sect. 1. Section 2 introduces various graph models used to model different kinds of information. We then present three kinds of scenarios and related graph-based approaches for fake news detection in Sects. 3 and 4. Section 5 concludes the chapter.

1   Characteristics and Challenges

We briefly elaborate the major challenges in anomaly detection, based on which several metrics are further built up to evaluate different approaches.

- **Adversarial scenario.** As fake news is deliberately produced to mislead ordinary users, it is difficult to distinguish it from truths. Moreover, smart fraudsters try to evade detection by conducting various camouflages. For example, spammers can vary their fake content to avoid detection based on textual similarity. As the information diffusion procedures are represented as edges in the graph model, diffusing fake news would inevitably yield a huge number of edges in a graph. Hence, unexpected cohesive or dense subgraphs occur. Graph-based approaches are more robust for detecting these subgraphs. However, there are still some strategies to evade such detection, while these camouflages require more than simply changing content. For instance, zombie followers, selling following services on Twitter, also follow many popular celebrities as normal users do. In addition, they can hijack normal accounts to produce fake news. Therefore, detection approaches are required to be *robust* in such an adversarial scenario.
- **High rate and large volumes.** Due to the low cost of publishing fake contents and enormous numbers of software-controlled profiles, e.g., social bots [8], nowadays fake news on social media is produced and diffused at large volumes and high rates [35]. This makes the fake news detection problem more challenging, particularly requiring the solutions to be *scalable* and *efficient*. Moreover, in terms of the fast propagations and damaging effects of fake news, it is important to develop approaches to the *early detection* of false news to prevent it from spreading on social media [22].

- **Limitation of datasets.** Existing algorithms for fake news detection have to be performed on limited datasets, where not all kinds of information related to anomalies are included. To be specific, there are three major limitations:

  (a) Scarcity. Most of the available datasets with "real" and "fake" labels are small due to the high costs of human annotation. To verify news in a rigorous way, the annotator needs to be equipped with the related knowledge. Therefore, it is hard to employ and manage enough human annotators to label news from different fields.

  (b) Imbalanced class. Normally, the number of "real" labels is much larger than that of "fake" labels due to the current strategies and policies on social media. It is challenging to address these highly imbalanced classes.

  (c) Noise. The quality of date collections is not always high, in terms of human factors and social media policies. For example, annotators may mislabel real news as fake in the datasets. In addition, some personal information may be lost during the collection of data, due to the privacy policies.

To tackle these challenges, the approaches to solveing the fake news detection problem should consider its robustness, scalability, efficiency, and accuracy.

2 Graph Models in Fake News Detection

In this section, we present various graphs used to model different kinds of information on social media. Existing approaches can be divided into different categories in terms of the information used to detect anomalies, e.g., unimodal approaches and multi-modal approaches, which will be explored in the following sections.

2.1 Information

To build graph models, we first elaborate the following different types of information.

- **Content.** One of the most important kinds of information is the content of news, which takes various forms, e.g., texts, images, and videos. As fake news is often produced by the same group of fraudsters or social bots, its content would share some similarities by which it can be efficiently detected. To better measure the similarities, different features are extracted from content in recent studies, e.g., stances of news [13], textual features (speaker affiliation or the source newspaper) [42], psychological or stylistic features [29], and linguistic features (punctuation, syntax, or readability) [30]. However, these methods can be easily evaded by varying the writing styles of content or simulating the real news.

- **Context.** Context-based (structural) information refers to the social interactions between users on social media, e.g., likes on Facebook, (re)tweets on Twitter, and reviews on Amazon, to detect fake news. Although fraudsters can carefully design the content to avoid detection, they will inevitably interact with others to diffuse news. Therefore, the context-based approaches can find some fake news escaping from the content-based detection. Moreover, the context-based information can be easily modeled as a graph where vertices denote the entities and edges represent the social interactions. The major challenge is how to extract useful features to characterize anomalies.
- **Temporal propagation.** The temporal propagation information tells us how news is spread on social media, by which we classify it into categories of interest, e.g., fake or real news. Note that there exists some overlap, as the context-based information can be regarded as a snapshot of the temporal propagation. As spreaders tend to diffuse the news that caters to their interests, analogous kinds of information would be diffused among similar traces on social media, e.g., they are diffused from the same sources, by the same people and in the same sequences [6]. Many studies have proposed to detect fake news based on the similarity of resulted spreading trajectories [43]. In addition, fake news, also sometimes called rumor, often arises and spreads faster [23]. These bursting features can be detected by the temporal information, i.e., the timestamps of diffusions.
- **Others.** There is also some other information that can be used to find anomalies, e.g., equipment information and IP address. Most of these messages are private and usually unavailable to publish in terms of policies of various social media platforms.

2.2 Graph Models

We discuss how to model the different kinds of information presented above as a graph. Existing graph-based approaches perform on the resulted graph models. We present the following two categories of graph models based on whether or not they contain temporal information.

- **Static graph model.** A static graph model consists of a set of fixed vertices and edges, where vertices denote various entities, e.g., users on Twitter or products on Amazon, and edges represent the social interaction, e.g., user-follows-user on Twitter or costumer-reviews-product on Amazon. Based on the types of entities, static graphs can be further divided into two categories: homogeneous graph only consisting of one type of entity and heterogeneous graph containing various types of entities, e.g., bipartite graph used to model costumer-product relationship on Amazon. Static graph models naturally contain context-based information. Other information, e.g., similarity of content-based features and temporal features, can be attached as the attributes of vertices and edges, e.g., weights and timestamps.

– **Dynamic graph model.** A dynamic graph model contains a set of vertices and edges that can be dynamically inserted and deleted. Various online applications are performed on such a dynamic graph where vertices denote entities and edges represent social interactions. It generalizes the static graph, as the static model can be regarded as a snapshot of a dynamic graph. Thus it not only models the context-based information but also temporal diffusion features. Intuitively, an edge will be generated when the social interaction occurs.

3 Unimodal Scenario: Static Graph-Based Methods

Considering the privacy policies imposed on the collection of data from social media platforms, it is hard to obtain all the anomaly information. A unimodal scenario refers to such a situation where only one kind of information, e.g., content-only or context-only, can be assessed. To tackle this scenario, various unimodal approaches are proposed to detect anomaly entities by focusing only on the available textual part or the structural part. The majority of graph-based approaches exploit the network connections to detect the abnormal groups. In this section, we first summarize some applications in this scenario and then present two kinds of graph-based methods.

3.1 Graph Statistics Detection

Many recent fraud detection algorithms are designed based on graph statistics, e.g., distribution of degrees [17], coreness [36], ranks, and eigenvalues [1]. The abnormal activities would inevitably be reflected in these graph statistics, while the fraud users try to evade detection by varying the content. Moreover, these graph statistics can be efficiently obtained by employing graph analysis tools or using existing algorithms, e.g., core decomposition and SVD.

CatchSync CatchSync [17] is a unimodal approach which models context information as a directed graph $G(V, E)$, to cluster suspicious vertices by detecting synchronized and rare behaviors based on the graph statistics, i.e., in-degree, out-degree, hubness, and authoritativeness. Here, two characteristics are used to detect anomalous vertex groups: abnormal vertices are usually *synchronized* to perform same tasks, which are quite *rare* compared with the majority.

We briefly recall the definition of related graph statistics and discuss how they characterize *synchronized* and *abnormal* patterns.

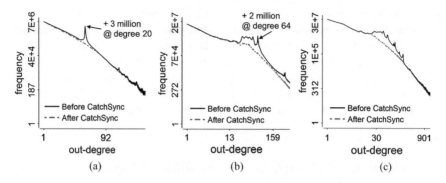

Fig. 1 Out-degree distributions and synchronized behaviors [17]. (**a**) Twitter (July 2009). (**b**) Tencent Weibo (January 2011). (**c**) Tencent Weibo (November 2011)

- **In-degree and out-degree.** Given a directed $G(V, E)$ with the vertex set V and the edge set E, the in-degree (resp., out-degree) of a vertex $v \in V$ is the number of edges incident to (resp., incident from) v. On Twitter, the abnormal users who buy fake followers usually have similar in-degrees that are much larger than normal users. In addition, the zombie followers, controlled by the same organization or company, would have a similar out-degree, as they offer the following services to similar customers. Therefore, the above synchronized behaviors are related to the degree values, which is further illustrated in Fig. 1. To be specific, a spike occurs at the out-degree 20 where more than 3 million followers follow exactly 20 users in Fig. 1a. The distribution becomes smoother and closer to a power law after removing the suspicious vertices.
- **Hubness and authoritativeness.** For a directed graph $G(V, E)$, the hubness and authoritativeness of a vertex $v \in V$, also known as HITS, are proposed in [18] to measure the importance of web pages. Specifically, the vertex with large authoritativeness is connected by many other vertices. The vertex that connects to "authoritative" nodes has a larger hubness than that which connects to ordinary nodes. On Twitter, fake followers often have a smaller hubness than the ordinary users with the same out-degree, as their customers are not as famous as some popular celebrities followed by ordinary users. In addition, abnormal users have smaller authoritativeness than the normal users with the same in-degree, as most of their followers are fake and not important.

Based on the above features, the authors of [17] propose an efficient and effective clustering algorithm to detect the fake users, the details on which we omit here.

Core Patterns The authors of [36], modeling context information as an undirected unweighted graph, proposed to detect two types of anomalies, i.e., "loner-star" and "lockstep behavior," based on the degree and coreness. Specifically, "loner-star" refers to the situation where vertices are mostly connected to others with a small degree denoted by "loners." "Lockstep behavior" is when a group of vertices behave similarly.

We briefly define some graph statistics and then elaborate how they are used to detect the mentioned anomalies.

- k-**core.** Given a graph $G(V, E)$ and an integer $k \geq 0$, k-core is a maximal subgraph of G with each vertex having a degree at least k.
- **Coreness.** The coreness or core number of a vertex v is the maximum k such that v belongs to k-core.

The k-core and coreness can be efficiently obtained by iteratively removing the vertex with the smallest degree, known as *core decomposition*. The authors of [36] propose to detect anomalies based on the observation that vertices with high coreness tend to have high degree and vice versa, as shown in Fig. 2a. We then present various anomalous patterns in real applications. To be specific, Fig. 2b marks a group of vertices with the highest degree 1383 but lower coreness 12. This "loner-star" pattern corresponds to the email account of the company's CEO, which is only used to receive emails. Among the received emails, 99.6% of corresponding email accounts are outside the company and have small coreness in the graph. These "loner-star" patterns also occur in other situations, as fake users tend to have small coreness in terms of the costs. Figure 2c presents an anomalous group corresponding to follower-boosting service on Twitter. The zombie users often follow the same groups of customers, which results in the highest coreness but low degrees. The similar lockstep behaviors called "copy-and-paste" can be found in the citation graph, where vertices represent different patents and edges represent citing relationships, as shown in Fig. 2d. Nearly 88% of vertices in the marked anomalous

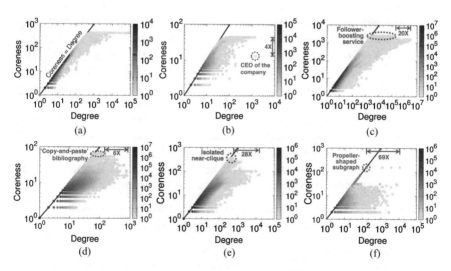

Fig. 2 Coreness and anomaly behaviors [36]. $\rho \in [-1, 1]$ denotes Spearman's rank correlation coefficient, and colors of heatmap represent the point density. (**a**) Catster ($\rho = 0.95$). (**b**) Email ($\rho = 0.99$). (**c**) Twitter ($\rho = 0.95$). (**d**) Patent ($\rho = 0.56$). (**e**) LiveJournal ($\rho = 0.93$). (**f**) NotreDame ($\rho = 0.99$)

group are patents from the same company. This is because the bibliography in previous patents has been reused in a "copy-and-paste" way, which leads to a dense subgraph with high coreness but low degrees. Figures 2g and h mark the similar patterns with high coreness but low degrees in the LiveJournal dataset and the NotreDame datasets, respectively.

3.2 Dense Subgraph Mining

Recent studies on anomaly detection proposed to exploit unique structures by bringing some insights from parallel research, known as *dense subgraph mining*, which aims to extract cohesive or dense subgraphs from the whole graph. Intuitively, the following selling services on Twitter often result in a cohesive or dense subgraph where zombie followers serve a huge but similar group of customers.

Dense Graph Structures in Fake News To capture such a cohesive or dense subgraph, various definitions of dense structures have been well studied.

(1) **Density-based structures.** Many studies on dense subgraph mining propose to find a subgraph (resp. k subgraphs) such that its density is highest (resp. top k highest). Different types of subgraphs would be yielded by varying the definitions of density.

 - **Edge density.** The edge density of an undirected graph $G(V, E)$ is defined as $\rho_e = \frac{|E|}{|V|}$, where $|E|$ and V respectively represent the number of edges and vertices [9, 11] . Intuitively, the target densest subgraph would have a high average degree. However, it fails to find the large near-clique subgraphs.
 - k-**Clique density.** Given an undirected graph and an integer $k > 0$, the k-clique density is defined as $\rho_c = \frac{|\mathcal{C}(k,G)|}{|V|}$, where $\mathcal{C}(k, G)$ is the set of all k-cliques in G [25, 39]. It is a kind of generalization and can discover the large near-clique subgraphs, as $k = 2$ reduces to the edge density.
 - **Pattern density.** Given an undirected graph G and a pattern Φ, also known as graphlet or motif, which is a small graph that consists of few vertices, the pattern density is defined as $\rho_p = \frac{\mathcal{C}(\Phi,G)}{|V|}$, where $\mathcal{C}(\Phi, G)$ is the set of all patterns Φ in G [7]. The pattern density is a more general model, as edges and k-cliques can be regarded as different input patterns.

(2) **Clique-based structures.** Another thread of studies tends to find the maximal subgraphs with respect to different target properties.

 - **Clique (biclique).** A clique is a subgraph of the general graph where each pair of vertices is adjacent. Similarly, a biclique refers to a subgraph of the bipartite graph such that every two vertices from the same side are connected [44].

- **Quasi-clique (quasi-biclique).** Quasi-cliques (resp. quasi-bicliques) can be regarded as a generalization of cliques (resp. bicliques) in terms of fault-tolerant scenarios. There are mainly two types of definitions: proportional error-tolerant structures and constant error-tolerant structures. To be specific, a γ-quasi-clique $H(V_H, E_H)$ is a subgraph of the general graph G where $\gamma \in (0, 1]$ and each vertex disconnects at most $(1 - \gamma)(|V_H| - 1)$ vertices [34]. Another kind of definition, also known as k-plex [3], refers to a subgraph H such that each vertex disconnects up to $k - 1$ vertices, $k \in \mathbb{N}^+$. We have the similar definitions of γ-quasi-bicliques [21] and ϵ-quasi-bicliques [37] where $\gamma \in (0, 1]$ and $\epsilon \in \mathbb{N}^+$.

FRAUDAR As existing algorithms, e.g., SPOKEN [31] and NETPROBE [26], tend to find an abnormal dense region, various camouflages are conducted to evade such detection. For example, as shown in Fig. 3b, the fake followers on Twitter also follow many popular celebrities as normal users do. Intuitively, they try to add enormous edges to result in a larger dense region consisting of many real entities, which the mentioned density definitions cannot tackle. For example, three examples of possible camouflages are shown in Fig. 3 where (a) fraudsters randomly add edges to ordinary users, (b) fraudsters mainly add edges to the vertices with high degrees, and (c) fraudsters hijack normal accounts to create fake edges.

FRAUDAR [15], modeling the structural information as a bipartite graph, proposes to discover anomalies in such an adversarial scenario. To find the most suspicious subgraph $S(V_S, E_S)$, it defines a new class of measurements $g(S)$ with the density form $g(S) = f(S)/|V_S|$, where $|V_S|$ is the number of vertices in S and $f(S) = f_V(S) + f_E(S)$ is the total suspiciousness of S with node suspiciousness $f_V(S)$ and edge suspiciousness $f_E(S)$. Moreover, it defined the following desired properties which are robust to the edge bursting:

- **Node suspiciousness.** Given other fixed conditions, a subgraph containing higher suspiciousness vertices is more anomalous than the one containing lower suspiciousness vertices. Formally, $|S| = |S'|$, $f_E(S) = f_E(S')$, $f_V(S) > f_V(S') \rightarrow g(S) > g(S')$.

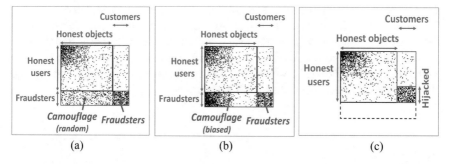

Fig. 3 Examples of possible camouflages [15]. (**a**) Random camouflage. (**b**) Biased camouflage. (**c**) Hijacked accounts

- **Edge Suspiciousness.** Fixing other conditions, adding edges to a subgraph increases its suspiciousness. Formally, $e \notin E_S \rightarrow g(S(V_S, E_S \cup \{e\})) > g(S(V_S, E_S))$.
- **Size.** Supposing vertices and edges weights are equal, larger subgraphs are more suspicious than smaller subsets with the same edge density. Formally, $|S| > |S'|, S \supset S', \rho_e(S) = \rho_e(S') \rightarrow g(S) > g(S')$.
- **Concentration.** A subgraph with fewer vertices is more anomalous than the one with the same total suspiciousness but more vertices. Formally, $|S| < |S'|, f(S) = f(S') \rightarrow g(S) > g(S')$.

Based on the proposed metrics, it proposes an approximation algorithm to find the subgraph S with maximum $g(S)$.

TellTail The density-based algorithms prefer the larger subgraphs, although they finally yield the densest subgraph. As shown in Fig. 4a, the densest subgraph based on edge density cannot find the injected clique, while it seems more surprising. To solve the size-biased issue, the authors of [14], modeling the context information as an undirected graph, develop a new model to measure the surprisingness of a subgraph based on the *extreme value theory*, as shown in Fig. 4b.

Intuitively, the surprisingness of a given subgraph is defined as the probability of generating a same size but denser subgraph under the random graph model. Various distributions can be employed to build the random graph model, such as the maximum likelihood to estimate unknown parameters. For example, the authors of [16] assume an Erdos–Renyi model which actually cannot model density regions of real graphs, e.g., the clustering structures and dense subgraphs, as shown in Fig. 4c and d. Moreover, the Gaussian distribution is also not suitable, as it decays too fast as shown in Fig. 4c and d. To better fit the real graph, the Generalized Pareto (GP) distribution is proposed in [14] .

3.3 Benefits and Issues

Graph-based approaches under unimodal scenarios only consider the structural information, which makes them easy to understand and implement. The structural information is almost always available in the published datasets, compared with the textual content. Moreover, the solutions are often efficient by employing existing graph mining algorithms. However, they mainly focus on static graphs and ignore the temporal features, which lead to a low accuracy.

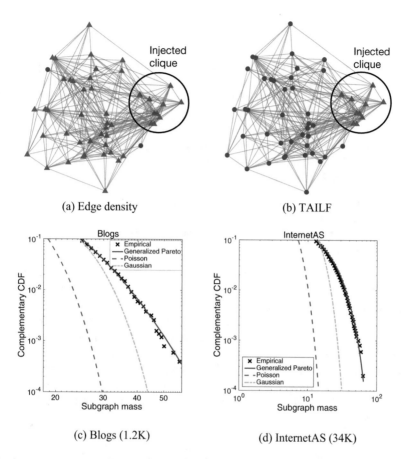

Fig. 4 The comparison of edge density and TAILF on the anomaly detection is given in (**a**) and (**b**) where red triangles show the subgraph detected using each measure. (**c**) and (**d**) show that different distributions fit mass distribution of two real datasets. The block crosses denote the empirical CCDF (i.e., 1-CDF) of the mass of 5000 random subgraphs [14]

4 Multi-modal Scenario

As unimodal approaches suffer from a low accuracy, recent studies devote more efforts to designing multi-modal methods which incorporate various types of information. However, the complicated models and algorithms impose more challenges on fake news detection. In this section, we discuss how to handle these issues under two different scenarios.

4.1   Dynamic Graph-Based Approaches

In this subsection, we discuss the graph-based fake news detection approaches under the dual-modal scenarios where the temporal propagation information will further assist the detection. We refer to dual-modal scenarios when considering both context-based information and temporal propagation. Since dynamic graphs are employed to model the mentioned two types of information, we also denote them as *dynamic scenarios* in this chapter. We first explore several temporal structures on the modeled dynamic graph and then present how existing approaches enable the detection by using these structures.

Temporal Graph Structures in Fake News   The following two kinds of temporal structures have received a lot of interest from recent studies. In particular, they are used to characterize the temporal properties, e.g., bursting events [33], periodic activities [32], and persistent groups [20].

– **Temporal density.** To capture a temporally dense structure, the traditional definitions of the densities have been extended to temporal graphs [24, 33]. In particular, the l-segment density is proposed in [33] to detect the bursting events. Intuitively, a vertex u in the given temporal graph $G(V, E, T = [t_s, t_e])$ has a sequence of varying degrees with respect to different snapshots, denoted as $\{N_u(G_i)\}_{i \in T}$ where G_i is the snapshot of temporal graph G at time i and $N_u(G_i)$ is the set of u's neighbors in G_i. The l-segment density actually indicates the average degree over the given sequence. To be specific, the l-segment density of vertex u, with respect to a subgraph induced by the vertex set S, is defined as follows:

$$SD(u, G_S(T)) = \frac{\sum_{i=t_s}^{t_e} |N_u(G_i) \cap S|}{t_e - t_s + 1}, \ t_e - t_s + 1 \geq l \qquad (1)$$

where $l \in \mathbb{N}^+$ is the minimum subsequence size. Clearly, a larger l-segment density means u interacts with enormous vertices in a short time, which corresponds to the bursting behaviors in fake news detection.

– **Temporal subgraph.** Most of the temporal graph structures are defined as the generalizations of standard structures. For example, Δ-cliques are introduced in [41] to generalize the classical notion of cliques. Given a temporal graph $G(V, E, T)$ with a set of vertices V, a set of times-edges $E \subseteq \binom{V}{2} \times T$, and a time interval $T = [\alpha, \beta], \alpha, \beta \in \mathbb{N}$, a Δ-clique $H(V', E', T' = [a, b])$ is a subgraph of G such that

  – $\forall u, v \in V', u \neq v \to ((u, v), t_{u,v}) \in E$ (*structural coherency*)
  – $\forall t \in [a, b - \Delta], \forall u, v \in V', u \neq v \to ((u, v), t_{u,v}) \in E, t_{u,v} \in [t, t + \Delta]$ (*temporal coherency*)

Basically, it is a classic clique within the time span. Moreover, it further requires that the fully connected structure always persists in every Δ time interval, by which it guarantees the bursty property. However, this definition does not reflect the hierarchical substructures, as it only requires that each pair of vertices have at least one edge within the time interval. Therefore, the denser subgraph, where every pair of vertices interacts frequently, cannot be discovered. To tackle this issue, the authors of [2] extend the Δ-cliques to (Δ, γ)-cliques where each pair of vertices has at least $\gamma \in \mathbb{R}^+$ edges within Δ time interval. Therefore, the Δ-clique can be further decomposed by setting various γ. In addition, the σ-periodic clique is proposed in [32] to identify the periodic events. To be specific, a σ-periodic clique S is a subgraph of a de-temporal graph $G(V, E)$ such that (1) S is a clique; (2) S is periodically formed at σ different snapshots, formally $\pi_\sigma(S) = \{t_1, t_2, \ldots, t_\sigma\}$ and $t_{i+1} - t_i = p$ for $i = 1, 2, \ldots, \sigma - 1$ with a constant p. It decomposes the whole graph based on these periodic patterns. However, to the best of our knowledge, the relationship between periodic behaviors and fake news is still vague. We roughly acknowledge that a group of fake users, controlled by the same organizations or social bots, tend to behave coordinately and periodically.

We then elaborate various studies to show how these temporal structures are used to detect fake news.

CopyCatch The authors of [4], modeling the temporal propagation information as a temporal dynamic bipartite graph, propose to discover anomalies by spotting lockstep behaviors. To be specific, the suspicious groups of users often enact similar behaviors together. For example, the fake customers on Amazon, controlled by a social bot, generally review the same groups of products around the same time.

To capture this behavior, a temporal structure, called $[n, m, \Delta t]$-temporally coherent bipartite core, is defined. Specifically, given a temporal bipartite graph $G(U, V, E, T)$ with two vertex sets U and V, an edge set E and the edge timestamps set T, a $[n, m, \Delta t]$-temporally coherent bipartite core is a subgraph $H(U', V', E', T')$ of G such that

- $(u, v) \in E', \forall u \in U', v \in V'$ (Structural coherency)
- $\exists t_v \in \mathbb{R}$ s.t. $|t_v - t_{u,v}| \leq \Delta t, \forall u \in U', v \in V'$ (Temporal coherency)
- $|U'| \geq n, |V'| \geq m$ (Size constraint)

Guaranteed by the structural coherency, the resulting subgraph is actually a biclique. It further requires that the edges adjacent to the same vertex have similar timestamps. Instead of computing a maximal subgraph, it incorporates size constraints to filter out small patterns. In terms of the low quality of data, the definition is relaxed as $[n, m, \Delta t, \rho]$-temporally near the bipartite core which is a subgraph $H(U', V', E', T')$ of G such that

- $|N(u, V)| \geq \rho|V|, \forall u \in U'$ (Structural coherency)
- $\exists t_v \in \mathbb{R}$ s.t. $|t_v - t_{u,v}| \leq \Delta t, \forall u \in U', v \in V'$ (Temporal coherency)
- $|U'| \geq n, |V'| \geq m$ (Size constraint)

where $N(u, V) \subseteq V$ denotes the neighbors of u. Intuitively, it relaxes the biclique structures as ρ-quasi-bicliques.

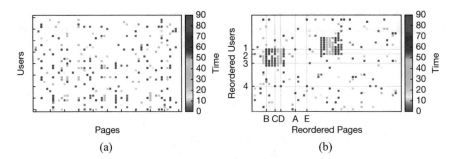

Fig. 5 User-likes-page behaviors over time on Facebook [4]. (**a**) Without CopyCatch. (**b**) With CopyCatch

CopyCatch is designed to detect these anomalies based on the temporally near-bipartite core. For example, Fig. 5 shows the performance of CopyCatch in detecting ill-gotten page likes. The users and pages are modeled as two sets of vertices U and V, and the page-like behaviors are represented by the set of edges E. Moreover, each edge attaches a timestamp to denote the time that corresponding behavior occurs. It is clear that groups of users like the same groups of pages at the same time.

4.2   Graph-Assisted Learning Approaches

To achieve a higher accuracy, most recent studies on fake news detection pay attention to the multi-modal scenarios where various kinds of information are available. However, it is quite challenging to integrate them into a graph model. There are also rare graph mining algorithms designed to solve such a complicated mining task. As learning-based methods make great progress in other fields, many researchers turn to employing machine learning or deep learning techniques to detect anomalies instead of traditional graph mining algorithms. To enable these learning frameworks, graphs still play an important role in modeling information and extracting features. In this section, we focus on learning-based approaches and discuss how graph features are extracted and used to support different learning frameworks.

Graph Features in Multi-modal Approaches Although we cannot construct one graph model to summarize all kinds of information, graph-based features can be extracted from various graph models. To this end, we briefly elaborate different types of graph-based features used in recent studies.

– **Vertex-level features.** In terms of the various entities, they are also referred to article-level or user-level features which are extracted to characterize the properties of each vertex by using the "local" information, e.g., degrees, neighbors, and content. For example, the syntactic features of each user can be extracted from the articles posted by the user. Inspired by node-level graph embedding methods, e.g., node2vec [12], DeepWalk [27], and LINE [38], recent studies turn to extracting vertex feature vectors based on deep representation learning.

- **Graph-level features.** Another kind of graph-based feature is extracted from the whole graph structure. In particular, many studies propose to learn the representation of cohesive subgraphs, e.g., bicliques [10] and diffusion patterns [43], based on the existing graph-level graph embedding methods such as subgraph2vec [5] and PV-DBOW [19]. As most of the mentioned cohesive subgraphs have vague meanings in fake news detection, graph-level features need to be further explored.

GTUT The authors of [10] propose a multi-modal fake news detection approach which operates in three phases as shown in Fig. 6. To integrate various kinds of information such as textual, user, and temporal data, it deploys an unsupervised learning framework and employs graph-based methods to extract features, e.g., biclique enumeration, graph-based feature vector learning, and label spreading.

Datasets and Graph Models To be specific, they consider a dataset consisting of articles $\mathcal{A}$, users $\mathcal{U}$, and posts $\mathcal{P}$. Each post $P_i \in \mathcal{P}$ is produced by a user $U_i \in \mathcal{U}$ with textual content c_i mentioning news article $A_i \in \mathcal{A}$ at time t_i, formally denoted as $P_i = [U_i, A_i, t_i, c_i]$. The task is to label articles as either fake or truthful based on $\{\mathcal{A}, \mathcal{U}, \mathcal{P}\}$. To tackle this problem, they model a bipartite graph G where vertices represent articles $\mathcal{A}$ and users $\mathcal{U}$, and an edge (A, U) between a user U and an article A represents that the user published a social media posting mentioning the article. Moreover, each edge (A, U) has a label or attribute $Attr(E_{AU})$ that contains a set of posts involving the article U. Formally, $Attr(E_{AU}) = \{P_i = [U_i, A_i, t_i, c_i] | P_i \in \mathcal{P}, U_i = U, A_i = A\}$.

Unsupervised Learning Approach Based on the datasets and graph models, GTUT consists of the following three phases, where it identifies seed fake and truthful articles in the first phase, and expands the labels to cover all articles in the next two phases:

- **Phase 1.** As fake news on social media often diffuses in a synchronized manner, a group of fraudsters create textually similar posts for same articles at similar times. This phase proposes to discover synchronous posting behavior based on the structural, temporal, and textual coherences. To extract structural features where groups of users comment on the same articles, they first enumerate all bicliques $\mathcal{B}$ with sizes larger than a threshold. For each biclique $B \in \mathcal{B}$ with articles B_A and users B_U, they estimate the coherence of each article in terms of

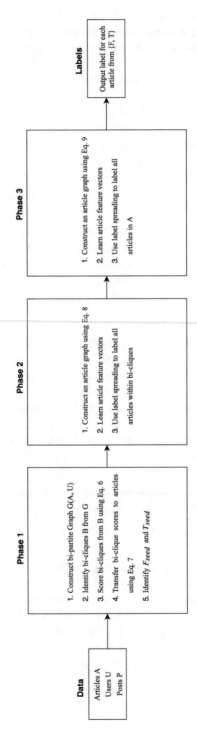

Fig. 6 Block diagram of GTUT [10]

the temporal bursty and the textual similarity (vertex-level features) as follows:

$$Temporal(A \in B_A, B) = \max\left(1 - \frac{BAS(A, B)}{T_{max}}, 0\right), \quad (2)$$

$$Textual(A \in B_A, B) = \frac{\sum_{(P_x, P_y) \in Posts(B,A)} sim(rep(c_x), rep(c_y))}{(|Posts(B, A)|) \times (|Posts(B, A)| - 1)} \quad (3)$$

where $BAS(A, B)$ denotes the bursty attention span covering most of the posts of A, T_{max} is a threshold, $rep(.)$ is the word2vec representation, and $sim(.)$ denotes the cosine similarity function. The larger the value, the more coherence it has. For example, if all the posts about A are created at a similar time, the time span $BAS(A, B)$ would be small, which leads to a large temporal score. Then they integrate above article-level features into subgraph-level features. For each biclique, we can obtain

$$Temporal(B) = \frac{\sum_{A \in B_A} Temporal(A, B)}{|B_A|}, \quad (4)$$

$$Textual(B) = \frac{\sum_{A \in B_A} Textual(A, B)}{|B_A|}, \quad (5)$$

$$TTScore(B) = \lambda \times Temporal(B) + (1 - \lambda) \times Textual(B) \quad (6)$$

where $\lambda \in (0, 1)$ is a parameter to indicate the weights of two types of features. Finally, the total score of an article is estimated based on the subgraph-level features as follows:

$$TTScore(A \in \mathcal{A}) = \frac{\sum_{B \in Bicliques(A,\mathcal{B})} TTScore(B)}{|Bicliques(A, \mathcal{B})|} \quad (7)$$

where $Bicliques(A, \mathcal{B}) \subseteq \mathcal{B}$ represents the subset of bicliques containing A. Basically, the higher the score an article has, the more suspicious it tends to be. Therefore, they select τ articles with the highest scores as the fake seed set and select τ articles with the lowest scores as the real seed set.

- **Phase 2.** As the seed articles have been identified, they continue to spread these labels to all articles within bicliques. To this end, they model a general weighted graph where vertices represent the articles within bicliques and each pair of vertices has an edge being weighted as follows:

$$Weight(E(A, A')) = \alpha \times \frac{|Bicliques(A) \cap Bicliques(A')|}{|Bicliques(A) \cup Bicliques(A')|} +$$

$$\beta \times \frac{|User(A) \cap User(A')|}{|User(A) \cup User(A')|} + (1 - \alpha - \beta) \times Sim(A, A') \quad (8)$$

where $Bicliques(A)$ denotes the set of bicliques containing A and $User(A)$ is the set of users who tweet article A. Intuitively, articles tend to share the same label, when they are within the same bicliques (term 1), are tweeted by the same users (term 2), and have similar textual content (term 3). Based on the constructed graph, they extract feature vectors of each article by node2vec [12] and then spread the labels over these features [45].

– **Phase 3.** They continue to spread labels to other articles outside the bicliques by employing the following weighting function to build a weighted graph and repeating the similar procedures in Phase 3:

$$Weight(E(A, A')) = \gamma \times \frac{|User(A) \cap User(A')|}{|User(A) \cup User(A')|} + (1 - \gamma) \times Sim(A, A') \quad (9)$$

In summary, they consider both vertex-level features and subgraph-level features to detect fake news. Various graph models are built during the feature extraction and the label spread, which shows their critical and unique roles in fake news detection.

4.3 Benefits and Issues

As an article with vivid content such as videos and images is more attractive, fraudsters tend to produce a multi-modal content. Therefore, the multi-modal cases are more common in real applications, and the multi-modal approaches are more general. In addition, these methods often achieve a higher accuracy compared with the unimodal ones, since the unimodal solutions only consider one kind of information but ignore others.

However, the multi-modal situation also introduces the extra costs of designing graph-based algorithms. For example, it is challenging to integrate various types of information into a graph model. Although we can alleviate these issues by employing dynamic graphs, the most mentioned temporal structures have not been well explored in fake news detection. In addition, another line of solutions based on learning-based approaches mainly focuses on vertex-level features. The effects of graph-level features on fake news detection need to be further explored.

5 Summary of the Chapter

This chapter presented recent studies on graph-based fake news detection. We first introduced its characteristics and corresponding unique challenges. We then summarized different types of information used to assist the detection and described how to model fake news as a graph. We discussed how graph structures are used to characterize features of fake news, such as summarizing various methods in terms of unimodal, dynamic (dual-modal), and multi-modal scenarios.

References

1. Akoglu, L., McGlohon, M., Faloutsos, C.: oddball: spotting anomalies in weighted graphs. In: Zaki, M.J., Yu, J.X., Ravindran, B., Pudi, V. (eds.) Advances in Knowledge Discovery and Data Mining, 14th Pacific-Asia Conference, PAKDD 2010, Hyderabad, June 21–24, 2010. Proceedings. Part II. Lecture Notes in Computer Science, vol. 6119, pp. 410–421. Springer, New York (2010). https://doi.org/10.1007/978-3-642-13672-6_40
2. Banerjee, S., Pal, B.: On the enumeration of maximal (Δ, γ)-cliques of a temporal network. In: Krishnapuram, R., Singla, P. (eds.) Proceedings of the ACM India Joint International Conference on Data Science and Management of Data, COMAD/CODS 2019, Kolkata, January 3–5, 2019, pp. 112–120. ACM, New York (2019). https://doi.org/10.1145/3297001.3297015
3. Berlowitz, D., Cohen, S., Kimelfeld, B.: Efficient enumeration of maximal k-plexes. In: Proceedings of the 2015 ACM SIGMOD International Conference on Management of Data, Melbourne, Victoria, May 31–June 4, 2015, pp. 431–444 (2015). https://doi.org/10.1145/2723372.2746478
4. Beutel, A., Xu, W., Guruswami, V., Palow, C., Faloutsos, C.: Copycatch: stopping group attacks by spotting lockstep behavior in social networks. In: 22nd International World Wide Web Conference, WWW '13, Rio de Janeiro, May 13–17, 2013, pp. 119–130 (2013). https://doi.org/10.1145/2488388.2488400
5. Chen, L., Li, J., Sahinalp, C., Marathe, M., Vullikanti, A., Nikolaev, A., Smirnov, E., Israfilov, R., Qiu, J.: Subgraph2vec: highly-vectorized tree-like subgraph counting. In: 2019 IEEE International Conference on Big Data (Big Data), Los Angeles, CA, December 9–12, 2019, pp. 483–492. IEEE, New York (2019). https://doi.org/10.1109/BigData47090.2019.9006037
6. Del Vicario, M., Bessi, A., Zollo, F., Petroni, F., Scala, A., Caldarelli, G., Stanley, H.E., Quattrociocchi, W.: The spreading of misinformation online. Proc. Natl. Acad. Sci. **113**(3), 554–559 (2016)
7. Fang, Y., Yu, K., Cheng, R., Lakshmanan, L.V.S., Lin, X.: Efficient algorithms for densest subgraph discovery. Proc. VLDB Endow. **12**(11), 1719–1732 (2019). https://doi.org/10.14778/3342263.3342645. http://www.vldb.org/pvldb/vol12/p1719-fang.pdf
8. Ferrara, E., Varol, O., Davis, C., Menczer, F., Flammini, A.: The rise of social bots. Commun. ACM **59**(7), 96–104 (2016)
9. Gallo, G., Grigoriadis, M.D., Tarjan, R.E.: A fast parametric maximum flow algorithm and applications. SIAM J. Comput. **18**(1), 30–55 (1989)
10. Gangireddy, S.C.R., P, D., Long, C., Chakraborty, T.: Unsupervised fake news detection: a graph-based approach. In: HT '20: 31st ACM Conference on Hypertext and Social Media, Virtual Event, July 13–15, 2020, pp. 75–83 (2020). https://doi.org/10.1145/3372923.3404783
11. Goldberg, A.V.: Finding a Maximum Density Subgraph. University of California, Berkeley (1984)
12. Grover, A., Leskovec, J.: node2vec: scalable feature learning for networks. In: Krishnapuram, B., Shah, M., Smola, A.J., Aggarwal, C.C., Shen, D., Rastogi, R. (eds.) Proceedings of the 22nd ACM SIGKDD International Conference on Knowledge Discovery and Data Mining, San Francisco, CA, August 13–17, 2016, pp. 855–864. ACM, New York (2016). https://doi.org/10.1145/2939672.2939754
13. Hanselowski, A., S., A.P.V., Schiller, B., Caspelherr, F., Chaudhuri, D., Meyer, C.M., Gurevych, I.: A retrospective analysis of the fake news challenge stance-detection task. In: Proceedings of the 27th International Conference on Computational Linguistics, COLING 2018, Santa Fe, New Mexico, August 20–26, 2018, pp. 1859–1874 (2018). https://www.aclweb.org/anthology/C18-1158/
14. Hooi, B., Shin, K., Lamba, H., Faloutsos, C.: Telltail: fast scoring and detection of dense subgraphs. In: The Thirty-Fourth AAAI Conference on Artificial Intelligence, AAAI 2020, The Thirty-Second Innovative Applications of Artificial Intelligence Conference, IAAI 2020, The Tenth AAAI Symposium on Educational Advances in Artificial Intelligence, EAAI 2020,

New York, NY, February 7–12, 2020, pp. 4150–4157. AAAI Press, Cambridge (2020). https://aaai.org/ojs/index.php/AAAI/article/view/5835

15. Hooi, B., Song, H.A., Beutel, A., Shah, N., Shin, K., Faloutsos, C.: FRAUDAR: bounding graph fraud in the face of camouflage. In: Proceedings of the 22nd ACM SIGKDD International Conference on Knowledge Discovery and Data Mining, San Francisco, CA, August 13–17, 2016, pp. 895–904 (2016). https://doi.org/10.1145/2939672.2939747

16. Jiang, M., Beutel, A., Cui, P., Hooi, B., Yang, S., Faloutsos, C.: Spotting suspicious behaviors in multimodal data: a general metric and algorithms. IEEE Trans. Knowl. Data Eng. **28**(8), 2187–2200 (2016). https://doi.org/10.1109/TKDE.2016.2555310

17. Jiang, M., Cui, P., Beutel, A., Faloutsos, C., Yang, S.: Catching synchronized behaviors in large networks: a graph mining approach. ACM Trans. Knowl. Discov. Data **10**(4), 35:1–35:27 (2016). https://doi.org/10.1145/2746403

18. Kleinberg, J.M.: Authoritative sources in a hyperlinked environment. J. ACM **46**(5), 604–632 (1999). http://doi.acm.org/10.1145/324133.324140

19. Le, Q.V., Mikolov, T.: Distributed representations of sentences and documents. In: Proceedings of the 31th International Conference on Machine Learning, ICML 2014, Beijing, 21–26 June 2014. JMLR Workshop and Conference Proceedings, vol. 32, pp. 1188–1196. JMLR.org (2014). http://proceedings.mlr.press/v32/le14.html

20. Li, R., Su, J., Qin, L., Yu, J.X., Dai, Q.: Persistent community search in temporal networks. In: 34th IEEE International Conference on Data Engineering, ICDE 2018, Paris, April 16–19, 2018, pp. 797–808 (2018). https://doi.org/10.1109/ICDE.2018.00077

21. Liu, X., Li, J., Wang, L.: Modeling protein interacting groups by quasi-bicliques: complexity, algorithm, and application. IEEE/ACM Trans. Comput. Biology Bioinform. **7**(2), 354–364 (2010). http://doi.acm.org/10.1145/1791396.1791412

22. Liu, Y., Wu, Y.F.B.: Early detection of fake news on social media through propagation path classification with recurrent and convolutional networks. In: Thirty-Second AAAI Conference on Artificial Intelligence (2018)

23. Ma, J., Gao, W., Wong, K.: Detect rumors in microblog posts using propagation structure via kernel learning. In: Proceedings of the 55th Annual Meeting of the Association for Computational Linguistics, ACL 2017, Vancouver, July 30–August 4, Volume 1: Long Papers, pp. 708–717 (2017). https://doi.org/10.18653/v1/P17-1066

24. Ma, S., Hu, R., Wang, L., Lin, X., Huai, J.: An efficient approach to finding dense temporal subgraphs. IEEE Trans. Knowl. Data Eng. **32**(4), 645–658 (2020). https://doi.org/10.1109/TKDE.2019.2891604

25. Mitzenmacher, M., Pachocki, J., Peng, R., Tsourakakis, C., Xu, S.C.: Scalable large near-clique detection in large-scale networks via sampling. In: Proceedings of the 21th ACM SIGKDD International Conference on Knowledge Discovery and Data Mining, pp. 815–824 (2015)

26. Pandit, S., Chau, D.H., Wang, S., Faloutsos, C.: Netprobe: a fast and scalable system for fraud detection in online auction networks. In: Williamson, C.L., Zurko, M.E., Patel-Schneider, P.F., Shenoy, P.J. (eds.) Proceedings of the 16th International Conference on World Wide Web, WWW 2007, Banff, Alberta, May 8–12, 2007. pp. 201–210. ACM, New York (2007). https://doi.org/10.1145/1242572.1242600

27. Perozzi, B., Al-Rfou, R., Skiena, S.: Deepwalk: online learning of social representations. In: Macskassy, S.A., Perlich, C., Leskovec, J., Wang, W., Ghani, R. (eds.) The 20th ACM SIGKDD International Conference on Knowledge Discovery and Data Mining, KDD '14, New York, NY, August 24–27, 2014, pp. 701–710. ACM, New York (2014). https://doi.org/10.1145/2623330.2623732

28. Pierri, F., Ceri, S.: False news on social media: a data-driven survey. SIGMOD Rec. **48**(2), 18–27 (2019). https://doi.org/10.1145/3377330.3377334

29. Popat, K., Mukherjee, S., Yates, A., Weikum, G.: Declare: debunking fake news and false claims using evidence-aware deep learning. In: Proceedings of the 2018 Conference on Empirical Methods in Natural Language Processing, Brussels, October 31–November 4, 2018, pp. 22–32 (2018). https://doi.org/10.18653/v1/d18-1003

30. Potthast, M., Kiesel, J., Reinartz, K., Bevendorff, J., Stein, B.: A stylometric inquiry into hyperpartisan and fake news. In: Proceedings of the 56th Annual Meeting of the Association for Computational Linguistics, ACL 2018, Melbourne, July 15–20, 2018, Volume 1: Long Papers, pp. 231–240 (2018). https://doi.org/10.18653/v1/P18-1022, https://www.aclweb.org/anthology/P18-1022/

31. Prakash, B.A., Sridharan, A., Seshadri, M., Machiraju, S., Faloutsos, C.: Eigenspokes: surprising patterns and scalable community chipping in large graphs. In: Zaki, M.J., Yu, J.X., Ravindran, B., Pudi, V. (eds.) Advances in Knowledge Discovery and Data Mining, 14th Pacific-Asia Conference, PAKDD 2010, Hyderabad, June 21–24, 2010. Proceedings. Part II. Lecture Notes in Computer Science, vol. 6119, pp. 435–448. Springer, New York (2010). https://doi.org/10.1007/978-3-642-13672-6_42

32. Qin, H., Li, R., Wang, G., Qin, L., Cheng, Y., Yuan, Y.: Mining periodic cliques in temporal networks. In: 35th IEEE International Conference on Data Engineering, ICDE 2019, Macao, April 8–11, 2019, pp. 1130–1141 (2019). https://doi.org/10.1109/ICDE.2019.00104

33. Qin, H., Li, R., Wang, G., Qin, L., Yuan, Y., Zhang, Z.: Mining bursting communities in temporal graphs. CoRR **abs/1911.02780** (2019). http://arxiv.org/abs/1911.02780

34. Ribeiro, C.C., Riveaux, J.A.: An exact algorithm for the maximum quasi-clique problem. Int. Trans. Oper. Res. **26**(6), 2199–2229 (2019). https://doi.org/10.1111/itor.12637

35. Shao, C., Ciampaglia, G.L., Varol, O., Yang, K.C., Flammini, A., Menczer, F.: The spread of low-credibility content by social bots. Nat. Commun. **9**(1), 1–9 (2018)

36. Shin, K., Eliassi-Rad, T., Faloutsos, C.: Patterns and anomalies in k-cores of real-world graphs with applications. Knowl. Inf. Syst. **54**(3), 677–710 (2018). https://doi.org/10.1007/s10115-017-1077-6

37. Sim, K., Li, J., Gopalkrishnan, V., Liu, G.: Mining maximal quasi-bicliques: Novel algorithm and applications in the stock market and protein networks. Stat. Anal. Data Mining **2**(4), 255–273 (2009). https://doi.org/10.1002/sam.10051

38. Tang, J., Qu, M., Wang, M., Zhang, M., Yan, J., Mei, Q.: LINE: large-scale information network embedding. In: Gangemi, A., Leonardi, S., Panconesi, A. (eds.) Proceedings of the 24th International Conference on World Wide Web, WWW 2015, Florence, May 18–22, 2015, pp. 1067–1077. ACM, New York (2015). https://doi.org/10.1145/2736277.2741093

39. Tsourakakis, C.: The k-clique densest subgraph problem. In: Proceedings of the 24th International Conference on World Wide Web, pp. 1122–1132 (2015)

40. Uno, T., Kiyomi, M., Arimura, H.: LCM ver. 2: Efficient mining algorithms for frequent/closed/maximal itemsets. In: Goethals Jr., R.J.B., Zaki, M.J. (eds.) FIMI '04, Proceedings of the IEEE ICDM Workshop on Frequent Itemset Mining Implementations, Brighton, November 1, 2004. CEUR Workshop Proceedings, vol. 126. CEUR-WS.org (2004). http://ceur-ws.org/Vol-126/uno.pdf

41. Viard, T., Latapy, M., Magnien, C.: Computing maximal cliques in link streams. Theor. Comput. Sci. **609**, 245–252 (2016). https://doi.org/10.1016/j.tcs.2015.09.030

42. Wang, W.Y.: "liar, liar pants on fire": a new benchmark dataset for fake news detection. In: Proceedings of the 55th Annual Meeting of the Association for Computational Linguistics, ACL 2017, Vancouver, July 30–August 4, Volume 2: Short Papers, pp. 422–426 (2017). https://doi.org/10.18653/v1/P17-2067

43. Wu, L., Liu, H.: Tracing fake-news footprints: characterizing social media messages by how they propagate. In: Proceedings of the eleventh ACM international conference on Web Search and Data Mining, pp. 637–645 (2018)

44. Zhang, Y., Phillips, C.A., Rogers, G.L., Baker, E.J., Chesler, E.J., Langston, M.A.: On finding bicliques in bipartite graphs: a novel algorithm and its application to the integration of diverse biological data types. BMC Bioinform. **15**, 110 (2014). https://doi.org/10.1186/1471-2105-15-110

45. Zhou, D., Bousquet, O., Lal, T.N., Weston, J., Schölkopf, B.: Learning with local and global consistency. In: Thrun, S., Saul, L.K., Schölkopf, B. (eds.) Advances in Neural Information Processing Systems 16 [Neural Information Processing Systems, NIPS 2003, December 8–13, 2003, Vancouver and Whistler, British Columbia], pp. 321–328. MIT Press, Cambridge (2003). http://papers.nips.cc/paper/2506-learning-with-local-and-global-consistency

Part II
Perspectives

Fake News in Health and Medicine

Ninu Poulose

Abstract With the rise of social media, the world is faced with the challenge of increasing health-related fake news more than ever before. We are constantly flooded with health-related information through various online platforms, many of which turn out to be inaccurate and misleading. This chapter provides an overview of various health fake news and related studies which have been reported in various news articles and scientific journals. Some of the studies conducted on health misinformation identified a prominence of vaccine- and cancer-related fake news. The popularity of so-called unproven natural cures for cancer and other diseases is alarming. The chapter also highlights the importance of maintaining accurate and effective scientific communication in this COVID-19 pandemic-hit world to safeguard public health. The current pandemic has also proved fertile ground for spreading misinformation. The chapter brings the audience's attention to the consequences of health misinformation, ranging from giving false hope to patients to the hurdles it poses to effective medical care. Finally, the chapter addresses some of the possible strategies to keep health misinformation in check.

Keywords Health misinformation · Natural cures · Vaccine · Pandemic

A lie can travel half way around the world while the truth is putting on its shoes.—Mark Twain

The earliest social media platform launched was SixDegrees in 1997 [1]. Since the launching of the modern social media app Friendster in 2002 and the many that followed, the twenty-first century has witnessed a social media boom, which includes famous platforms such as Facebook, Twitter, YouTube, and Instagram [1]. There are millions of people using these platforms, which has significantly changed the nature of human interactions and relationships. A parallel boom in the smartphone industry has enabled people to access these sites and information

N. Poulose
University of Oxford, Oxford, UK
e-mail: ninu.poulose@nds.ox.ac.uk

© Springer Nature Switzerland AG 2021
Deepak P et al., *Data Science for Fake News*, The Information
Retrieval Series 42, https://doi.org/10.1007/978-3-030-62696-9_9

just at their fingertips. This digital empowerment has not only improved our lives in a multitude of ways but also paved the way for building several online communities that bring people from various fronts together. Now that the world is more connected than ever, individuals are able to share news and views, disseminate knowledge, exchange culture and experiences, market business, participate in games and entertainment, and also engage in dialogues with an unknown person on the other side of the globe.

The medium through which people access news has also changed significantly over these years. According to a Pew Research Center survey conducted in 2018, around two-thirds of US adults (68%) get news on social media sites, at least occasionally, with around 43% getting news from Facebook [2]. However, we often turn a blind eye to the elephant in the room that is "fake news". Social media sites are a major platform for fake news providers to increase the web traffic of such news. The magnitude of this was evident in the US presidential elections in 2016 where fake news stories from hoax sites and hyperpartisan blogs outperformed real ones from major news outlets on Facebook [3]. There was also a preponderance of pro-Trump (anti-Clinton) fake news over pro-Clinton (anti-Trump) fake news, with 115 pro-Trump fake stories shared on Facebook a total of 30 million times and 41 pro-Clinton fake stories shared a total of 7.6 million times [4]. Reports show that people aged 65 and older tend to share fake news on social media more than younger people, as in the case of the 2016 presidential elections [5]. Another report by the Stanford History Education Group in 2019 provided evidence that high school students had difficulty discerning fact from fiction online, reiterating the importance of promoting digital media literacy among these students [6] who would be future voters in a few years.

Accurate and effective scientific communication is imperative to keep the public informed about the latest scientific developments and empower them to fully benefit from these advancements. Social media has facilitated the health industry in reaching out to the public by providing platforms to share health information, build patient-to-patient support networks, and allow the public to provide useful information and feedback. However, we are living in an era of information overload with a million websites on health issues, inaccurate scientific reporting, and health misinformation. Though health misinformation has always existed, with the surge in social media use, dissemination of fake news related to health and medicine is on an ever-alarming rise than ever before. Such fake news spreads like wildfire in our hyperconnected world, reaching a wide audience. Without proper third-party filtering, fact-checking, or editorial reviews, false information circulated through social media can cause confusion and mislead the audience. The sad truth is that an individual with no proven medical/scientific background can sometimes reach more readers than a credible news channel or journal. With the global media operating 24 h/day and with the increasing demand for news, even some trusted media outlets have misreported scientific facts (Blue Latitude Health). In order to safeguard public health and people's trust in the health industry in general, discriminating between scientifically proven facts and fake news has become the need of the hour.

There are different types of fake news. One such is inadvertent reporting of mistakes that occur due to errors and gaps in the editorial process. Rumours turning into fake news form the second category. Conspiracy theories are the third type, which are typically created by people who believe them to be true, and are difficult to verify as true or false. Fourth, fake news can also originate from satire that is likely misconstrued as factual. Fifth, false statements by pharmaceutical companies or medical professionals for making profits also contribute to fake news. The last category is the most problematic, where reports are not outright false but are slanted to mislead the public [7]. Fake news related to health is mostly associated with vaccines, cancer cures, conspiracy theories, and the recent COVID-19 pandemic. A number of studies conducted across the world have shed some light onto the landscape of fake health news stories. Some of these studies and interesting fake news articles are discussed in this chapter.

1   Polish Health Misinformation Study

A pilot study conducted by the Medical University of Gdansk sought to measure the extent of health misinformation stories shared on Polish language social media [8]. This study assessed the top shared health web links between 2012 and 2017 employing the BuzzSumo application, using keywords related to the most common diseases and causes of death. Surprisingly, 40% of the most frequently shared links fell into the category of fake news which was shared more than 450,000 times. It also found that the majority of these were vaccine related. This study came to the conclusion that analysing top shared news in social media could help in identifying medical misinformation.

2   Stanford University Study: Cannabis, a Cure for Cancer

Another study conducted by Stanford University researchers evaluated the growing online interest in using cannabis to cure cancer [9]. By employing the Google Trends' relative search volume (RSV) tool, they compared the online search activity over time (from January 2011 through July 2018) for "cannabis and cancer" versus standard cancer therapies. They arrived at some interesting results and conclusions. Whereas the RSV of "cannabis and cancer" queries increased nearly twofold, the RSV for "standard cancer therapy" queries changed only a little over the duration. The rate of increase in RSV of "cannabis and cancer" queries was ten times faster than that of standard cancer therapies' queries. Cannabis legalisation also had an impact on RSV growth, with higher growth in states where medical or recreational cannabis was legalised before 2011. Using the BuzzSumo social media analyser, they found that 51 of 136 high-impact news stories (37.5%) referencing "cancer cure/therapy/treatment" were the ones claiming a cancer cure with alternative treatments, of which 12 (23.5%) proposed cannabis as a cancer cure. It is disheartening that the top fake news proposing cannabis as a cancer cure generated 4.26 million engagements, whereas top accurate news stories discrediting this false news generated 0.036 million engagements. This again reminds me of the famous quote by Mark Twain. This study also highlights the importance of physicians and cancer organisations clarifying such misinformation.

3   NBC News Study

NBC News conducted an analysis in 2019 following the same methodology used in the previous two studies [10]. Using the BuzzSumo tool, they searched for keywords related to common diseases and causes of death in the USA. They also extended the search criteria to include topics often targeted by misinformation campaigns such as vaccines, fluoride, and natural cures. Articles with more than 25,000 engagements were considered for the study, with 80 articles in the final study. They found that articles on cancer, unproven cures, and vaccines were the most circulated health misinformation. On specified topics like cancer and fluoride, fake news dominated the overall news. Some of the most shared fake news articles are detailed below.

(a) "Big Pharma" *hiding cure for cancer*: The cancer-related article of highest engagement in 2019 pushed a medical conspiracy that "Big Pharma", consisting of a group of doctors and federal health organisations, is hiding a cure for cancer. The article, "Cancer industry not looking for a cure; they're too busy making money", gathered 5.4 million engagements on Natural News, a website owned and operated by Mike Adams, a dietary supplement purveyor who goes by the moniker "The Health Ranger". Facebook facilitated the highest engagement for the article where Natural News had nearly three million followers until it was banned later.

(b) *Natural cures for cancer and other diseases:* Cancer was the most popular topic of health misinformation, marijuana being one of the most popular alleged cures. Ranking among the top engaged articles were those advocating a fear of processed foods and a change to so-called natural cures without evidence. One article that titled "Scientists Warn People to Stop Eating Instant Noodles Due to Cancer and Stroke Risks" garnered 300,000 engagements. Another article that generated over 800,000 engagements claimed that "Ginger is 10,000x more effective at killing cancer than chemo". Other natural products which were falsely claimed to be cures for cancer, diabetes, asthma, and the flu included papaya leaf juice, elderberry, dates, thyme, garlic, jasmine, limes, okra, and other herbs, vegetables, and exotic fruits.

(c) *Vaccine, the villain:* Vaccines are one of the greatest discoveries in the medical field and provide a safe and effective means to fight and eradicate infectious diseases. Though limitations to their effectiveness exist, there are some well-funded anti-vaccination activists who actively work to promote the false claim that vaccines cause harm and death. This analysis revealed the identity of the most popular anti-vaccine news creators of 2019: Adams' Natural News; Children's Health Defense, an organisation led by the anti-vaccine activist

Robert Kennedy Jr.; and Stop Mandatory Vaccination, a website led by the self-described social media activist Larry Cook. Such anti-vaccine news was received well, with over a million engagements. Many of these articles that posit vaccines to be dangerous for children and pregnant women often misinterpret research and even claim vaccines to be the cause of death in some babies [10]. Another article highlighted by NBC News discussed how anti-vaxxers target women who have lost babies unexpectedly to death and turn them into crusaders against vaccines [11].

4 Dandelion, the Magical Weed

Dandelion weed was yet another popular alleged cure for cancer and an immunity booster, which was claimed to work better than chemotherapy [12]. According to CBC News, a 72-year-old leukaemia patient's cancer went into remission 4 months after he had dandelion root tea [13]. The article with the headline "Dandelion weed can boost your immune system and cure cancer" received more than 1.4 million shares, likes, and comments, according to two separate web analysis tools [7]. Although dandelion extracts have been shown to suppress different types of cancer cell proliferation in vitro [14, 15], there is no clinical evidence so far to support its miraculous properties in curing cancer or boosting immunity.

5 Polarised Facts

Polarised facts which are often heard in a political context are not entirely foreign to the medical field. A study by Hofmann sought to investigate how polarised research, where researchers hold radically opposite views on the same issue, produces polarised facts [16]. Using mammography screening for breast cancer as an example, a widely debated topic, he demonstrated a strong polarisation of the results. The biggest advantage of the screening is reduced breast cancer mortality, while the major disadvantage is overdiagnosis and overtreatment. Hence, overdiagnosis to mortality reduction ratio (OMRR) is an estimate of the risk-benefit ratio for mammography screening. Some researchers are proponents of high reduction in mortality and low rate of overdiagnosis with screening, while others claim mortality reduction to be moderate and overdiagnosis high. Analysing 8 published studies on OMRR revealed a huge difference among the ratios, up to 25-fold, from 0.4 to 10. Interestingly, a strong correlation existed between the OMRR and the authors' attitudes to screening ($R = 0.9$). This analysis sheds some light on how strong professional interests can polarise research and potentially influence important health policy decisions and therefore proposes that researchers disclose professional interests along with financial interests when submitting research articles.

6 Fake News During the Pandemic

Health agencies have the added responsibility of managing misinformation during a health crisis like a pandemic. It is important that health agencies and related organisations are equipped with a social media management plan to counter misinformation during crises. Two great examples are the 2014 Ebola crisis and the current COVID-19 crisis.

(a) Ebola

During the 2014 Ebola crisis, there was a constant stream of inaccurate claims circulating through social media. After the first patient was diagnosed in the USA, the number of virus-related tweets per minute skyrocketed. People continued tweeting and spreading fear despite the fact that all the potential cases tested came out negative in Newark, Miami Beach, and Washington, DC. A statement was issued by the Department of Public Health in Iowa dispelling social media rumours that Ebola had arrived in the state. In order to curb the spread of misinformation, the Centers for Disease Control and Prevention (CDC) sent out constant updates on Ebola on its website and social media accounts. The CDC swung into action fairly quickly by sending a tweet illustrating how people can and cannot contract the virus within less than 3 h after confirming the Ebola case in Dallas. This was retweeted more than 4000 times. The CDC's "Facts About Ebola" image was tweeted by another account with one million followers and was retweeted almost 12,000 times, spreading the message much further than it did through the original CDC tweet. The CDC also hosted a Twitter chat answering questions about Ebola. All these measures helped to spread the right information to people in a timely manner and stem fearmongering on the web to some extent [17].

(b) COVID-19

An excellent example of fake news in the medical/health field is the current COVID-19 pandemic that has forced many countries into lockdown to prevent further spread. On a daily basis, we read and hear many claims related to the pandemic through various social media platforms, many of which are not true. Many of us tend to believe these without verifying their authenticity and participate in propagating this false information. As rightly said, a little knowledge is a dangerous thing. Ali Therani, founder of Astroscreen, a London-based start-up which uses artificial intelligence to seek out disinformation on social media, says that coronavirus fake news does not appear to be a targeted campaign; instead people spread fake cures or conspiracy theories themselves [18].

There are many conspiracy theories on the origin and spread of the virus. One such conspiracy theory claims that 5G masts are the true cause of the coronavirus outbreak. Unfortunately, celebrities with huge followings were also part of spreading the story. NHS England's national medical director Stephen Powis dismissed the claims as "rubbish" and the worst kind of fake news. To make matters worse, a 5G mast was set on fire in Birmingham, UK, on 3 April 2020 in a suspected arson attack, after the technology was linked online to the spread of coronavirus [19]. Many other theories regarding the virus' origin have been circulating on social media. Some of them claim that it was conceived as a bioweapon, while others believe it was accidently released from a lab in Wuhan, the city where the coronavirus outbreak was first detected [20]. Despite rampant speculation, however, currently there is no evidence to prove either of these claims. A statement released by the US national intelligence director's office on 30 April 2020 dismissed the claims of its origin as a bioweapon. It also said the intelligence community is rigorously examining "whether the outbreak began through contact with infected animals or if it was the result of an accident at a laboratory in Wuhan" [20]. Moreover, a study of the coronavirus genome published in March 2020 in *Nature Medicine* from Scripps Research in California concluded that SARS-CoV-2 is not a laboratory construct or a purposefully manipulated virus. Instead, they proposed two plausible modes of origin: natural selection in an animal host before zoonotic transfer and natural selection in humans following zoonotic transfer [21]. Other baseless claims include that coronavirus was sent to Wuhan by a Canadian-Chinese spy team and that it might have originated in the USA and been brought to Wuhan by the US Army [22].

News about unproven cures for COVID-19 has been rampant on social media since the beginning of the pandemic. Unfortunately, in Iran, over 700 people died from ingesting toxic methanol following the spread of rumours that it helps cure the coronavirus [23]. In India, a cow urine drinking party of around 200 people was conducted by a Hindu group (Akhil Bharat Hindu Mahasabha) based on the belief that it would ward off the deadly virus [24]. Developed nations are also not immune to receiving such misinformation. US president Donald Trump, during the White House coronavirus task force briefing, suggested scientists explore the possibility of injecting bleach to treat COVID-19. Thankfully, a prompt response was given by medical doctors and disinfectant firms warning people about the danger of ingesting or injecting disinfectants [25]. During the initial stage of the pandemic, Trump also touted the malaria medication hydroxychloroquine as a possible treatment for coronavirus, although the claims had no adequate clinical backing. Hospital admissions from hydroxychloroquine poisoning have been reported in different parts of the world [26]. All this false information comes as a real blow to the tremendous efforts made by scientists and healthcare professionals around the world to tackle the pandemic.

Facebook-owned WhatsApp is a popular messaging service, where multiple groups can be created, and is yet another major platform for spreading misinformation. Since the pandemic started tightening its grip on the world, I have been receiving a constant flow of messages in different WhatsApp groups, on topics varying from home remedies to prevention of COVID-19 and videos from people including doctors and nurses from pandemic-affected and other countries and snapshots of NHS messages (in a UK WhatsApp group) and news updates. Some of the fake health messages which I have received through WhatsApp are shown in Fig. 1. One of the messages recommends eating alkaline foods as a way to beat the coronavirus. I received the same message in different WhatsApp groups. Though lemon, due to its citric acid content, has a pH around 2.2, the picture claims its pH to

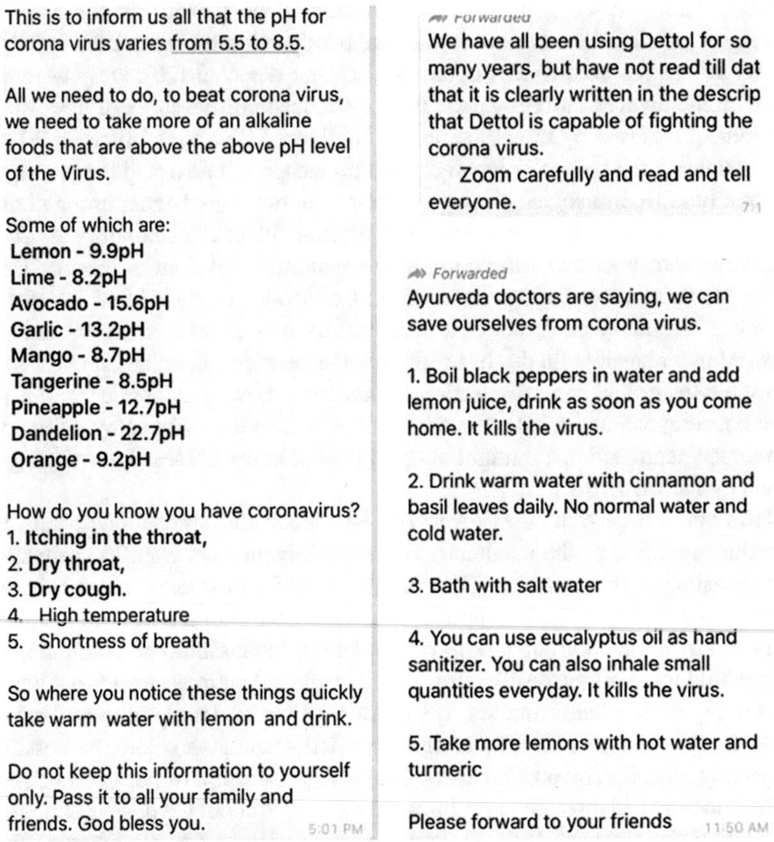

This is to inform us all that the pH for corona virus varies from 5.5 to 8.5.

All we need to do, to beat corona virus, we need to take more of an alkaline foods that are above the above pH level of the Virus.

Some of which are:
Lemon - 9.9pH
Lime - 8.2pH
Avocado - 15.6pH
Garlic - 13.2pH
Mango - 8.7pH
Tangerine - 8.5pH
Pineapple - 12.7pH
Dandelion - 22.7pH
Orange - 9.2pH

How do you know you have coronavirus?
1. Itching in the throat,
2. Dry throat,
3. Dry cough.
4. High temperature
5. Shortness of breath

So where you notice these things quickly take warm water with lemon and drink.

Do not keep this information to yourself only. Pass it to all your family and friends. God bless you.   5:01 PM

↩ Forwarded
We have all been using Dettol for so many years, but have not read till dat that it is clearly written in the descrip that Dettol is capable of fighting the corona virus.
    Zoom carefully and read and tell everyone.   7:1

↪ Forwarded
Ayurveda doctors are saying, we can save ourselves from corona virus.

1. Boil black peppers in water and add lemon juice, drink as soon as you come home. It kills the virus.

2. Drink warm water with cinnamon and basil leaves daily. No normal water and cold water.

3. Bath with salt water

4. You can use eucalyptus oil as hand sanitizer. You can also inhale small quantities everyday. It kills the virus.

5. Take more lemons with hot water and turmeric

Please forward to your friends   11:50 AM

Fig. 1 Sample of COVID-19 fake news

be 9.9 and claims dandelion has an extremely high pH of 22.7. It's a known fact that all the food we ingest, despite its pH, is exposed to the strongly acidic gastric fluid in the stomach. Being a respiratory virus, linking the coronavirus's viability to the pH of food is quite illogical as well as scientifically unproven. In an effort to prevent the public from falling for such a hoax, the World Health Organization (WHO) has an official page called Mythbusters with a brief description and pictorial representation of myths and facts. The National Health Service (NHS) is also constantly updating their websites regarding the latest information on COVID-19.

7   Consequences of Health Misinformation

Health misinformation has a huge impact on society. According to a study by leading health economists from Kingston University, London, more than 60% of online fake news about healthcare issues is considered not credible. More importantly, people's trust in fake news seems to increase with increased exposure. In other words, repetition counts: "the more someone sees something, the more they believe it" [27]. This study also revealed that warnings about potentially inaccurate information had a limited impact on users' behaviour in terms of believing or sharing information. An article by David N Rapp discusses how reading inaccurate information is likely to influence subsequent decision-making processes, even when a person is better informed. This is explained as a predictable consequence of the routine cognitive processes [28].

One of the most vulnerable groups affected by fake news is those patients who suffer a debilitating condition and seek quick relief. Dr. Shilpi Agarwal, a board-certified family medicine physician in the Washington, DC, area, pointed out that "False medical information and news makes patients scared unnecessarily and can often delay necessary medical care and attention" [29]. Fake news misleads many patients to pursue unproven cures for life-threatening diseases, disregarding the approved medical treatments. This makes the doctor's job harder, and the patient may also develop trust issues with the doctor. "We often spend a good amount of a medical visit correcting misinformation and re-educating the patient", says Agarwal. Health misinformation about vaccines is a major threat to global health, as it can lead to lower vaccination levels below herd immunity and put minors at risk.

Giving false hope to patients is a major consequence of health misinformation. A perfect example is an article published in the *Telegraph* entitled "Gene editing could end HIV, scientists hope, after second patient is 'cured' using rare mutation" [30]. The title somewhat misleads the reader to assume that some sort of gene editing was used to cure HIV. However, the main content of the article talks about using stem cell transplantation from a donor with a mutation in an HIV co-receptor CCR5 gene to cure the disease, and gene editing is proposed only as a possibility in the future. For an undiscerning public, especially those living with HIV and requiring a lifetime of medication, such headlines could lead to false expectations, which have to be then managed by healthcare professionals. Loss of valuable time for health professionals is another undesired outcome.

"Misinformation is being weaponised against vulnerable communities in a particular place at a particular time", says Dr. Shakuntala Banaji, Professor of Media, Culture, and Social Change at the London School of Economics (LSE). For instance, in 2018 an explosion of misinformation on WhatsApp about child kidnappers fuelled gruesome mob violence in an Indian village. Similarly, misinformation about coronavirus is promoting violent reactions across the world, from abuse levelled at Asian Americans in the USA to blaming Muslims for the virus in India [18].

Incidents like the 5G mast being set on fire following linking of the baseless conspiracy theory to the pandemic is yet another example of a dangerous act instigated by misinformation.

8 Managing Health Misinformation

It's a tough battle ahead to manage and reduce misinformation in health care. Social media platforms, healthcare providers, scientists, and the public all need to work hand in hand to fight this fake news pandemic which has crippled the virtual world. An article by Brady et al. discusses the grounds for propagation of misinformation from a physician's perspective. The article indicates that whereas a large proportion of misinformation can be traced back to computer-generated "bots" (automated programs designed to perform a specific task), credible sources like medical professionals have also contributed to spreading misinformation [31]. The article cautions medical authors about using online "quick shots", such as the visual abstract often used in journals now which can potentially cause oversimplification and omission of small but critical details of a medical study, which can therefore lead to misrepresentation by readers. Given that this new format is more Twitter-friendly than the traditional abstract, there is a risk of spreading inaccurate information. Another point raised was the existence of non-overlapping "social media bubbles" of physicians and patients, with the physicians probably unaware of the false information exposed by the patients. The article also highlights the importance of a greater presence of physicians on social media platforms to combat the spread of misinformation by actively engaging in discussions, critically evaluating posted information, and extending their social media bubble to include patients. It's reassuring to see that the WHO has an online page called Mythbusters, with a brief description and pictorial representation of various myths and facts, in order to prevent the public from falling for the common fake news related to COVID-19. The National Health Service (NHS) is also constantly updating their websites regarding the latest information on COVID-19.

As readers, we have a great deal of responsibility to filter the kind of information we receive, apply caution, and handle the information diligently. One way to do that is to seek information from reliable sources like the NHS or Cancer Research UK for information related to cancer or the British Heart Foundation for a new heart disease study. Instead of simply sharing Facebook news, retweeting, or forwarding a WhatsApp message, it's advisable to take time to verify if the content is worth sharing. If the authenticity cannot be verified by one's limited medical background, it may be better not to share such information. If you are a patient, it's advisable to talk to your doctor about a new treatment or drug that has captured your attention. It's always better to get information from multiple reliable sources and apply one's own logical reasoning to it, rather than just seeing and believing. According to an article published in the journal *Science*, one explanation for the faster and broader reach of falsity over truth is that false rumours are significantly more novel than the truth across all novelty metrics [32]. Hence readers have to be wary of the sensational headlines of many fake news stories that grab their attention. Promoting media literacy in schools would be an important step to prepare teenagers who may have difficulty discerning facts from fiction online.

Social media platforms have also taken measures to curb the spread of misinformation around coronavirus. On WhatsApp, one is now only able to send frequently forwarded messages in a single chat at a time. This has brought down message forwarding by 70%. Meanwhile, the Google-owned platform YouTube is removing anything that contradicts advice from the WHO, while Facebook users who have read, watched, or shared false information about the virus will now receive a pop-up alert urging them to visit the WHO's website. Facebook, which was heavily criticised following the 2016 elections, has now partnered with third-party fact-checkers to rate and review the content on the platform. Together, technology and human experience combined with media literacy will be instrumental in combating the issue.

References

1. Allen, J. https://www.future-marketing.co.uk/the-history-of-social-media/ (2017)
2. Shearer, E., Eva Matsa, K.: News use across social media platforms 2018. Pew Research Center, September 10. https://www.journalism.org/2018/09/10/news-use-across-social-media-platforms-2018/ (2018)
3. Silverman, C.: This analysis shows how fake election news stories outperformed real news on Facebook. BuzzFeed News, November 16. https://www.buzzfeednews.com/article/craigsilverman/viral-fake-election-news-outperformed-real-news-on-facebook (2016)
4. Allcott, H., Gentzkow, M.: Social media and fake news in the 2016 election. J. Econ. Perspect. **31**(2), 211–236 (2017)
5. Lardieri, A.: https://www.usnews.com/news/politics/articles/2019-01-09/study-older-people-are-more-susceptible-to-fake-news-more-likely-to-share-it (2019)
6. https://news.stanford.edu/2019/11/18/high-school-students-unequipped-spot-fake-news/
7. Kanekar, A.S., Thombre, A.: Fake medical news: avoiding pitfalls and perils https://fmch.bmj.com/content/7/4/e000142 (2019)
8. Waszak, P.M., Kasprzycka-Waszak, W., Kubanek, A.: The spread of medical fake news in social media – The pilot quantitative study. Health Policy Technol. **7**, 115–118 (2018)
9. Shi, S., Brant, A.R., Sabolch, A., Pollom, E.: False news of a cannabis cancer cure. Cureus. **11**, e3918 (2019)
10. Zadrozny, B.: https://www.nbcnews.com/news/us-news/social-media-hosted-lot-fake-health-news-year-here-s-n1107466 (2019)
11. Zadrozny, B., Nadi, A.: https://www.nbcnews.com/tech/social-media/how-anti-vaxxers-target-grieving-moms-turn-them-crusaders-n1057566 (2019)
12. Caylor, M.: http://bestimmunebooster.com/dandelion-weed-can-boost-your-immune-system-and-cure-cancer/ (2016)
13. CBC News. https://www.cbc.ca/news/canada/windsor/dandelion-tea-touted-as-possible-cancer-killer-1.1129321 (2012)
14. Ovadje, P., Ammar, S., Guerrero, J.-A., Arnason, J.T., Pandey, S.: Dandelion root extract affects colorectal cancer proliferation and survival through the activation of multiple death signaling pathways. Oncotarget. **7**(45), 73080–73100 (2016 Nov 8)
15. Zhu, H., Zhao, H., Zhang, L., Xu, J., Zhu, C., Zhao, H., Lv, G.: Dandelion root extract suppressed gastric cancer cells proliferation and migration through targeting lncRNA-CCAT1. Biomed. Pharmacother. **93**, 1010–1017 (2017 Sep)
16. Hofmann, B.: Fake facts and alternative truths in medical research. BMC Med. Ethics. **19**, 4 (2018)

17. Luckerson, V.: Fear, misinformation, and social media complicate ebola fight. https://time.com/3479254/ebola-social-media/ (2014)
18. Heathman, A.: https://www.standard.co.uk/tech/fake-news-coronavirus-spreading-online-a4432601.html (2020)
19. Martin, A.: https://news.sky.com/story/coronavirus-celebrities-criticised-for-fanning-the-flames-of-5g-conspiracies-11968570 (2020)
20. Rincon, P.: https://www.bbc.co.uk/news/science-environment-52318539 (2020)
21. Andersen, K.G., Rambaut, A., Lipkin, W.I., Holmes, E.C., Garry, R.F.: The proximal origin of SARS-CoV-2. Nat. Med. **26**, 450–452 (2020)
22. Sardarizadeh, S., Robinson, O.: https://www.bbc.co.uk/news/world-52224331 (2020)
23. https://www.independent.co.uk/news/world/middle-east/coronavirus-iran-deaths-toxic-methanol-alcohol-fake-news-rumours-a9487801.html
24. https://www.independent.co.uk/news/world/asia/coronavirus-news-hindu-cow-urine-drinking-party-india-a9402491.html
25. BBC Trump. https://www.bbc.co.uk/news/world-us-canada-52407177
26. Spring, M.: Coronavirus: the human cost of virus misinformation. https://www.bbc.co.uk/news/stories-52731624 (2020)
27. Kingston University. https://phys.org/news/2019-11-fake-news-healthcare-online-major.html (2019)
28. Rapp, D.N.: The consequences of reading inaccurate information. Curr. Direct Psychol. Sci. **25**, 281–285 (2016)
29. Booth, S.: https://www.healthline.com/health-news/how-fake-health-news-may-be-influencing-you-to-make-dangerous-decisions (2018)
30. Knapton, S.: https://www.telegraph.co.uk/science/2019/03/06/gene-editing-could-end-hiv-scientists-hope-second-patient-cured/ (2019)
31. Brady, J.T., et al.: The Trump effect: with no peer review, how do we know what to really believe on social media? Clin. Colon Rectal Surg. **30**(4), 270–276 (2017)
32. Vosoughi, S., Roy, D., Ara, S.: The spread of true and false news online. Science. **359**, 1146–1151 (2018)

Ethical Considerations in Data-Driven Fake News Detection

Deepak P

Abstract Data-driven and AI-based detection of fake news has seen much recent interest. The focus of research on data-driven fake news detection has been on developing novel and effective machine learning pipelines. The field has flourished with the rapid advances in deep learning methodologies and the availability of several labelled datasets to benchmark methods. While treating fake news detection as yet another data analytics problem, there has been little work on analyzing the ethical and normative considerations within such a task. This work, in a first-of-its-kind effort, analyzes ethical and normative considerations in using data-driven automation for fake news detection. We first consider the ethical dimensions of importance within the task context, followed by a detailed discussion on adhering to fairness and democratic values while combating fake news through data-driven AI-based automation. Throughout this chapter, we place emphasis on acknowledging the nuances of the digital media domain and also attempt to outline technologically grounded recommendations on how fake news detection algorithms could evolve while preserving and deepening democratic values within society.

Keywords Ethics · Fairness · Fake news detection · Data science

1 Introduction

Data-driven fake news detection involves the usage of machine learning and data analytics methods in order to combat fake news. It is still early days in this discipline, and thus algorithms in this space have largely explored supervised methods for the task, with some limited work on unsupervised fake news detection. Active learning, transfer learning, and reinforcement learning are not yet popular for fake news detection. With the growing ecosystem of fake news and a widespread recognition of the pervasiveness of fake news or disinformation through buzzwords like *post-truth*, fake news is arguably something we would need to live with in the

© Springer Nature Switzerland AG 2021
Deepak P et al., *Data Science for Fake News*, The Information
Retrieval Series 42, https://doi.org/10.1007/978-3-030-62696-9_10

long run. *Collins Dictionary* chose *fake news* as the word of the year in 2017,[1] in the aftermath of the 2016 US Presidential Elections during which the phrase was used heavily. Given these trends, one could envisage fake news detection as being embedded by default in various information delivery platforms in the future, much in the same way that spam detection has become a default feature offered by most email service providers. As this chapter is being authored, Microsoft has started including its *NewsGuard* plugin, which rates website credibility, in mobile versions of its Edge browser[2] as a feature turned on by default; Fig. 1 shows the NewsGuard plugin in action.

In this chapter, we consider the ethical aspects of data-driven fake news detection (DFND). As a first work in this topic, we endeavor to consider a broad set of ethical dimensions in DFND. We place emphasis on ensuring that this chapter is understandable for a broad audience much beyond technologists working in the area, providing abundant context wherever necessary. We outline several ethical dimensions that are pertinent for DFND in Sect. 2. This is then followed by a discussion of fairness in DFND in Sect. 3. We consider aspects around the uptake of DFND in Sect. 4, especially from the perspective of how democratic values could be presented during the course of such uptake; in this section, we also endeavor to provide some concrete recommendations that could help guide AI approaches to fake news detection. We then conclude the chapter in Sect. 5.

2 Ethical Dimensions of DFND

The ethical considerations in DFND fall under the broad umbrella of ethical considerations of any data-driven optimization task but are confounded greatly by the societal importance of the task. Thus, the domain of fake news poses some unique ethical considerations in that it operates in a domain where it seeks to make judgments on news and could thus influence opinions and substantive decisions made by humans. In order to illustrate the contrast with other domains, consider product recommendation, the task of determining whether a user may like a product or not. This task is relatively benign in moral terms in that a bad prediction may only result in users receiving bad product suggestions. For example, a chocolate ice cream lover may be sent offers pertaining to vanilla ice creams due to an inaccurate decision, or a beer lover could be sent wine recommendations. While these are evidently problematic, their effects are limited to creating user frustration and/or leading them into bad choices but (arguably) have limited impact beyond the

---

[1] https://www.independent.co.uk/news/uk/home-news/fake-news-word-of-the-year-2017-collins-dictionary-donald-trump-kellyanne-conway-antifa-corbynmania-a8032751.html.

[2] https://www.cnet.com/news/microsofts-edge-browser-warns-you-about-fake-news/.

Fig. 1 Edge NewsGuard
Plugin displaying a warning

purchase and consumption of the product. On the other hand, fake news on climate change being labelled as *non-fake* has serious ramifications. It could sway individual users' and public opinion away from green policies and could be harmful to society as a whole. Similarly, xenophobic fake news has been increasingly used as a tool by certain political parties to sway public opinion toward themselves.

We analyze the space of ethical considerations of DFND across three dimensions, which are briefly outlined herein:

– **Mismatch of Values:** A core ethical consideration in the context of data and AI technologies whose growth has been fuelled by automation efficiency and other values of the market is the tension between the values embedded in them and the values held in society. This includes the tension between *accuracy* and *fairness* as well as that between *convenience* and *dignity* and others considered in various contexts [21]. This conflict is also relatable to positions in the political spectrum in reasonably unambiguous ways.

- **Nature of Data-Driven Learning:** An important ethical consideration comes from the nature of data-driven algorithms themselves. Data-driven algorithms look to build statistical models from the past (past encoded in historical datasets used for training) and attempt to use such models for the future. While this is done explicitly during the training process in the case of supervised learning, assumptions based on the past are implicitly encoded within the design of both supervised and unsupervised algorithms. This involves an implicit assumption of a static nature of the high-level data and application scenario, which causes ethical ramifications in the domain of fake news.

- **Domain Properties:** There are certain properties of the domain that spawn ethical and normative considerations. As a simple example, unlike ad recommendations where the same ad could be relevant for a user and irrelevant for another, a news article judged to be fake needs to be judged fake for all users. Further, certain other inconsistencies may be inadmissible. As an example, the veracity decision should not depend on who is quoted in the article; in other words, fake news should be judged as fake regardless of who is quoted as relaying it.

We will delve into such ethical considerations in detail in the following subsections. We do not claim that this covers the full spectrum of ethical considerations but do hope to cover many important ones.

2.1 Mismatch of Values

We will now consider ethical ramifications from the mismatch of values for which ML algorithms are designed to optimize and those expected in the application domains such as fake news detection. We will consider the historical context of ML, and how things have changed from thereon, and outline various ethical facets of the value mismatch.

Historical Context of ML It helps to consider a historical perspective of machine learning in order to understand the context of the ethical considerations that emanate from the *mismatch of values*. The initial efforts of machine learning were targeted toward automating tasks that were inappropriate or difficult to be automated by means of rule-based methodologies. As an oft-quoted example, consider the case of handwriting recognition. It is extremely hard, if not impossible, to come up with a set of rules that would fit together to form a system to effectively recognize handwritten text. Tasks such as handwriting recognition are, by nature, tasks that humans are quite good at but often perceive as quite mundane. Data entry tasks that involve handwriting recognition were considered within the lowest rung of IT-enabled jobs in terms of skills required. Thus, machine learning, in its early days, aimed to automate such mundane tasks using abundant historical traces of human performance over such tasks to learn statistical models that would help replace or reduce manual labor spent on the task. Most early advances in machine learning were around tasks of a similar nature such as image recognition, text to speech translation, and automation of search. All of these target the optimization of mundane tasks for which the natural metric of success is *amount of labor automated*. Such is the case with another application realm for machine learning, that of robotics, where physical labor was sought to be automated. With optimization of manual labor being a priority for businesses who wanted to improve their competitive advantage in an emerging IT-oriented marketplace, investment in machine learning was aligned with market priorities. Cumulative metrics such as *precision*, *recall*, and *accuracy* were the natural targets for optimization, since they are easily translatable into automated cumulative labor. In semi-automated machine learning pipelines where humans were always available to correct errors, such as a supervisor who would gloss over handwriting recognition outputs to correct any apparent errors, the quantum of manual effort is the obvious area to be minimized. Such settings, and their more automated counterparts, did not offer any incentive to consider the *distribution of errors*. As a hypothetical example, a system which always misidentified a particular rare word, say a complex one such as *xylophone*, would be acceptable if that misidentification helped the model move toward such directions that ensure correct identification of a number of other common words. Within the historical context of machine learning envisaged as a minor and passive player that seeks to automate a set of mundane tasks within a sophisticated ecosystem, such market-driven and efficiency-oriented considerations being the sole or primary consideration made natural sense. These automation-oriented metrics, in the landscape of political philosophy, align with the schools of utilitarianism [19]. This school of thought seeks to maximize cumulative good, which translates well into automation being the target of maximization in the case of ML.

Current ML Applications Of late, one may observe that two major changes have taken place. First, the realization of the power of data-driven learning began to accelerate the uptake of machine learning quite dramatically. Machine learning algorithms started to become *major players* (as against minor ones) in an increasingly IT-enabled ecosystem and consequently started playing a more *active role* than a merely passive one. Second, machine learning started to be applied for tasks much beyond the original limited remit of *mundane tasks* that are worthy of automation. As Narayanan [15] opines, ML has been moving from the domain of *automating perception* (e.g., tasks such as handwriting recognition and face recognition) to domains of *automating judgment* (e.g., spam detection, fake news detection) and *predicting social outcomes* (e.g., predictive policing, predicting criminality from a face! [8]). Narayanan argues that the task of predicting social outcomes can be regarded as fundamentally dubious from an ethical perspective. The interplay between the first factor (pervasive use of ML) and the second factor (usage for predicting social outcomes) leads to serious issues that may not be apparent when considering them separately. As an example from an ML use case from predictive

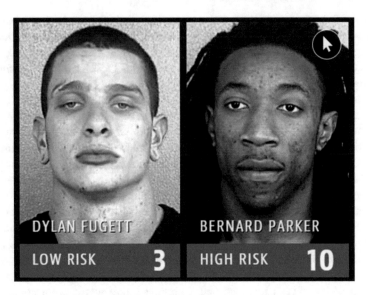

Fig. 2 The risk scores from COMPAS for two individuals detained for drug possession, illustrating racial bias (image Source: ProPublica)

policing, an initial preference for labelling minority areas as crime prone can be imbibed by an algorithm that aims to identify areas for higher surveillance. Crimes are caught only when *committed* as well as *observed*; higher surveillance in minority areas increases the observation rate of committed crimes, reinforcing the bias. Thus, not only does ML enable institutionalizing bias inherent within historical data, it creates even more biased data for the next generation of tools to work on, compounding the problem. Thus, machine learning algorithms are today employed in making decisions that significantly affect human lives. In what has become a widely cited example of bias, COMPAS, a software tool to predict recidivism in the USA, has been widely criticized for being biased against blacks; Fig. 2 shows the *risk scores* assigned by COMPAS to two individuals detained on account of possession of drugs.

An evolutionary perspective predicts that most diets and fitness programs will fail, as they do, because we still do not know how to counter once-adaptive primal instincts to eat donuts and take the elevator.
—Daniel E. Lieberman in "The Story of the Human Body: Evolution, Health, and Disease" [13]

The Facets of the Mismatch While the historical context of automation of mundane tasks made cumulative efficiency-oriented market-aligned metrics the natural ones to optimize for, the new application scenarios make them least suited due to their conflicts with the values of society. ML algorithms that have historically been advanced along a certain direction (guided by cumulative efficiency-oriented metrics) now need to be steered in a different direction! Such *mismatches* are hardly unique to machine learning; the most studied mismatch is that of *evolutionary mismatch* [14], which refers to evolved traits that were once advantageous but became maladaptive due to changes in the environment. An oft-quoted example is that of human diet, where humans, having evolved for long durations in the African savanna, developed a penchant for rare foods that contain both sugar and fat, as Daniel Lieberman states in the quote cited above. This evolutionary liking encourages us to seek out foods high in fat and sugar, which the market has overtly exploited through abundant placement in supermarket shelves, leading to a pervasive obesity problem in the population. While the analogy does not go the whole way, since ML algorithms are different in being actively designed by humans rather than evolving through natural selection, it does suggest that adapting to the needs of the new tasks is likely to require radical reimagination as opposed to patchwork fixes. The mismatch of values between those *from markets* and those *in society* has several facets, some of which we examine below, within the context of fake news detection:

– **Utility vs. Fairness:** While fake news algorithms should rightly aim to develop the capability to debunk as much fake news as possible (i.e., high utility in terms of fraction of fake news debunked), this should not come at the cost of an asymmetry along facets that matter. For example, even if fake news about *tapeworm* is only 2% of medical fake news, a method that is totally unable to capture that space of fake news would not be acceptable. In other words, the cumulative accuracy/inaccuracy should not have a high distributional skew along facets that are reasonably important.
– **Problematic Features:** Typically, ML algorithms are designed by making use of all features that can potentially tell something about the target variable, since it would help the ML algorithm achieve better accuracy. Thus, if a particular user handle, U, is largely used to share fake news, an ML algorithm may learn that pattern. This could, for example, be through a high value of conditional probability $P(fake|U)$ or more sophisticated mechanisms. However, such a feature could be problematic to use, since it undermines the user's (or, for that matter, any human's) ability to evolve, and if such estimates are used widely and the user's posts are blocked more often than not, it disenfranchises the user's voice in the media. It may be argued that something that has a systematic bias against a particular user cannot be construed as being part of a *fair process*. A similar argument could be used, though less compellingly, against using news source IDs as a feature. It may also be noted that simply not using a particular feature may not be enough, since there could be other proxy features. For example, a user may be identifiable through a distinctive language style, and thus the user's correlation with fake news may be learnt indirectly by an ML algorithm.
– **Responsibility and Accountability:** In applications of ML which fall under the *automating perception* category (in Narayanan's categorization [15]), it

was plausible to make an argument that *the further we go, the better*. In other words, it was possible to argue that some amount of automation is better than no automation, and more is better than less. However, when it comes to tasks such as fake news detection, the fact that ML is being used or claimed to be used in this regard implicitly involves much more responsibility. This means that deepening of automation may need to be held off until there is capacity to shoulder the responsibility that comes with such higher levels of automation. There are at least two fronts of responsibility and accountability that come from functioning in a democratic society:

- To the media sources whose news stories are being labelled as fake or non-fake
- To the user who is expected to consume the decision made by the algorithm

It is still an open question as to how these responsibilities may be fulfilled. One possibility could be that a trail or explanation is generated to support the decision, which can be made public, so as to be challenged or debated upon. Then again, should these be subject to legal regulations? If a legal framework needs to be instituted, it would require that the process of ensuring compliance with the legal regulations be laid out clearly. It could also be argued that such enforcements should not be made by legal frameworks but through voluntary compliance with ethical standards developed in the community.

2.2   Nature of Data-Driven Learning

Big Data processes codify the past. They do not invent the future. Doing that requires moral imagination, and that's something only humans can provide.
—Cathy O' Neil in "Weapons of Math Destruction" [16]

We now consider ethical issues that emanate from the very nature of ML or data-driven learning. The broad task in data-driven machine learning is to make use of historical/past data (in conjunction with several other constraints coming from an understanding of the domain) in order to make meaningful decisions about the future. Any perspective that is historically rooted would pooh-pooh a proposal that aims to make an assumption that the past is predictive of the future *when it comes to making decisions on substantive societal issues*. So, how did we come about to even attempting to use ML for such tasks? The answer once again lies in the historical context in which ML developed.

Following on from the narrative in the previous section, we can see that the simple hypothesis of *past predicts the future*—more technically, that training and testing data come from the same distribution—works exceedingly well for tasks such as speech recognition, or characterizing supermarket purchase patterns, especially in the short term. Handwriting styles remain quite static for an individual, and the nature of errors made by ML algorithm handwriting recognition does

Fig. 3 A post from the Discussion Board */pol/* from *4chan*, a discussion board often noted for extremist political ideologies

not influence how the person would change his or her handwriting. Similarly, people generally have some amount of periodicity in purchasing regularly used FMCG products, and stores organizing products based on purchase patterns, while enhancing convenience, are not likely to affect consumer purchasing behavior. However, when one considers other domains of activity and the long term, people do evolve substantially. Peddlers of fake news work in a highly dynamic ecosystem of social media platforms, where certain features are more useful than others for propagating fake news. For example, anonymous posting functionalities provided by social networks like *4chan* have been regarded as being exploited heavily by agents that drive fake news.[3] Figure 3 shows a post from *4chan*'s */pol/* discussion board which has been noticed for extremist political ideology as well as alternative facts. WhatsApp recently restricted its forwarding functionality in view of fake news.[4] Such measures lead to a gamification between fake news peddlers and social media platforms, in turn leading to an ever-changing character of fake news, limiting the ability of using historical data in predicting the future. Viewed from another perspective, naively learning from historical data without accounting for the dynamics of the space would lead to techniques that would be biased in being able to discover certain kinds of fake news more than others.

The dynamic environment that exists in the misinformation space escalates in volatility even further with the presence of ML-based fake news detection as an active player. When certain techniques for fake news detection gain prominence and get widely applied, incentives to devise workarounds also emerge along with it. The resultant gamification between fake news detection techniques and fake news itself would lead to a perpetual race by each party to stay one step ahead of the curve. In such a scenario, the nature of fake news detection techniques will also decide the future nature of fake news, and vice versa. This could result in fake news detection mechanisms employing highly complex decision surfaces to stay current

---

[3]https://news.sky.com/story/research-examines-fake-news-hate-speech-and-4chan-10910915.
[4]https://www.theverge.com/2020/4/7/21211371/whatsapp-message-forwarding-limits-misinformation-coronavirus-india.

and usable. The change in the behavior of fake news detection methods need not be due to conscious engineering by data scientists. The same algorithms when fed with newer labelled data encoding the changes in character of fake news will itself result in changes in the nature of the models built by the same learning methods. In a way, the same ML working as a meta-model using inductive learning will produce different models in response to different labelled datasets.

The highly volatile landscape with multiple actors trying to outpace one another is not quite new. It exists in the case of other domains, a very relatable one being *antivirus* software and, to a lesser extent, *spam detection* software. The makers of antivirus software and the makers of viruses are always in a relationship similar to that between fake news debunking software and fake news peddlers. The difference is that while virus makers are keen on finding new ways to squeeze *self-replicating code* into machines, fake news creators operate in ways to *sway the user's thoughts* in directions that suit their political or economic interests. While the former may be argued to be *morally neutral*, the latter definitely is not so. The intent of fake news differs in ways in which computer viruses do not; political fake news that is "useful" for one party would be "harmful" for another. Since fake news operates in the space of *swaying user opinions and thoughts*, one which has plentiful moral dimensions, care needs to be accorded to how detection algorithms are built.

Staying Updated in a Volatile Ecosystem An ML-based fake news detection method that is out of sync with the configuration of the ecosystem over which it would be used for fake news detection could result in a plethora of inaccurate decisions. Reactively adjusting to discovered errors may not be sufficient. This is so since some errors may never be discovered, or are less likely to be discovered; for example, news labelled as fake by a detection method may be hidden from view (depending on how the method is embedded within a software tool), and thus there may not be an opportunity to identify such false positives. Consistently making erroneous decisions that curtail the propagation and visibility of certain opinions can be argued to stand against the spirit of democracy and compromise *reasonable pluralism* [6] in public discourse; this aspect makes this issue distinct from analogous scenarios within antivirus and spam detection software. Continuously procuring a current set of labelled data followed by extensive benchmarking and method refinement may not be feasible due to resource and economic considerations. Nevertheless, a continuously updated conceptual picture of the ecosystem within which the technique would be embedded needs to be maintained, and the technique needs to be periodically contrasted against it in order to ensure that it is current. In particular, the ML method may need to be refined in two distinct dimensions to remain updated: by varying the training data and by varying the method. We consider important questions in this space, the answers to which may point to directions in which the techniques should be refined.

- **Training Data Curation:** The training data, in the case of supervised methods, determines the capabilities that will be infused into the fake news detection model that is eventually learnt. This makes curation of training data an important consideration in ensuring that the fake news detection technique stays current. This involves aspects such as the following:

- *How old can the training data be?* Very old training data may be inappropriate to use since they may be obsolete artifacts from an ecosystem that has substantively changed.
- *What is the relevance of training data elements?* Even temporally recent training data elements may be of limited relevance if they are associated with aspects of a media ecosystem that no longer exists. For example, one could argue that a social media post that is sparse in content and rich in emojis may be of limited relevance if that was soon followed by a radical change in the affordances with respect to emojis on the social media platform where it is situated.

- **Technique Design:** Every ML method, implicitly or explicitly, makes use of some assumptions about the domain in order to carry out the learning process. Some of these assumptions may be violated with changes in the media ecosystem that happen due to ML or non-ML actors as outlined above. Within unsupervised learning methods that do not have the luxury of being guided by training data, technique design considerations are more central and worthy of more attention. As an example, a *truth discovery* approach [22] makes an implicit assumption that fake news is represented on a minority of websites and that fake narratives diverge from facts in different directions. The presence of a widespread orchestrated fake news campaign could easily upturn such an assumption and lead the technique to discovering fake news as real and vice versa. Similarly, behavioral heuristics such as assumptions on synchronous user activity employed by recent methods (e.g., [7]) could also be invalidated by novel strategies by fake news peddlers.

2.3 Domain Properties

Fake news detection operates in the space of media, often referred to as the *fourth estate*, a space where actors have significant indirect influence in the political ecosystem. Further, the nature of the media domain entails some unique considerations for AI interventions within it. We outline some such unique aspects below:

- **Veracity Decisions as Impersonal and Universal:** This is an era of personalization, where ML algorithms routinely make use of user profiles to tailor decisions to them. We compared fake news detection with spam detection many times over, and within spam there is an element of personalization that could be legitimately brought in by making use of the inputs from the user on what is spam for them and what is not. Indeed, most ML-based personalized algorithms operate at two levels: one that makes use of cross-user data to learn general trends across a large dataset and another that makes use of user-specific data to learn specific likings of the user. The decision for a user, such as whether an email is spam, is one that blends both these factors. Thus, an email that goes into the spam folder for a user may legitimately need to land in the inbox for another. However, such personalization is inherently incompatible with the task of fake news detection, since there is no reason why a news is *fake* for one and *legitimate* for another. This aspect needs to be seen normatively, rather than in terms of utility. For example, *fake news* on vaccines and autism may be comforting for an *anti-vax* activist, and thus personalization that does not flag the news as fake may be better for improving user satisfaction for him/her, the utilitarian metric that most such methods aim to optimize. Despite such factors, the veracity decision needs to be consistent across users, from a normative standpoint.

- **Decision Timeliness, Reversals, and Accountability:** The emerging understanding of fake news involves a finding that exposure is hard to correct [18]; in other words, a person exposed to a news is still influenced by it long after it is exposed as fake to the same person. A news delivery system which claims to have a fake news detection functionality thus needs to ensure timely decisions to reduce exposure to fake news, in view of the accountability considerations discussed in an earlier section. It may also be argued that there is value in deferring dissemination of news articles until they are verified, especially if the fake news detection is implemented on a news delivery platform such as Google News. If that is not done (and it may be infeasible to do so in cases where the fake news identification is embedded within a browser plugin, where the user acts independently of the service), it may be argued that the service may be considered accountable to those who read a news in their system which was later labelled as fake. Does the system have an implicit obligation to proactively inform such users about the finding of fakeness? ML systems make decisions on the basis of data. As new data emerges, decisions may have to be reversed, or the confidence in a particular decision may deteriorate to an ambiguous range. It is interesting to analyze, from the perspective of accountability, as to how systems should handle such decision reversals. A somewhat similar case exists in online media where it is considered a good practice to make all edits to a published article public. In any case, there is a higher degree of accountability toward users who viewed an earlier decision that was reversed, as compared to somebody who viewed the article prior to any decision from the fake news detection method. It may also be seen here that these dimensions of accountability around decision timeliness and reversals do not apply to the earlier generation of tasks such as handwriting recognition (at least, not anywhere close to the same extent).

- **Veracity of Reporting or Reported Information:** Consider a case where a famous person, say X, makes a verifiably fake claim, such as *turmeric water can cure COVID-19*. What would be the veracity label attached to a news article, or a tweet such as that shown in Fig. 4, that carries the statement: *X says that turmeric water can cure COVID-19?* There are arguments on two sides. First, that the news piece is *non-fake* since X did actually make the claim. Second, that the news piece is *fake* since it contains a verifiably false claim. By treating fake

Fig. 4 A tweet reports that a famous person claimed a COVID-19 cure. Do we verify whether the claim was correct or whether the reporting was indeed factual? These choices lead to different veracity decisions

news detection as a data science problem, such important nuances could easily be brushed under the carpet by relegating them to the ways in which they are labelled, which in turn may depend on how individual labellers think about them. However, it is important to consider whether fake news detection should restrict itself to superficial verification (e.g., whether the statement reported was actually made) and whether it needs to go a level deeper (i.e., whether the statement is actually true). This could lead to different kinds of fake news detection systems.

3 Fairness and DFND

We now consider aspects of fairness and how they apply to the task of DFND. Fairness is used variously and is interpreted as related to a number of other concepts such as *equity*, *impartiality*, *unbiased*, and *equality*. Fairness has a long tradition within philosophy over the centuries, and the most widely accepted usage today could be in the context of *justice as fairness*, Rawls' pioneering work [17] in 1971.

While a broad discussion of fairness is well beyond the scope of this work, we will consider fairness in the way it has been used in machine learning literature, fair ML being a very active area since an early work [5] in 2012.

Streams of Fairness in Machine Learning Fairness in machine learning has been studied under two distinct streams: *individual fairness* and *group fairness*. While this distinction has come under recent criticism [3], we will use it as it provides a conceptual distinction between routes for deepening fairness. Individual fairness is commonly interpreted as being related to application of *fair procedure*, in that the task is done without partiality to the individual and in full sincerity to the aspects that matter to the task. A fair job selection process should thus only make use of attributes or features of a candidate that matter to the job and nothing else. In most analytics tasks, this would mean that *similar objects get assigned similar outcomes* and that similarity is assessed in a task-relevant manner. While all of this should come across as natural, what it keeps out of scope is important to analyze. It does not consider the *historical context* of data and interprets *data as given*. Thus, an individually fair or procedurally correct method discards any historical context of entrenched oppression that has caused some ethnicities to be disadvantaged with respect to access to quality education or if the metrics that measure *future productivity* in the job are set up in a way that is advantageous to certain ethnicities. This means that a method that agrees to tenets of individual fairness could produce *unequal outcomes* on dimensions such as ethnicity or other dimensions such as gender within which historical asymmetries exist. On the other hand, *group fairness* algorithms interpret fairness as a property of outcomes. It usually works by designating some attributes as *sensitive*; these are typically attributes that an individual usually does not have much role in determining for herself, or on whom asymmetries in societies usually function. Thus, these could include *gender*, *ethnicity*, *nationality*, and *religion*. Group fair algorithms try to ensure that parity is maintained across such specified sensitive dimensions. For example, if blacks have a one-seventh representation in a population, as is roughly the case in the USA, group fairness would be violated when the proportion of blacks among successful hires deviates much from one-seventh. These constraints are enshrined, though not to the fullest extent, within legal provisions such as the *Uniform Guidelines on Employee Selection Procedures (1978)* [2] in the USA, and provisions for affirmative action by way of quotas (commonly called *reservations*) within the Indian Constitution, viz., Article 15(4).[5]

Fairness and Impersonal Data The above notions of fairness are well motivated when making decisions about human beings based on their data. Indeed, the notion of *equity* and *equality* is most supported within a society when it comes to treatment of individuals. This would also naturally extend to cases where certain other attributes are correlated with individuals' *sensitive attributes*. As an example, we may argue that predictive policing methods should enforce some

---

[5]https://indiankanoon.org/doc/251667/.

kind of parity between minority and non-minority neighborhoods to ensure that societal anti-minority stereotypes are not reinforced by heavily policing minority neighborhoods. Such fairness arguments may also be extended toward geographical regions, where we may expect a *decent* level of public infrastructure across regions. For example, we may want to ensure that road works are not unduly delayed in rural neighborhoods even if the roads are less heavily used there as compared to urban neighborhoods. These issues are particularly of concern in cases where crowdsourcing is used to collect reports on issues such as *report a pothole* services that are being deployed[6] by governments worldwide. Solely relying on such IT-enabled crowdsourcing mechanisms could reinforce existing asymmetries. Rural roads are likely both less busy and residents may be less tech-savvy, both of which could cause underreporting of issues from rural localities. This may be seen as a notion of group fairness when treating *geographical region* as a sensitive attribute. Thus, the applications of principles of fairness could extend beyond personal data and could be carefully and meaningfully extended to data that does not pertain to human beings.

Fairness and Fake News Detection How would we go about thinking about the usage of fairness principles and their applications in the task of fake news detection? One possibility is to first consider violations of fairness we would necessarily want to avoid. We discuss some examples here:

– **Political Alignment:** Consider an example where a political party enters the fake news detection business and provides a plugin that explicitly states that *it debunks fake news from its political opponents*. In certain other cases, the political alignment may be less explicit than this but may serve a similar function. Would we want to permit such a fake news detection method even if it truthfully admits the bias? Such a tool may work either by keeping news sources that it favors completely out of the detection remit or by ensuring they are labelled as non-fake through other means. It may be argued that such tools reinforce the echo chamber effects that personalized news is often criticized for [20]. Due to such reasons, such politically aligned fake news detection that is unfair in being biased toward particular political positions may be considered undesirable. Should we then consider political leaning as a *sensitive attribute* and shoot for group fairness? We will consider such options soon.
– **Different Standards:** We may also want to avoid fake news detection tools that apply different standards to different parts of the news domain. Such different standards could emerge from seemingly legitimate reasons. As an example, a fake news detection engine may decide that news from a particular country may be fact-checked against authoritative sources within that country, as a maxim of procedural fairness. This would entail that disparities in authoritativeness between reference sources across various nations would naturally manifest as different standards. As an example, the same news could be assigned different

---

[6]https://www.nidirect.gov.uk/services/report-pothole.

veracity labels based on which country it stems from. However, in this case, unlike the case of political affiliations, we may choose to allow the possibility of a fake news detection engine that admits upfront that news from certain countries is likely to be judged with higher confidence than news from other countries.

The above examples suggest that extreme violations of *procedural fairness* along dimensions of political affiliation and regions should be considered as unacceptable. By way of *procedural fairness*, one would mean that the procedure for determining veracity should not be biased to favor some over others.

Procedural Fairness in DFND The desirability of procedural fairness places significant constraints on what kinds of DFND algorithms would be acceptable and which ones would be unacceptable. In other words, the notion of procedural fairness places constraints on how to go about building DFND methods. We will consider political affiliation as the dimension of consideration for fairness for this discussion; however, the ensuing discussion is equally applicable to any choice of dimension over which fairness is desired, such as gender, ethnicity, or geographical region. First, consider a purely data-driven DFND approach that is trained over historical labelled data. The notion of procedural fairness translates into fair representation of different political positions within each label in the training data. For example, if most fake news were from the right wing and most legitimate news were from the left wing, it would be easy for a learner to learn the (undesirable) mapping from political positions to a fake/real label. Even when representational parity is ensured, there is a possibility of algorithmic steps encoding some bias. Consider an example of a case where fake news from the left wing is more dispersed than fake news from the right wing. When regularizers are applied during the learning process for parsimonious model learning, the compact model may be inherently incapable of learning an accurate model to characterize the dispersed left wing fake news and thus would be able to deliver higher accuracies in detecting the more coherent right wing fake news. Thus, algorithmic steps including the usage of regularizers should be carefully scrutinized from the perspective of fairness. Second, for DFND methods that additionally incorporate external knowledge sources to inform decision-making, such sources should also be well distributed across political positions, with attention being paid to dispersion considerations as in the previous case. While the above checks do not yield a comprehensive fairness auditing method for DFND algorithms, a procedurally fair DFND method should necessarily align with the above principles.

Impact Fairness in DFND While we have seen that some forms of violations of procedural fairness would not be agreeable, it is interesting to consider what that entails for impact fairness. In general cases of using machine learning over person-level data, procedural and impact fairness are often in conflict. For example, if historical legacies of unfairness (such as racial unfairness) have resulted in significantly altered standing in terms of social, educational, and economic achievements across various categories (e.g., racial categories), a procedurally fair method would necessarily result in reflecting the biases. In other words, a race-agnostic and procedurally fair selection process could potentially result in much higher selection rates for whites than blacks if the former have a lower educational (and consequently, skill and achievement) profile, with race being correlated to the selection criterion of achievement level due to historical discrimination. Thus, in cases of handling person-level data, it is often argued, at least within progressive political circles, that impact fairness should take precedence over procedural fairness to counter historically entrenched discrimination manifesting as socioeconomic inequalities across dimensions over which fairness is desired. Streams of political philosophy such as Rawls [17], while stopping short of stating that impact fairness should trump procedural fairness, do prefer configurations where the inequality in impact is kept to as low levels as possible. With that background, the first consideration could be to ask whether impact fairness is in conflict with procedural fairness within the context of DFND. In other words, are there intrinsic or entrenched reasons as to why fake news is more abundant within a political position as opposed to another? In fact, while studies have generally been cautious about asserting a political correlation in fake news, there is increasing evidence that conservatives have historically played a much larger part in propagating fake news than liberals [10]. The study finds that extremely conservative people are almost twice as likely to spread fake news than extremely liberal people (these are self-reported labels, so need to be taken with a pinch of salt) on Twitter. Similar skewed distributions are potentially likely to be found when analyzing the kind of gender and ethnic stereotypes used in fake news authoring; for example, misogynistic fake news may be more prevalent than misandrist fake news. Evidence of such skewed distributions takes us back to the discussion on representation parity that we alluded to in the previous section. If indeed pro-conservative fake news is more prevalent than pro-liberal fake news, achieving representational parity requires us to sub-sample from available pro-conservative fake news, in order to construct a balanced training dataset. Even if such a balanced dataset is created, the method might still produce a significantly larger number of *fake* verdicts for conservative articles, since the skew along political alignment would exist in the unseen data over which these algorithms are used. The training data curation considerations are confounded in the presence of biases in user judgments (some bias mitigation strategies appear at [11]), especially when crowdworkers may have a different distribution of political positions as compared to that of fake news.

Summary Remarks on DFND Fairness We considered the two streams of machine learning fairness, those of individual fairness and group fairness. We observed that these are well motivated within the context of data about people but would still apply in some way in DFND. We considered concrete scenarios where DFND that is skewed on political and other positions may legitimately be considered as unacceptable. Building upon this, we observed that attention to procedural fairness along such dimensions would be necessary, and this places constraints and obligations on the design of DFND methods in myriad dimensions such as training data curation, algorithm design, and selection of external knowledge sources to use. Turning our attention to group fairness, we observed that such attention to

procedural fairness (that correlates to individual fairness) could lead to violations
of group fairness which may be unavoidable against the backdrop of the observed
skew of fake news distribution across dimensions such as political positions and
gender. In summary, we may wish to place more emphasis on procedural fairness in
DFND and pay significant attention to training data bias by ensuring well-designed
and debiased data collection paradigms.

4   Democratic Values and Uptake of DFND

DFND, much like any classification task, is a computational labelling task. The
ultimate goal is to pronounce a decision on each news article, which could be
either a binary label of *fake* or *legitimate* or a veracity scoring on an ordinal
scale. In typical data science scenarios, the predicted label is often used in a very
straightforward way, that of associating the label with the data. In an automated job
shortlisting scenario over received applications, the predicted label, which could be
one of *shortlist*, *reject*, or *unsure*, could lead to concrete actions, such as sending a
shortlisting or decline letter, or channelizing for more manual perusal of the job
application. Guided by such commonly encountered settings, typical fake news
detection software also uses a crisp verdict in the form of a predicted label. We
saw this in the case of the NewsGuard application in Fig. 1 where the label is
shown prominently at the top of the news article. We first take a look at the
variety of ways in which fake news-related information (or in general, any kind
of information veracity assessments) is presented to users, following which we will
assess normative considerations within a liberal democratic framework and how
they may favor some forms of presentations more than others.

Current DFND-Based User Engagement   We will now consider a set of currently
available software tools and algorithmic techniques with a focus on how they aid
in tackling fake news and how they present veracity information to users, along
with any forms of analysis they promote by way of the presentation modalities they
use. One of the first efforts at developing a plugin for fake news detection was as
early as 2016, leading to a tool called *BS Detector*.[7] The plugin has since been
taken down and few traces of vivid details of its workings and resulting presentation
modalities are available in the public domain. However, a screenshot available on
the web seems to suggest that it places a veracity label on news articles and, at
least in the case illustrated in Fig. 5, places an emphasis on the website from which
the news is sourced. A similar software, *Stop the Bullshit*, illustrated in Fig. 6,
squarely informs the user that the reasoning is based on the website, rather than the
content that is attempted to be accessed. To illustrate the pervasiveness of website-
level reasoning used in veracity result presentation, we may observe that Wikipedia

---

[7]https://www.bustle.com/articles/195638-the-bs-detector-chrome-extension-flags-the-sites-
containing-a-load-of-political-bs.

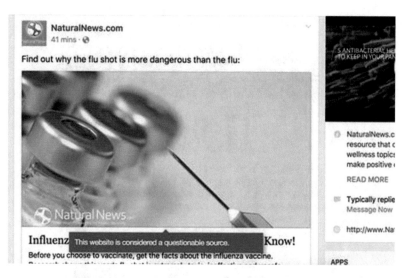

Fig. 5 BS Detector Plugin labelling a news piece as "This website is considered a questionable source" (picture courtesy: Bustle.com)

Fig. 6 *Stop The Bullshit* software (picture courtesy: ProductHunt.com)

has a web page on *List of Fake News Websites*.[8] *Know News*, a veracity detection plugin by Media Monitoring Africa, which has been widely reported on the web as employing veracity detection on content, also predominantly uses website-level reasoning in presenting results (Fig. 7).

The website-level reasoning relentlessly expressed and promoted in veracity assessment presentation, we will see, may be critiqued from the perspective of alignment with democratic values. For now, it may be noted that such reasoning does not allow to recognize that the same website may host content of varying veracities. Despite the pervasiveness of website-level reasoning, there have been a few efforts that focus on the content and allow for veracity to be checked without using any

---

[8]https://en.wikipedia.org/wiki/List_of_fake_news_websites.

Fig. 7 *Know News* plugin, a veracity assessment software, presenting veracity results basing the reasoning on the website (picture courtesy: Chrome Web Store)

kind of information about the source of the content. *ClaimBuster* [9], unlike the software that we saw so far, places an emphasis on automating the fact-checking of claims. While it calls itself an *end-of-end fact-checking system*, it limits itself to presenting pertinent information to a claim (mostly related claims) along with veracity information associated with such pertinent information. The user could then consider such information in arriving at a veracity judgment herself. Figure 8 illustrates the veracity results presented over a manually entered claim. By stopping short of providing a concrete and crisp veracity judgment, ClaimBuster places a higher cognitive burden on the user since the user has to digest and assimilate the related information presented in order to decide whether or not to trust the claim. However, in doing that, it also allows acknowledging the nuanced nature of veracity determination.

There have been several other "indirect" methods of tackling the fake news problem used by several stakeholders in the media ecosystem. These include enhancing *findability* of credible news sources. Facebook, the social media giant, explicitly admits using credibility/veracity information in the ranking of stories within news feeds. On one of its help pages,[9] it says: "Showing false stories lower in News Feed: if a fact-checker rates a story as false, it will appear lower in News Feed. This significantly reduces the number of people who see it." While this sounds quite reasonable, such credibility adjustments are notably done without

---

[9]https://www.facebook.com/help/1952307158131536.

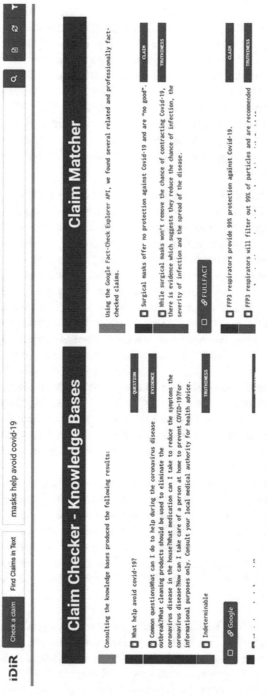

Fig. 8 ClaimBuster in action; when a user enters a claim in free text, they find related claims and veracity information associated with them

user engagement. In particular, reducing the chances of seeing a lower credibility news story is sharply different from displaying a warning (like NewsGuard, BS Detector, and other examples seen before) since it marks a shift of agency (in the decision-making process of whether the article is to be read) from user to algorithm. The Facebook veracity/credibility judgments are on the basis of feedback from both users and what are called *third-party fact-checkers*. The same help page referenced above says: "Identifying false news: we identify news that may be false using signs like feedback from people on Facebook. Fact-checkers may also identify stories to review on their own." In particular, it says precious little on how an aggregate score on veracity is arrived at when there are conflicting signals from across users and fact-checkers and how a weighting is determined to balance user feedback and fact-checkers' judgments.

How Should DFND Be Used? Having looked at how DFND has been used, we now consider how DFND *should* be used. In this discussion, we draw heavily from the EU HLEG report on Disinformation [4] and look at high-level principles for DFND usage. Any way of ensuring that the results of DFND are put to use would result in some kind of barrier or constraint on the free consumption of all forms of information within society. Much like the institution of some binding norms could benefit everyone (e.g., traffic discipline helps everybody get to their destinations faster), it could be argued that enforcing binding norms that prevent creation and consumption of fake news could deepen democratic discourse. However, unlike traffic signal violations, there is an enormous amount of subjectivity in enforcing norms on media, to the extent that it would almost be impractical. The UN joint declaration on fake news [12], on the other hand, underlines the potential of fake news to mislead and interfere with the public's right to seek and receive, as well as impart information and ideas of all kinds. It also highlights the *positive obligation* of states to create an enabling environment for freedom of expression. This perspective, in contrast to the one illustrated earlier, puts forward a rights-based need for intervention. The EU report suggests that any disinformation interventions should be focused on two general objectives: (i) *increase long-term resilience* and (ii) *ensure that disinformation responses are up-to-date*. Let us now consider how these high-level and long-term objectives translate into the design of DFND usage within digital interfaces as well as elsewhere within society. The EU HLEG report also stresses on the importance of fake news responses to abide by five pillars, viz., *transparency, media and information literacy, empowering users and journalists, safeguarding the diversity and sustainability of the media ecosystem*, and *promoting continued research on the impact of disinformation*. These are summarized in Table 1.

How Do Current DFND Methods Fare? We now assess, qualitatively, as to how current DFND user uptake methods fare against the normative principles outlined earlier. The mainstream method of DFND adoption, that of assigning source-level (i.e., at the level of the website of the top-level domain) verdicts and presenting these to users, evidently does *not* conform to the *transparency* and *empowerment*

Table 1 Recommendations from the EU HLEG report on disinformation [4]

General objectives	Increase long-term resilience
	Keep disinformation responses up-to-date
Normative principles	Transparency
	Media and information literacy
	Empowering users and journalists
	Safeguard the diversity and sustainability of the media ecosystem
	Promote continued research

criteria above and also may be seen as agnostic to safeguarding the diversity and sustainability of the media ecosystem. By explicitly indicating the verdict and allowing the user to disregard a *fake news* warning and continue reading, it may be said that there is some regard to media and information literacy, as well as user and journalist empowerment, within that model. The source/website-level verdict would need to be revised over time in order to satisfy the general objective of ensuring that disinformation responses are up-to-date; this would require that a website that has stopped sharing dubious content not be disadvantaged even after the change in character. In contrast to this analysis, reducing the findability of fake news by taking veracity into account in generating the ranking for the news feed, as used within Facebook, is quite weak in adherence to the normative objectives and principles laid out above. It may be argued that they neither satisfy the general objectives nor the five principles. In particular, such *under-the-cover-*type fake news exposure reduction methods, while sounding attractive in terms of offering a seamless integration into current systems, are quite poor when it comes to adherence to democratic and liberal values that have motivated most of the normative recommendations for fake news responses. The ClaimBuster approach, which starts with a claim and then presents related claims along with their veracity information, enabling a user to arrive at a judgment, can be seen as most amenable to the considerations seen above. However, as it was developed as a standalone tool which is not meant to intercept user-media interaction, its impact could be limited. The system, when augmented with a claim detection method, could well be packaged as a plugin which searches for relevant claims to the core claim on the web page attempted to be perused. That said, another drawback is that of the detailed nature of the presentation (i.e., related claims and their veracities), which makes it hard to be presented within a plugin format without significant detriment to user experience.

Improving Adherence to Normative Principles in DFND Uptake We now discuss how we could potentially improve the adherence to the principles from Table 1 in DFND uptake. Our attempt is not intended to outline a concrete and novel exemplary DFND approach, since the development of such a framework would naturally take several years of research effort. However, we will attempt to outline recommendations based on currently available technologies in order to translate the high-level principles into a language that is better understood by technologists in machine learning and data science. In particular, the aspects of *transparency* and *user empowerment* are quite interesting to analyze in terms of how they could be realized computationally. We outline some high-level components of a roadmap toward enhancing adherence to the principles outlined earlier:

- **Relaxing (Implicit) Obligations of Showing a Crisp Decision:** Virtually all veracity-oriented software and tools do pronounce a crisp decision on what is evaluated. This is likely almost construed as entailing from the task undertaken by the tool. However, we may argue that there is no need to show a crisp decision as such. Showing critical information that would empower the user to arrive at a decision for herself could be considered as *enough*. We realize that such a tool may lose out on *user appeal*, and the ability to show a crisp and clear decision is often part of the *hard sell* marketing that DFND tools may use; thus, market forces are likely not conducive toward relaxing the paradigm of showing a crisp decision. Relaxing the paradigm of crisp decision-making would require the interested user to engage better with the information being presented and thus could empower users. Furthermore, crisp decisions that are mostly informed by source-level (i.e., website-level) cues could be also seen as a soft censorship and are thus not well aligned with the goal of safeguarding the diversity and sustainability of the media ecosystem.

- **Confidence Scores with Decisions:** The attractive feature of offering a crisp decision is often used without an understanding of whether such decision-making is valid; in other words, we seldom ask the question *have we designed the algorithm to tell us when it does not know enough* to make a decision. Often times, the decision-making is made in a comparative manner, based on which is the best choice among available decisions. This obscures information on whether the algorithm is indeed confident about the decision it is making or whether it is the best effort choice made under considerable ambiguity. DFND methods, given the importance of the domain of operation, need to have both a mechanism of reporting the *error bar* in some intelligible manner and a probability score (perhaps expressed as a percentage). Such confidence scores help deepen the adherence to the normative principles in Table 1.

- **Explaining Decisions:** There has been much recent interest in explainability in AI and machine learning (e.g., [1]). Enhancing explainability in fake news detection is a direct way of enhancing transparency, as well as media and information literacy. Neural network-based models have often been criticized for lack of understandability of the decision-making process, and enforcing a condition of explainability within the learning process may cause them to operate at lower levels of accuracy. Thus, explainability could stand in the way of achieving the best possible accuracies in decision-making, creating an interesting trade-off between accuracy and democratic values. Another way of tackling the problem is to do post hoc explanations, whereby the explanation is made after the decision-making. Cases when a post hoc explanation cannot be derived could be indicated explicitly, so the user may treat that as an indicator of having to take the decision with a pinch of salt.

- **Showing Pertinent Credible Information:** For systems that make use of sources of credible information (e.g., PubMed articles in the health domain, ontologies for science, and so on), a straightforward way to use them would be to display *pertinent credible information that would enable manual verification* directly to the user. This would enhance user engagement while also implicitly training users to exercise own judgment and analysis, something that could be deemed critical for long-term resilience. Showing information from known credible source is related to, but different from, the ClaimBuster approach of showing related claims in the dataset along with their labels.
- **Showing Pertinent Non-Credible Information Marked Clearly:** A complementary approach to showing credible information would be to show fallacious/fake information marked so, as long as it is related enough to the article upon which a decision is to be made. This paradigm, one may argue, might nudge the user to engage in fact-checking or verification, by showing that there is quite similar content that is known to be fake. However, this paradigm needs to be used with abundant caution due to several reasons. First, showing non-credible information to users enhances user familiarity with such content, and this heightened familiarity along with the psychological bias called *illusion of truth effect*[10] could eventually lead to an enhanced belief in such fake news. Second, showing such non-credible information in an easily accessible manner might risk the DFND method being perceived as an easy-access channel for fake news. One way to mitigate such risks could be to mark such non-credible information very clearly, as illustrated in Fig. 9.
- **Encouraging User Engagement and Deeper Analysis:** In addition to the above, DFND methods may explore interactive tools in order to enhance user engagement and empowerment. For example, the suggestions above of showing credible information, explanations, and confidence scores could all be operationalized using a mouse hover paradigm. For example, hovering the mouse over a particular sentence could bring up a tooltip with information localized to the sentence. DFND methods could also employ force-directed graph-based interfaces[11] which show the interrelationships between segments of the article in an interactive manner to aid user-specific explorations to enable deeper understanding of the veracity judgments presented.
- **Ability to Provide Feedback and Other Information:** DFND methods, much like any machine learning method, are not designed to achieve perfect decision making, and could make wrong decisions. Thus, it would be in the interest of the DFND method to continuously improve the decision-making processes through crowdsourcing feedback from own users. In addition, the facility to provide feedback and other kinds of information (e.g., credible information pertinent to the article in question) will hopefully enhance user confidence in the system and promote media literacy and user engagement.

---

[10]https://en.wikipedia.org/wiki/Illusory_truth_effect.

[11]https://en.wikipedia.org/wiki/Force-directed_graph_drawing.

Fig. 9 Abundant caution is necessary while displaying fake news to users, even if it may be to debunk it. The picture shows how AltNews, an India-based fake news detection engine, uses several ways of indicating that the information is fake while displaying it to the user

The above recommendations are not meant to be comprehensive, but we hope this will enhance and sharpen the debate on how AI should go about tackling fake news while acknowledging, preserving, and deepening the democratic values within society.

5 Conclusions

In this chapter, we analyzed and discussed several ethical and normative considerations that are relevant to the context of data-driven and AI-based automation of fake news detection. We began by outlining the increasing pervasiveness of fake news detection and its societal and political importance, motivating the need for increased attention to the non-technological aspects of data-driven fake news detection (DFND). We also analyzed the main pillars of ethical considerations. First, we considered the historical context of machine learning, and argued that legacy considerations of optimizing for automation-era metrics are squarely inappropriate to drive its evolution within domains such as digital media within which DFND is situated. Second, we considered why the very nature of data-driven learning could be critiqued for usage in domains within volatile dynamics within which automation could be an active player. Third, we described that the domain of fake news detection has some unique features which make it unlike other analytics tasks with a similar structure such as spam detection and product recommendations. Following this, we delved deeper into fairness considerations in DFND. We analyzed DFND from the perspectives of the two streams of fairness concepts used within ML, viz., individual and group fairness. We argued that individual or procedural fairness may be considered as being more important for DFND and contrasted it within analytics involving person-level data where group fairness may legitimately be considered more critical. We then turned our attention to analyzing DFND uptake modalities and how they fare against recent recommendations on normative principles that DFND should align with. We observed that under-the-cover and seamless integration of DFND results, while sounding attractive, would fare significantly worse on such normative principles, as opposed to more explicit ways of delivering the results to the user. Based on such analyses, we outlined several technologically grounded recommendations that could inform the design and development of DFND methods that would be well aligned toward preservation and deepening of democratic values within society.

References

1. Arrieta, A.B., Díaz-Rodríguez, N., Del Ser, J., Bennetot, A., Tabik, S., Barbado, A., García, S., Gil-López, S., Molina, D., Benjamins, R., et al.: Explainable artificial intelligence (XAI): concepts, taxonomies, opportunities and challenges toward responsible AI. Inf. Fus. **58**, 82–115 (2020)
2. Bernardin, H.J., Beatty, R.W., Jensen Jr., W.: The new uniform guidelines on employee selection procedures in the context of university personnel decisions. Person. Psychol. **33**(2), 301–316 (1980)
3. Binns, R.: On the apparent conflict between individual and group fairness. In: Proceedings of the 2020 Conference on Fairness, Accountability, and Transparency, pp. 514–524 (2020)
4. Buning, M.D.C., et al.: A multidimensional approach to disinformation. EU Expert Group Reports (2018)
5. Dwork, C., Hardt, M., Pitassi, T., Reingold, O., Zemel, R.: Fairness through awareness. In: Proceedings of the 3rd Innovations in Theoretical Computer Science Conference, pp. 214–226 (2012)
6. Freyenhagen, F.: Taking reasonable pluralism seriously: an internal critique of political liberalism. Polit. Philos. Econ. **10**(3), 323–342 (2011)
7. Gangireddy, S.C.R., Long, C., Chakraborty, T.: Unsupervised fake news detection: a graph-based approach. In: Proceedings of the 31st ACM Conference on Hypertext and Social Media, pp. 75–83 (2020)
8. Hashemi, M., Hall, M.: Criminal tendency detection from facial images and the gender bias effect. J. Big Data **7**(1), 1–16 (2020)
9. Hassan, N., Zhang, G., Arslan, F., Caraballo, J., Jimenez, D., Gawsane, S., Hasan, S., Joseph, M., Kulkarni, A., Nayak, A.K., et al.: Claimbuster: the first-ever end-to-end fact-checking system. Proc. VLDB Endowm. **10**(12), 1945–1948 (2017)

10. Hopp, T., Ferrucci, P., Vargo, C.J.: Why do people share ideologically extreme, false, and misleading content on social media? A self-report and trace data-based analysis of countermedia content dissemination on Facebook and twitter. In: Human Communication Research (2020)
11. Hube, C., Fetahu, B., Gadiraju, U.: Understanding and mitigating worker biases in the crowdsourced collection of subjective judgments. In: Proceedings of the 2019 CHI Conference on Human Factors in Computing Systems, pp. 1–12 (2019)
12. Huff, M.: Joint declaration on freedom of expression and "fake news," disinformation, and propaganda. Secrecy Soc. **1**(2), 7 (2018)
13. Hysolli, E.: The story of the human body: Evolution, health, and disease. Yale J. Biol. Med. **87**(2), 223 (2014)
14. Lloyd, E., Wilson, D.S., Sober, E.: Evolutionary mismatch and what to do about it: a basic tutorial. Evolut. Appl. 2–4 (2011)
15. Narayanan, A.: *How to Recognize AI Snake Oil.* Arthur Miller Lecture on Science and Ethics. MIT, Cambridge (2019)
16. O'neil, C.: Weapons of Math Destruction: How Big Data Increases Inequality and Threatens Democracy. Broadway Books, New York (2016)
17. Rawls, J.: A Theory of Justice. Harvard University Press, Cambridge (1971)
18. Roets, A., et al.: 'Fake News': incorrect, but hard to correct. The role of cognitive ability on the impact of false information on social impressions. Intelligence **65**, 107–110 (2017)
19. Smart, J.J.C., Williams, B.: Utilitarianism: For and Against. Cambridge University Press, Cambridge (1973)
20. Thurman, N.: Personalization of news. Int. Encycl. Journal. Stud. **2019**, 1–6 (2019)
21. Whittlestone, J., Nyrup, R., Alexandrova, A., Dihal, K., Cave, S.: Ethical and Societal Implications of Algorithms, Data, and Artificial Intelligence: A Roadmap for Research. Nuffield Foundation, London (2019)
22. Yin, X., Han, J., Philip, S.Y.: Truth discovery with multiple conflicting information providers on the web. IEEE Trans. Knowl. Data Eng. **20**(6), 796–808 (2008)

A Political Science Perspective on Fake News

Muiris MacCarthaigh ⓘ **and Connel McKeown** ⓘ

Abstract Contemporary concerns about "fake news" are typically framed around the need for factual accuracy, accountability, and transparency in public life at both national and international level. These are long-standing concerns within political science, but the problem of "fake news" and its associated impact on the fundamental political questions about who governs and how have taken on new potency in the digital age. In this chapter, we begin by considering what is meant by fake news before examining the issue in a historical political context. The chapter then turns to more recent manifestations of fake news and the real-world challenges it presents. A final section considers how fake news has attracted interest in the study of elections and voting behaviour, international relations and strategic narratives, and transparency and trust in government.

Keywords Fake news · Political science · International relations · Elections · Transparency

1 Introduction

Access to information about the activities and decisions of rulers by those being ruled is generally believed to be an important underlying condition for the functioning and continued legitimacy of virtually all systems of government. In theory at least, governments govern with the continued consent of the people who have, by a variety of means, appointed them to office, and with the expectation that once in office, those people require information about what actions are being taken on their behalf. Governments are also expected to inform citizens about what is happening in the world beyond national borders.

In the case of liberal Western democracy, there is an expectation (with roots tracing to fifth-century Athenian democracy) that citizens cannot leave it to

M. MacCarthaigh · C. McKeown
Queen's University Belfast, Belfast, UK
e-mail: M.MacCarthaigh@qub.ac.uk; CMcKeown32@qub.ac.uk

© Springer Nature Switzerland AG 2021
Deepak P et al., *Data Science for Fake News*, The Information
Retrieval Series 42, https://doi.org/10.1007/978-3-030-62696-9_11

governments alone to provide this information, but rather they have a duty "to learn about the social and political world, exchange information and opinions with fellow citizens and arrive at considered judgements about public affairs" [1]. In this endeavour, the popular media have, since the late eighteenth century, had a vital role to play. It was in this century that the British Parliament led the way in allowing newspaper writers access to their proceedings so as to inform the public of their deliberations. Referring to them as the "fourth estate" (after the concept of parliament consisting of three estates representing the clergy, nobility, and commoners), Irish parliamentarian and philosopher Edmund Burke recognised that parliamentary reporters were an increasingly powerful group in determining the success or otherwise of the government's political agenda, as well as that of their opponents.

Over the course of the nineteenth and twentieth centuries, the means and methods of gathering access to information about political and public affairs nationally – and internationally – grew steadily. Central to this was evolution in the range of print, radio, and subsequently television-based media, facilitated by new means of communicating across the globe and ever-greater access to the workings of public institutions and politicians themselves. And as is well established, the dawn of the digital and online age has resulted in a huge proliferation of information from a variety of state and non-state sources. The term "fifth estate" has been coined to describe the emergence of online-only news journalism and popular commentary.

However, any belief that increased access to greater amounts of information would bolster democratic accountability and transparency and the social contract between rulers and the ruled has been undermined by the phenomenon which has become loosely known as "fake news". An important moment for the idea of fake news was the 2016 US Presidential Election, when the term entered the popular lexicon as it was used extensively by Republican candidate Donald Trump to portray news that was not politically supportive as being factually incorrect or inaccurate. Indeed, he went as far as to identify certain media outlets as "the true enemy of the people" [2]. As a result of this, the term "fake news" has itself become the subject of official and political contestation, perhaps best captured by it being revealed as the *Collins Dictionary* word of the year in 2017.

A general interpretation of fake news is "fictitious accounts made to look like news reports" [3], but there is now greater acceptance that what is meant by fake news encompasses a wide range of activities. Chadwick et al. suggest that fake news can range from the "outright fabrications created by online news 'factories' that exploit advertising syndication systems for financial gain" to "online production and circulation of information that is exaggerated, sensationalized, selective, or assembled from a web of partial truths in hybrid networks of reputable and less reputable sources" [1]. The term has also experienced misuse and contradictory use – at times it has been ascribed to factual sources of information or even opinion pieces.

There are other, competing, definitions of fake news which seek to explain several different varieties of disinformation and misinformation. This distinction between "disinformation" and "misinformation" is an important one. Wardle proposes that disinformation can be understood as "the deliberate creation and sharing of information known to be false", whereas misinformation is the "the inadvertent sharing of false information" [4]. A British parliamentary committee inquiry published in 2019 asserted that "definitions in this field matter" [2], and after their preliminary research concluded, they resolved to change the name of this inquiry from simply "Fake News" to "Fake News and Disinformation". The committee defined fake news not only according to its characteristics but also its purpose, describing it as being "created for profit or other gain, disseminated through state-sponsored programmes, or spread through the deliberate distortion of facts, by groups with a particular agenda, including the desire to affect political elections" [5].

Edson et al. [3] created a typology based on an examination of 34 articles which made use of the term between 2003 and 2017, drawing attention to the fact that the term has an older lineage than is often perceived. They identified six types of fake news: news satire, news parody, fabrication, manipulation, advertising, and propaganda. News satire can be understood as mock news programmes which make use of humour and exaggeration to present audiences with news updates. News parody differs from satire in that it uses non-factual information "to inject humour" [3]. Fabrication "refers to articles which have no factual basis but are published in the style of news articles to create legitimacy" [3]. Manipulation refers to "the manipulation of real images or videos to create a false narrative" [3]. Native advertising is when "news may function as fulfilling both advertising and news goals" [3]. Finally, propaganda refers to "news stories which are created by a political entity to influence public perceptions" [3].

They cluster these types of fake news by facticity and intention. Facticity can be understood as "the degree to which fake news relies on facts", whereas intention is the "degree to which the creator of fake news intends to mislead" [3]. Native advertising and propaganda have both a high level of facticity and intention to deceive. Manipulation and fabrication have a low level of facticity and a high intention to deceive. News satire has a high level of facticity and a low intention to deceive. News parody has a low level of facticity and a low intention to deceive. Combined, they demonstrate the range of interpretations which may be applied to the concept of fake news.

As will be detailed below, the presentation of fictitious accounts as factual ones has a long pedigree in public affairs. However, the proliferation in household access to the Internet since the turn of the century, the existence of online versions of legacy media sources, and exponential growth in social media platforms and users present challenges that are unprecedented in human development and political life. The role of fake news and social media "echo chambers" [6], in which citizens only consume and share information that conforms to their worldview, has even been compared to an infectious disease [7]. It has also given rise to the philosophical concept of a "post-truth" age in which what were previously deemed to be accepted norms of scientific inquiry are questioned and non-scientific assertions (often based on emotion) are treated as of equal value to scientific findings.

In 2017, the world's online population grew to 3.8 billion people [8], effectively half of the world's population collectively consuming and exchanging enormous amounts of information. These developments have resulted in increased ease of access for the public to information and news. Much of this concerns politics and public affairs and includes information that is politically sensitive and significant. It also includes news which may be, intentionally or otherwise, factually incorrect. The civic and democratic implications of this are evident when one considers that social media giant Facebook claimed 2 billion followers, or roughly the same as the world's population of Christians. Similarly, over 1.8 billion people use YouTube, an equivalent figure to the followers of Islam [2].

A review by an EU-commissioned expert group published in 2018 preferred the term "disinformation" and recommended clear and unequivocal abandonment of using "fake news". The group argued that fake news did not adequately capture what is a complex range of print or digital information, some or all of which might not be factual, as well as the fact that the term is used in a partisan manner to dismiss arguments by perceived political opponents [9]. Whilst this is a valid appeal, for the purpose of inquiry and coherence, this chapter will use the term fake news throughout. In the next section, we consider more closely the origins and evolution of the concept before looking at its consequences for the practice and study of politics.

2   The Origins of Fake News

Edson et al. question the idea of fake news as a modern problem facing society, pointing out that "misinformation in the media is not new" [3]. Even ancient civilisations, with their formative writing systems, employed a mix of what Marcus refers to as "horizontal" and "vertical" propaganda [10]. Horizontal propaganda can be understood as propaganda used by "members of the elite in an attempt to influence other members of the elite". In contrast, vertical propaganda describes how "rulers attempt to influence the behaviour of the ruled" [10]. Although fake news connotes malign endeavours, the more benign and ancient literary canon of political satire is based upon inaccurate representations of politicians and political views.

What is widely perceived to be the first written history, Greek writer Herodotus's *Histories* (his account of the fifth-century Persian Wars), has long been recognised as riddled with inaccuracies and fantastical claims. More recently, the post-WWII Cold War involved an extensive proxy propaganda war, with the spread of inaccurate information used by all parties to delegitimise then dominant global political ideologies. Authoritarian and dictatorial regimes have in many respects always been characterised by the use of fake news to reinforce particular values, demonise outsiders, and secure the authority of their leaderships.

A much more contemporary historical and ongoing example of "fake news" can be seen in the form of "tabloid journalism". This can be understood as journalism primarily comprising sensationalised and subjective news stories, often involving openly partial political opinions and commentary. "New tabloid journalism" in the

UK can be traced back to the 1930s rebirth of the *Daily Mirror* which challenged the "journalistic norm of objectivity" [11]. In the USA, a tabloid press emerged in the late 1890s, being pejoratively described as "yellow journalism". As with the tabloid press, yellow journalism can similarly be understood as mass-produced newspapers which adopted "varying proportions of sensationalism, populism, and socialism to address the interests of new, urban, working-class, and immigrant readers" [12].

Many point to the tabloid press's exploitation of social media as the most prevalent and politically impactful source of viral disinformation and misinformation. Of course, that it might provide a fertile ground for fake news may not be unexpected given that for some the stock in trade of the tabloid press has always been news of questionable civic value. And so the idea that "fake news" represents a digitisation of the tabloid press has wide appeal. Chadwick et al. contend that there are "affinities between tabloid news and misinformation and disinformation behaviours on social media" and that "sharing tabloid news on social media is a significant predictor of democratically-dysfunctional misinformation and disinformation behaviours" [1]. However, fake news in its modern form goes beyond simply digitalisation of pre-existing forms of sensationalism and questionable assertions.

3   Fake News in the Twenty-First Century

As has been established, the contemporary idea of fake news is not new. Rather, what is new is the environment in which fake news now exists – an increasingly interconnected and digitised world with advanced information communication technologies. Fake news may be distinguished from traditional vertical and horizontal political propaganda in that it is not always elite led. Indeed, in contrast to these concepts, fake news can be disinformation which is produced by and/or circulated by members of the general public and non-state organisations, as well as by political elites. And it can be rapidly spread and legitimised by political elites and popular figures at a low cost.

As we enter the third decade of the twenty-first century, fake news (however defined) is generally believed to represent a fundamental challenge to liberal representative democracies globally and has become a subject which political institutions around the world have sought to address [1]. Reflecting the need for international cooperation on this, the UK Parliament's Digital, Culture, Media and Sport Committee conducted an inquiry into fake news and also established a "Grand International Committee" in 2018, involving parliamentary representatives from Argentina, Belgium, Brazil, Canada, France, Ireland, Latvia, and Singapore to examine the democratic challenge presented by fake news and disinformation. The increased choice in media has resulted in an environment, according to the Committee inquiry, where users are only presented with material "that reinforces their views, no matter how distorted or inaccurate while dismissing content they do not agree with as fake news" [2]. They further proposed that the potential ramifications in terms of accountability, transparency, and democratic government are evident as this "has a polarising effect and reduces the common ground on which reasoned debate, based on objective facts, can take place" [2].

An important problem identified by the Committee in the attempts to tackle fake news, by legislation or other means, is the issue of press freedom. It has long been argued that new media sources are not held to the same professional and ethical journalistic standards as more traditional news sources. Edson et al. contend that "most legacy news media are committed to truth and draw the line at altering images to create a misleading or inauthentic narrative" [3] and contrast this with extensive manipulation of images and interpretations on social media.

The continued use of fake news through the term of office held by US President Donald Trump and the suggestion that the media were the enemy of the people stand in stark contrast to more typical and historical views of political elites on the press. It has long been a convention that press freedom is an integral part of liberal representative democracy. In a much-quoted speech, British Prime Minister Winston Churchill remarked that:

> A free press is the unsleeping guardian of every other right that free men prize; it is the most dangerous foe of tyranny ... Under dictatorship the press is bound to languish ... But where free institutions are indigenous to the soil and men have the habit of liberty, the press will continue to be the Fourth Estate, the vigilant guardian of the rights of the ordinary citizen. [13]

The popularisation of fake news challenges this ideal as individuals and groups may apply the label to undermine information and commentary that is factual or legitimate. In another interpretation, it raises the "plausible risk of the substitution of the Fifth Estate for the Fourth Estate [and] the potential for audiences to be more selectively exposed to the news" [14]. This presents a challenge to the effective functioning of democratic accountability and transparency regimes as this "news" is "unmediated by editors and professional journalists, in ways that could lead also to less diversity and the reinforcement of prejudices" [14]. These concerns may be overstated. In a review of the French and Italian cases, Fletcher et al. [15] found that websites presenting fake news were far less engaged with than the websites of established news sites.

Traditional media sources such as TV, radio, and print are considered as "central to pluralist democratic processes" [16]. Online media also offers an opportunity to enhance civic engagement, communication between those holding public office and the public, and by virtue of this access to information the democratic quality of government. Indeed, the potential of social media to share political news, information, and opinion was initially touted to be an "essential raw material for good citizenship" [1]. This was, however, premised on the assumption that increased interconnectivity and exposure to social media would result in citizens being exposed to a plurality of alternative perspectives and reasoned, valid, and well-informed opinions. In practice, the proliferation of new media presents a substantial challenge for democracy because access to digital technologies and social media has brought with it a corresponding proliferation in dissemination of disinformation and misinformation.

Elaborating on the idea of echo chambers (above), Sunstein noted the ability of digital technology to increase people's ability to filter what they want to read, see, and hear such that "you need not come across topics that you have not sought out ... you are able to see exactly what you want to see, no more, no less" [6]. Some social media platforms make use of user data to algorithmically tailor posts and content which appear on their newsfeed, so they only see what corresponds to their interests. This is often done automatically and invisibly – users typically must "opt out" rather than "opt in" to such a scheme. The effect of these activities on citizen preferences and voting behaviour is increasingly contested however [17].

The proliferation of fake news has also resulted in "fact-checking" organisations and associated websites such as ClaimBuster and PolitiFact.com, with ratings for the veracity of claims made by politicians and governments [18]. As of April 2020, one of the most popular websites – reporterslab.org – claimed there were 237 fact-checkers in nearly 80 countries. However, keeping up with the volume and speed of transmission of disinformation and misinformation is a constant challenge, with a large number of scientific papers suggesting ways and means of improving this [15, 19, 20].

4 Fake News and the Study of Politics

For political scientists, fake news has application across a wide variety of issues in government and politics and the relationship between the citizen and the state. We consider here three sub-fields in political science where fake news has generated particular interest. These are elections and voting behaviour, international relations and strategic narratives, and transparency and trust in government.

That fake news has real-world implications for democratic accountability and governance, and electoral politics, is now well established. Prominent examples of how fake news infused the democratic process include the 2016 US Presidential Election (when, as noted, the term also entered popular discourse), the 2016 Brexit referendum and associated campaign activities, and the UK General Elections of 2017 and 2019. There are, however, earlier examples of social media-based fake news being part of electoral competition. For example, research has also been conducted into the use of social media in the 2015 Argentine presidential election [21] and also the 2012 US Presidential campaign [22]. The predispositions or otherwise of voters to endorse or reject political conspiracies and rumours have also attracted the attention of political scientists [23].

Fake news has also been strongly connected to the emergence of what is termed "populism", with the electoral success of individuals such as Jair Bolsonaro in Brazil [24], Rodrigo Duterte in the Philippines [25], and Narendra Modi in India [26] identified as prominent cases of populist leaders benefitting from incidences of fake news during their campaigns. Reflecting these developments, the UK parliamentary inquiry into fake news identified that "data has been and is still being used extensively by private companies to target people, often in a political context, in order to influence their decisions" [2].

There are also geopolitical power struggles at play, and the Russian Federation in particular has been implicated in this use of fake news via various digital channels to influence the outcome of popular votes. The inquiry by the UK's Digital, Culture, Media and Sport Committee found that Russia supports "organisations that create and disseminate disinformation, false and hyper-partisan content, with the purpose of undermining public confidence and of destabilising democratic states" [5]. In 2020, the US Senate published a report confirming that the Russian Government used fake social media accounts and bots to interfere in the 2016 US Presidential Election [27]. It proposed that this was done with the objective of boosting the candidacy of Donald Trump and harming the electoral prospects of Hillary Clinton.

Because of this, fake news is not only of interest to students of elections and voting behaviour but also of increasing interest to scholars of international relations and strategic narratives. Strategic narratives are those tools used by political actors to articulate a position on a specific issue and to shape perceptions and actions of domestic and international audiences [28]. For example, Khaldarova and Pantti [29] examined how the transmission of strategic narratives and counter-narratives through television and fact-checking websites respectively by the parties to the Russian-Ukraine conflict over Crimea was used to appeal to popular emotions and infuse reality with fiction. In South Korea, the need for government to manage potentially damaging "cyber-rumours" necessitates Internet surveillance systems to try and mitigate this [30].

The third and final area where we see interest among political scientists in the effects of fake news is in respect of transparency and trust in government. The UK Brexit campaign was heavily influenced by disinformation. In the lead-in to the referendum, many right-wing tabloid newspapers strongly advocated for Britain's exit from the European Union, and their online work was an important part of their strategy. For instance, *The Express* ran a story that a leaked document from the European Union indicated that they intended to force the privatisation of the NHS so as to remove an impediment to equal access to the European Single Market. Whilst this was a totally unfounded story, it "became the single most-shared news article on social media during the Brexit referendum campaign, with 464,000 shares, comments, and interactions on Facebook" [1].

With the use and spread of rumours and unsubstantiated claims, many of which elicit rapid and voluminous responses, much research in political science (and political communications) has increasingly focused on whether or not online activity is undermining the integrity of the political process and citizen trust in government. In their analysis of the 2012 US Presidential Election, Garret et al. [31] found that exposure to ideological media encouraged inaccurate beliefs, regardless of what consumers knew of the evidence presented to them. Insights from psychology about individual propensity to consume and believe fake news [23, 32, 33] and the use of political attitude profiling by social media platforms to target political messaging [34, 35] have also emerged as topics of importance to political scientists. Kreiss and Mcgregor [36] argue that such is their importance to political outcomes that scholars of political communication need to consider such firms as active rather than passive agents in the political process.

5   Conclusion

The existence and use of what might be termed fake news is not necessarily a new or novel phenomenon for political science. Rather, what is new is the environment in which it is disseminated. In this interpretation, fake news represents traditional forms of disinformation adapting to modern technologies and social media platforms. It is effectively a popular and catch-all term encompassing propaganda, misinformation, disinformation, and subjective journalism as they present in the digital age.

The availability of social media and the Internet offers extensive opportunities for individuals to easily access unprecedented amounts of information about the institutions of national and global governance and those in power. However, it also facilitates the rapid dissemination of information that is factually incorrect or mischievous. This can be damaging when this false information is political in nature, undermining public trust in institutions and political figures and influencing voting behaviour and the outcomes of elections. This problem has been compounded by the increasing commercialisation of social media, which has incentivised the production of fake news.

How democracies in particular respond to the challenges posed by fake news, disinformation, and misinformation is an evolving process. At the time of writing, the focus is on getting gargantuan social media and technology companies to adhere to rules allowing citizens more control over their personal data and its use. There is also a need to more easily identify the sources and veracity of information and scrutinise the financial activities and operations of technology companies, many of whose funding models are based on facilitating the rapid spread of unchecked information. In addressing the problems associated with the fake news phenomenon, political scientists have a distinctive role to play in helping to better understand the consequences of fake news on voting behaviour and electoral outcomes, inter-state relations, and trust in the institutions of government.

References

1. Chadwick, A., Vaccari, C., O'Loughlin, B.: Do tabloids poison the well of social media? Explaining democratically dysfunctional news sharing. New Media Soc. **20**(11), 4255–4274 (2018)
2. Digital, Culture, Media and Sport Committee: Disinformation and 'fake news': final report. House of Commons (HC1791), London (2019)
3. Tandoc, E.C., Lim, Z.W., Ling, R.: Defining "fake news". Digit. Journal. **6**(2), 137–153 (2018)
4. Wardle, C.: Fake news. It's complicated. First Draft News, 16 (2017)

5. Digital, Culture, Media and Sport Committee: Disinformation and 'fake news': interim report. House of Commons (HC363), London (2018)
6. Sunstein, C.: Republic.com. Princeton University Press, Princeton, NJ (2001)
7. Kucharski, A.: Post-truth: study epidemiology of fake news. Nature. **540**, 525 (2016)
8. Domos. Data Never Sleep 6.0. https://www.domo.com/assets/downloads/18_domo_data-never-sleeps-6+verticals.pdf (2018)
9. Buning, M.D.: A multi-dimensional approach to disinformation: report of the independent High-level Group on fake news and online disinformation. European Commission 2018. https://ec.europa.eu/digital-single-market/en/news/final-report-high-level-expert-group-fake-news-and-online-disinformation (2018)
10. Marcus, J.: Mesoamerican Writing Systems: Propaganda, Myth, and History in Four Ancient Civilizations. Princeton University Press, Princeton, NJ (1993)
11. Bromley, M.: Objectivity and the other Orwell: the tabloidism of the Daily Mirror and journalistic authenticity. Media Hist. **9**(2), 123–135 (2003)
12. Campbell, W.J.: Yellow journalism: why so maligned and misunderstood? In: Sensationalism, pp. 3–18. Routledge, New York (2017)
13. Churchill: The Leveson. In: An inquiry into the culture, practices and ethics of the press. Stationary Office, London (1949)
14. Newman, N., Dutton, W., Blank, G.: Social media in the changing ecology of news: the fourth and fifth estates in Britain. Int. J. Internet Sci. **7**(1), 6–22 (2013)
15. Fletcher, R., Schifferes, S., Thurman, N.: Building the 'Truthmeter': training algorithms to help journalists assess the credibility of social media sources. Converg. Int. J. Res. New Media Technol. **26**(1), 19–34 (2017)
16. Dutton, W.H.: The fifth estate emerging through the network of networks. Prometheus. **27**(1), 1–15 (2009)
17. Dubois, E., Blank, G.: The echo chamber is overstated: the moderating effect of political interest and diverse media. Inform Commun Soc. **21**(5), 729–745 (2018)
18. Graves, L.: Deciding What's True: The Rise of Political Fact-Checking in American Journalism. Columbia University Press, New York (2016)
19. Boididou, C., Middleton, S.E., Papadopoulos, S., Dany-Nguyen, D., Boato, G., Kompatsiaris, Y.: Verifying information with multimedia content on twitter: a comparative study of automated approaches. Multimedia Tools Appl. **77**, 15545–15571 (2017)
20. Ciampaglia, G.L.: Fighting fake news: a role for computational social science in the fight against digital misinformation. J. Comput. Soc. Sci. **1**, 147–153 (2018)
21. Filer, T., Fredheim, R.: Popular with the robots: accusation and automation in the Argentine Presidential Elections, 2015. Int. J. Polit. Cult. Soc. **30**, 259–274 (2017)
22. Kreiss, D.: Seizing the moment: the presidential campaigns' use of Twitter during the 2012 electoral cycle. New Media Soc. **8**(8), 1473–1490 (2014)
23. Miller, J.M., Saunders, K.L., Farhart, C.E.: Conspiracy endorsement as motivated reasoning: the moderating roles of political knowledge and trust. Am. J. Polit. Sci. **60**(4), 824–844 (2015)
24. Bracho-Polanco, E.: How Jair Bolsonaro used "fake news" to win power. The Conversation. https://theconversation.com/how-jair-bolsonaro-used-fake-news-to-win-power-109343 (2019)
25. Otto, B.: Facebook removes accounts linked to Duterte's former social-media manager. Wall Street J. https://www.wsj.com/articles/facebook-removes-accounts-linked-to-dutertes-former-social-media-manager-11553861357 (2019)
26. Poonam, S., Bansal, S.: Misinformation is endangering India's election. The Atlantic, 1 April. https://www.theatlantic.com/international/archive/2019/04/india-misinformation-election-fake-news/586123/ (2019)
27. Select Committee on Intelligence. Report of the Select Committee on Intelligence United States Senate on Russian Active Measures Campaigns and Interference in the 2016 U.S. Election (Vol. 4). Washington: Houses of Congress. https://www.intelligence.senate.gov/sites/default/files/documents/Report_Volume4.pdf (2019)
28. Miskimmon, A., O'Loughlin, B., Roselle, L.: Strategic Narratives: Communication Power and the New World Order. Routledge, New York (2014)

29. Khaldarova, I., Pantti, M.: Fake news. Journal. Pract. **10**(7), 891–901 (2016). https://doi.org/10.1080/17512786.2016.1163237

30. Kwon, K.H., Rao, H.R.: Cyber-rumor sharing under a homeland security threat in the context of government Internet surveillance: the case of South-North Korea conflict. Gov. Inf. Q. **34**(2), 307–316 (2017)

31. Garrett, R.K., Weeks, B.E., Neo, R.L.: Driving a wedge between evidence and beliefs: how online ideological news exposure promotes political misperceptions. J. Comput. Mediat. Commun. **21**(5), 331–348 (2016)

32. Dagnall, N., Drinkwater, K., Parker, A., Denovan, A., Parton, M.: Conspiracy theory and cognitive style: a worldview. Front. Psychol. **6**, 1–9 (2015)

33. De Keersmaecker, J., Roets, A.: Fake news': incorrect, but hard to correct. The role of cognitive ability on the impact of false information on social impressions. Intelligence. **65**, 107–110 (2017)

34. Dutton, W. H., Reisdorf, B. C., Dubois, E., Blank, G.: Social shaping of the politics of Internet search and networking: moving beyond filter bubbles, echo chambers, and fake news. SSRN Electronic Journal: Quello Centre Working Paper No. 2944191. https://papers.ssrn.com/sol3/papers.cfm?abstract_id=2944191 (2017)

35. Kreiss, D., Mcgregor, S.C.: The "arbiters of what our voters see": Facebook and Google's struggle with policy, process, and enforcement around political advertising. Polit. Commun. **36**(4), 499–522 (2019)

36. Kreiss, D., Mcgregor, S.C.: Technology firms shape political communication: the work of Microsoft, Facebook, Twitter, and Google with campaigns during the 2016 U.S. Presidential Cycle. Polit. Commun. **35**(2), 155–177 (2018)

Fake News and Social Processes: A Short Review

Girish Keshav Palshikar

Abstract The explosive growth in social media, social networking, and messaging platforms has seen the emergence of many undesirable social phenomena. A common thread among many of these social behaviors is *disinformation propagation*, through falsehoods of many shades and grades that are quickly propagated to millions of people. In this chapter, we focus on disinformation propagation mainly in the garb of *fake news*, which contains deceptive, distorted, malicious, biased, polarizing, inaccurate, unreliable, unsubstantiated, and unverified or completely false or fabricated information. We examine the literature related to the sociological analysis of the fake news phenomenon and its impact on social processes such as elections and vaccination. We also outline directions for further research.

Keywords Fake news on social media · Disinformation in elections · Disinformation in anti-vaccine propaganda · Disinformation propagation · Rumors

1 Introduction

The world is witnessing an explosive growth in social media, social networking, and messaging platforms (which we collectively call *information sharing channels*, or just *channels*) and their deepening reach into all strata of societies across the world. Along with many benefits, this has also led to the emergence of several types of undesirable online social behaviors, including rumors, fake news, fake reviews, fake images, fake videos, spam emails, identity theft, cyber-stalking, phishing, etc. [3].

A common thread among many of these online social phenomena is *disinformation propagation*, which consists of dynamic, distributed social processes for the creation and dissemination of deceptive, distorted, malicious, biased, polarizing, inaccurate, unreliable, unsubstantiated, unverified, or completely false or fabricated

G. K. Palshikar
TCS Research, Tata Consultancy Services Ltd., Pune, India
e-mail: gk.palshikar@tcs.com

© Springer Nature Switzerland AG 2021
Deepak P et al., *Data Science for Fake News*, The Information
Retrieval Series 42, https://doi.org/10.1007/978-3-030-62696-9_12

information, often as a coordinated *campaign* spread across multiple channels, targeting specific classes of users and achieving a specific impact or goal. Typical goals include promoting (or attacking) specific religious/political views, promoting (or attacking) specific people/organizations/products, spreading fear/hatred/ anger, causing confusion/suspicion, influencing social events such as elections/protests, and influencing strategic decisions. The disinformation is often expressed in the form of fake news, fake images, fake videos, fake documents, and fake textual messages or posts, although the expression could be much more subtle instead of being outright false, e.g., misleading information crafted around a core of true facts. Disinformation campaigns are typically created and launched by *campaign managers*, who, behind the scenes, coordinate, sustain, and manage the spread. Fake accounts and bots play an important role in initiating and sustaining the campaign. Still, a disinformation campaign continues due to the active participation in its dissemination of supporting or interested users on different channels [32], who are often unaware of the true intentions of the managers. Disinformation works because it appeals to simple factors in human nature: humans respond to emotional triggers; humans share disinformation if it appeals to or reinforces their existing beliefs and prejudices; and humans long to belong to groups sharing similar beliefs (*echo chambers*) [33]. The voluntary participants in disinformation campaigns add weightage and credibility to them.

In this chapter, we will focus on disinformation propagation involving fake news, although, in our opinion, the use of other channels should be considered to construct an integrated view of a disinformation campaign. After the advent of the Internet, the traditional sources of news—such as newspapers, news magazines, and television—are seeing a rapid decline in the access of their original print avatars, and most have switched to online digital formats. While many digital newspapers remain behind paywalls, news aggregators, such as Apple News, Google News, and Upday, are gaining importance as they provide personalized online news access to users. Platforms such as Facebook, Instagram, Twitter, and WhatsApp are becoming primary sources for getting, sharing, and discussing news [18].

Since it is now easy to create and quickly disseminate user-generated content at scale, as mentioned earlier, a new class of undesirable online social phenomena has emerged. Important among these is the phenomenon of fake news. *Fake news* can be defined as a text (or other media such as image, audio, or video) content, masquerading as real and authentic news, that intentionally and verifiably contains falsehoods and disinformation presented as true facts. A wider definition might allow for the presence of more subtle (not directly verifiable) falsehoods, such as insinuations, misinterpretations, etc. Fake news has the appearance of authentic news but whose source is not any well-known, official, trustworthy agency and which often contains targeted, usually negative, malicious, biased, polarizing, unverified, unsubstantiated, unreliable, inaccurate, and even completely false or fabricated information [5, 25, 32, 35]. The fake news phenomenon consists of dynamic, distributed processes of creation and dissemination of disinformation, often through online social media or messaging platforms. Fake news campaigns often evoke counter-processes, such as detection, control, and retaliations.

Fake news campaigns generally target influential persons, organizations, products, or countries in order to damage (or boost) their activities, finances, or reputations. Widespread prevalence of fake news has resulted in the reduction of trust in the news circulating on various platforms, despite efforts by these platforms to build public confidence. Across all countries, the average level of trust in the news is about 42%, and less than half of the users trust the news media they themselves use [18]. Facebook, WhatsApp, Twitter, Instagram, and other such platforms often allow dissemination of content within closed, opaque groups (*information black holes*) without any serious fact-checking validations, control, or judgment, making it difficult to control the dissemination of falsehoods. Several initiatives have been created to help in the detection of fake news, e.g., publicizing lists of known fake news publishers, public-domain datasets of fake news, associations such as the International Fact-Checking Network (poynter.org/ifcn/)/European Union Disinformation Lab (disinfo.eu), and fake news detection websites such as FactCheck.org, snopes.com, and altnews.in.

While fake news and other disinformation campaigns may benefit some in the short run, it will almost certainly have large social and political costs in the long run, if the falsehood remains persistent and widespread. Some of the long-term effects of fake news are fragmentation and polarization of societies, social unrest, distrust of democratic institutions, and misgivings about scientific temper. In individuals, fake news may result in feelings of alienation, confusion, suspicion, cynicism, and distrust of authorities.

In this chapter, we review fake news related to social and political processes. Since fake news affects many social processes and events, we further narrow down our review to elections and vaccine hesitancy. Fake news is being used to influence election outcomes and thereby wreak havoc with the foundations of democratic institutions [6, 12]. Fake news is also being used to create and fan social unrest, to push political/religious agenda, to create anger/hatred in society, to damage reputation/legitimate activities of political opposition, and to create fear/confusion in vulnerable sections of a society (e.g., vaccine hesitancy). Figure 1 shows two

Fig. 1 Examples of disinformation

examples of recent fake news. One is based on a fake letterhead that "shows" how the Delhi state government "appeased" Muslim citizens during recent communal riots [11]. The other is a claim that the Indian Prime Minister was invited to head a task force against Coronavirus. Interestingly, there was another related fake news which involved a claim by an Indian Union minister that Prince Charles was cured of COVID-19 with Ayurvedic treatment, which was promptly denied by the British authorities [20].

2 Sociological Studies of Disinformation

Research in disinformation (and fake news in particular) can be grouped into two broad (sometimes overlapping) categories: computational and sociological. Computational research into disinformation investigates several questions, as follows:

1. **Origin of misinformation:** While anyone can see isolated posts, messages, fake news stories, etc., across different channels, how can we collect them, link them, and group them into *campaigns*, having specific beginning and end points in time, geographical spread, etc.? Can we identify people or accounts (i.e., *managers*) that initiated, coordinated, and fueled a campaign? Which channels are more effective for fake news campaigns?
2. **Contents of disinformation:** What was the *level* of the disinformation, on a scale from 0 (no disinformation) to 1 (completely false information)? Can we identify the broad conceptual categories of disinformation? How was the disinformation expressed, e.g., as text, images, video?
3. **Spread of disinformation:** How long did a campaign last? How and why did it die? How many people did a campaign reach? How many of the receivers participated in its further spread? What were the strategies used to increase the *spread* of a campaign? Was there a serious opposition to the campaign?
4. **Participants:** How to identify the shared characteristics (e.g., demographics) of users who were the true targets of a campaign? What are the shared characteristics of the people who supported the campaign and participated in its dissemination? How persuasive was the campaign, i.e., what was the probability that a receiver would participate in its further spread after seeing n messages?
5. **Questions related to the impact of misinformation:** Was the campaign able to affect the responses (e.g., voting decision) of a significant number of receivers? How many receivers actually agreed with the common message of the campaign? How can the success or impact achieved by the campaign be measured?

Sociological studies attempt to understand fake news as a social behavior:

1. What is the extent of the prevalence of fake news in different news categories, e.g., health, politics, business, entertainment, etc.?
2. How well does fake news succeed in manipulating and influencing public opinion, decisions, and policies?

3. What factors affect the success of fake news campaigns? What are the demographic factors (sex, age, gender, education, income, ethnicity, political orientation) of the susceptible population segments targeted by fake news? Does fake news affect across countries?
4. What situational factors drive people to share fake news? How do people judge the reliability of the information (e.g., news) they receive on social media or messaging?
5. How many "views" of different fake news stories does a user typically need to see before they accept the common viewpoint of these fake news stories?
6. What should organizations and governments do to control fake news and reduce its impact?

In this chapter, we will largely focus on sociological research into the phenomenon of fake news.

3 Vaccine Hesitancy

Most modern governments have a compulsory vaccination program for children, which typically includes vaccines for measles–mumps–rubella (MMR), diphtheria–pertussis–tetanus (DPT), polio, and hepatitis B, among others. Fake news has played an important part in creating the social phenomenon of *vaccine hesitancy*, which refers to reluctance, confusion, distrust, suspicion, fear, anger, or hostility in the minds of parents toward vaccination of children. Vaccine hesitancy makes children susceptible to easily preventable diseases and thus is a major threat to the health of millions of children worldwide.

Fake news has been used to spread disinformation and conspiracy theories about vaccines. An example is a fake news that claimed autism as a potential side effect of MMR vaccines, often quoting a paper that was later retracted and proven to be fraudulent. Disinformation about vaccines often goes much beyond (false) claims of autism as a side effect. Other examples of disinformation about vaccines include [7] messages claiming that babies have died or suffered severe disabilities and other claims that sow seeds of doubts. For example, "so, a baby can handle 8–9 viruses all at once via vaccination, but cannot handle one single virus when it's wild caught?" Such disinformation has several logical fallacies; this example ignores the fact that the viruses in a vaccine have carefully attenuated virulence, unlike viruses in the wild. In general, such disinformation often mis-appropriates scientific terms, hides relevant facts, makes unacceptable assumptions, and has errors in its inferences.

Of course, no link between MMR vaccine and autism was found to exist [17]. Nevertheless, according to the WHO, measles cases increased by 30% globally in 2018, and in 2019 and a state emergency was declared in Washington, USA, due to a measles epidemic. While the controversy was specifically about MMR vaccine, researchers have noted that there were *spillover effects*, which led to hesitancy about other vaccines. This indicates the potential of fake news to create damage that goes beyond its prima facie goals. Moreover, such fake news creates feelings of disillusionment, powerlessness, and distrust in the authorities and in the minds of parents, which is a more dangerous consequence.

Smith et al. [28] demonstrated a significant increase in MMR non-receipt after the media coverage of the MMR-autism controversy, which was fueled by fake news. In the 1995 cohort, only 0.77% of children had not received MMR vaccine, which rose to 2.1% in 2000, coinciding with the emergence of the controversy. They also noted the return of the MMR vaccination rates to the pre-controversy levels, after the media coverage (and the associated fake news campaigns) had died down.

The uptake of MMR vaccines in the UK dropped by over 5%, before it rose again [2]. Rather surprisingly, by analyzing the data from local health authority areas, the authors showed that the uptake rate of MMR vaccines declined faster in areas where most parents had higher educational levels than in areas where most parents had relatively less education.

Jolley and Douglas [15] empirically established negative influence of anti-vaccine conspiracy theories on health-related behaviors. They conducted two surveys that (1) demonstrated negative correlation between beliefs in anti-vaccination conspiracies and vaccination intentions and (2) revealed that participants exposed to anti-vaccination conspiracy theories showed less intention to vaccinate than those in controls.

Chang [9] established that (1) in the USA, the MMR-autism controversy led to a decline in vaccination rates in the immediate years; (2) there were negative spillovers onto other vaccines; (3) more highly educated mothers had developed higher levels of vaccine hesitancy (refusal or delay of vaccination); and (4) the vaccine hesitancy was proportional to the media attention to the controversy. Point (3) was also noted in [2]. A possible explanation is the so-called *health allocative efficiency* hypothesis, which states that this education gradient in health outcomes is due to greater access, absorption, and response to online health information by more highly educated individuals. Clearly, this is not an adequate explanation, since higher education should instill better discerning and reasoning abilities.

Finally, we note that the disinformation and fake news campaigns related to vaccination, combined with other social factors such as prevalence of religious fundamentalism, have led to serious violence against and killings of health professionals delivering vaccinations to children. Such violence has persisted even in 2019 [16], indicating the alarmingly long life of the lies about vaccinations.

We should also analyze *why* people run vaccine misinformation campaigns. What are their motives? What do they want to achieve? Can we develop techniques to detect whether any communities are being formed on social media consisting of anti-vaccinationists or of people having vaccine hesitancy? What are the factors that convince people with vaccine hesitancy to change their health behavior (e.g., refuse or delay vaccine to a child)? Such drivers may include a general tendency to believe in conspiracy theories, beliefs in alternative healthcare systems, a general distrust of authorities, and so on [7].

Obviously, the usual informative campaigns explaining the real benefits of vaccinations should continue. Several social media platforms have responded positively [7]. Pinterest ensures that searches for vaccine-related topics will only show links to reputable public health organizations. Instagram has blocked hashtags that make patently false claims such as #vaccinescauseaids. YouTube has removed advertisements from anti-vaccination, videos so that their posters will not make any money. But such responses are clearly not enough because the platform owners probably worry about censorship or restricting the freedom of speech. Hence, AI, healthcare, and social scientists need to develop proactive detection, control, and retaliatory strategies that governments and other institutions can use to stop controversies and control the spread of disinformation campaigns, whenever they surface, and whatever channels they use—whether fake news, fake videos, or disinformation messages on messaging platforms.

4   Elections

Elections constitute an important political process in a democracy, and even in other institutions, for which factually well-informed electorate is an essential prerequisite. Since elections are a road to political power, it is not surprising that political parties, candidates, and their supporters would use all possible means, including fake news, to influence voters and win elections. In 2018, it was discovered that a company called Cambridge Analytica used the private account data of Facebook users in order to make targeted delivery of political campaign material and to manipulate political opinions [24]. As mentioned earlier, fake news was delivered not just through fake news websites, but also Twitter, Facebook, and WhatsApp, among others, were important channels for disinformation during elections.

The US Election in 2016 was among the first major elections in which a large number of fake news campaigns appeared; many other later elections (e.g., in Europe) also were victims of fake news. The role of fake news in the 2016 US election is well studied, and we will summarize some findings here. Examples of highly circulated fake news prior to the 2016 US elections are as follows: FBI agent suspected in Hillary email leaks found dead in apparent murder-suicide; Pope Francis shocks world, endorses Donald Trump for president; and Hillary sold weapons to ISIS [23]. Figure 2 shows examples of political disinformation of the kind seen during elections in India [4]. Broadly, election-related disinformation typically maligns political parties or leaders or tries to mislead voters, often through links to fake news websites, fabricated images of screen grabs of TV news, or doctored screenshots of newspapers.

The main motivations for the creation and dissemination of political fake news are economic or ideological. Teenagers in a Macedonian town flooded primarily pro-Trump fake news in the 2016 US elections, not for any ideological reasons but because that gave them much higher click-based advertising revenues [30]. Ideological motivation stems from the fact that fake news can serve as a propaganda vehicle to influence voters to support a particular political party, candidate, or party position about an issue.

Fig. 2 Examples of political disinformation

Grinberg et al. [12] studied 16,442 Twitter accounts in the USA from August to November 2016 and found that 5% of tweets received by them and 6.7% of the URLs in tweets generated by them came from fake news sources. Only 1% of them consumed 80% of the volume from fake news (*superconsumers* and *superspreaders*), showing that not everybody had an equal affinity for (or were targets of) fake news. Their demographic analysis showed that individuals most likely to "engage" with fake news sources were conservatives, older, and highly interested in political news. A 2016 survey by news and entertainment site BuzzFeed found that fake news fools American adults about 75% of the time [26].

Vosoughi et al. [32] found by analyzing 126,000 stories tweeted by 3 million people more than 4.5 million times that false stories diffused significantly farther, faster, deeper, and more broadly than the truth in all categories of information, but particularly so for political stories. Significantly more false cascades than true cascades exceeded a depth of 10. The top 1% of fake news cascades reached to between 1000 and 100,000 people, compared to true news that rarely reached more than 1000 people. They suggest that fake news is more novel, and hence people are more likely to share it.

Allcott and Gentzkow [1] showed that (1) pro-Trump fake news was shared on Facebook three times more often compared to pro-Clinton fake news and (2) 41.8% of web traffic to fake news sites came from social media compared to 10.1% for regular news sites. They found that 27% of people visited a fake news source a few weeks before the 2016 US presidential election, and visits to fake news sources were only 2.6% of total visits to reliable news sites. They conducted a post-2016 election survey and found that, on average, an adult received 1.14 fake news stories per person and 8% of the respondents believed them. Further analysis showed that Republicans were more credulous of fake news compared to Democrats. Among the demographic attributes, they show that the ability to discern fake news from true news weakly improves with education and age.

Bovet and Makse [6] analyzed 171 million tweets sent by 11 million users in the 5-month period preceding the 2016 US presidential election, among which 30.7 million tweets, from 2.2 million users, contained a URL to news outlets.

They found that 25% of these tweets spread either fake or highly biased news. The tweeting rate for fake news tweets was four times higher than that of normal news tweets. Furthermore, the top news spreaders for traditional news tweets were mostly known journalists or leaders, and those for fake news tweets were unknown or deleted accounts. Both of these suggest a role of bots. Users spreading fake and extremely biased news were smaller in number but were more active and also connected (through retweets) to more users on average than users in the traditional news networks. Among users, 64% and 8% were Clinton and Trump supporters, respectively, although Trump supporters were, on average, 1.5 times more active than Clinton supporters. Through causal analysis of tweet volume time series, the authors found that the dynamics of fake news spreaders was mostly governed by Trump supporters.

One key question is: did fake news *decisively* influence the 2016 US election? While there are some extreme positions [22], the broad consensus seems to be "No" [34], although, in our opinion, more work is needed in devising a general methodology for reliably answering such questions.

5 Other Social Processes

Fake news often indirectly promotes a particular political viewpoint or agenda, by promoting conspiracy theories, pseudo/anti-scientific narratives, and anti-media, anti-globalization, or anti-migration viewpoints. Conspiracy theories [8] usually claim that rich and powerful people or organizations clandestinely organize key events in order to protect their power. Examples are as follows: `Boston Marathon Bombings were perpetrated by U.S. Navy Seals;` the `2012 shootings at the Sandy Hook school were staged to motivate gun control legislation;` and `Orlando shooting was a hoax. Just like Sandy Hook, Boston Bombing, and San Bernardino. Keep believing Rothschild Zionist news companies.` Conspiracy theories offer an interesting vehicle to promote political agenda, as believers in them may be more likely to also believe in other unverifiable information. Starbird [29] performed a detailed analysis of fake news stories about mass shootings circulating on Twitter and showed that Twitter users who engaged with conspiracy theories liberally used fake news to support the theories.

With hundreds of millions of users in India alone, WhatsApp and Facebook have become important channels for information sharing, and, unfortunately, for disinformation campaigns, often appealing to people's prejudices and biases and leading to mob violence. Some lynchings took place in India because of disinformation about the presence of child abductors in an area [19]. Similarly, there were some incidents of lynching of Muslims in India, due to rumors about their consumption of beef or about them taking cows to the abattoir for slaughter [21]. Some disinformation campaigns attempt to boost cultural, nationalist, or religious supremacist positions;

for example, the chief minister of a state in India claimed that the Internet was invented by ancient Indians (Hindus) thousands of years ago [27].

Crisis and disaster situations often bring out disinformation tendencies. A number of fake images emerged during Hurricane Sandy [13]. The Ebola epidemic has generated many conspiracy theories, which were not just harmless scaremongering but led to real consequences such as social resistance, suspicion of authorities, noncompliance, and aggression and generally made outbreak prevention and control difficult [31]. The ongoing coronavirus pandemic has given rise to many disinformation campaigns (at least in India), some with interesting intersections with politics. For instance, a fake news claimed that Prime Minister Modi was selected to lead an international task force against the coronavirus [10]. Fake news can sometimes have disastrous economic consequences. A tweet falsely claiming that injury to President Obama in an explosion eroded $130 billion from the stock markets [32]. The use of a fake WMD dossier was instrumental in the 2003 US invasion of Iraq. A routine military exercise (Jade Helm 15) was misinterpreted as the beginning of a civil war in the USA.

6   Conclusions

In this chapter, we summarized the literature about how social processes (elections and vaccinations, among others) are affected by all kinds of disinformation, including fake news. Disinformation is here to stay and is flourishing due to the deep reach and high speeds of social media channels. The key question is, of course, what can be done to contain the production and dissemination of fake news and the damage they inflict on society. While statistical and AI techniques for detection of fake news will undoubtedly improve over time, more theoretical work is needed for mathematical (e.g., game-theoretical) modeling of fake news as a social phenomenon. Predictive models about what kinds of fake news campaigns will emerge, say, for the next election or the next pandemic, are clearly lacking. At the very least, we need techniques that can "pick up" and group signals from very recent social media posts in real time and predict whether one of these would become a fake news campaign. Effective automated intervention techniques for countering disinformation campaigns are yet to be developed. Clearly, much more work needs to be done to detect and control disinformation campaigns on social media channels. Sociologists, journalists, politicians, and social media platforms must come together with technologists and work toward an overhaul of the news industry [14]. To conclude, computational and AI technologies cannot form the final solution to what is essentially a social malady.

References

1. Allcott, H., Gentzkow, M.: Social media and fake news in the 2016 election. J. Econ. Perspect. **31**(2), 211–236 (2017)
2. Anderberg, D., Chevalier, A., Wadsworth, J.: Anatomy of a health scare: education, income and the MMR controversy in the UK. J. Health Econ. **30**(3), 515–530 (2011)
3. Apte, M., Palshikar, G.K., Baskaran, S.: Frauds in online social networks: A review. In: Ozyer, T., Bakshi, S., Alhajj, R. (eds.) Social Network and Surveillance for Society, pp. 1–18. Springer, New York (2019)
4. Bansal, S., Garimella, K.: Fighting fake news: Decoding 'fact-free' world of WhatsApp. In: Hindustan Times (2019)
5. Berghel, H.: Lies, damn lies, and fake news. Computer **50**(02), 80–85 (2017)
6. Bovet, A., Makse, H.A.: Influence of fake news in Twitter during the 2016 US presidential election. Nat. Commun. **10**(7) (2019).
7. Burki, T.: Vaccine misinformation and social media. Lancet Dig. Health **1**(6), E258–E259 (2019)
8. Byford, J.: Conspiracy Theories: A Critical Introduction. Palgrave Macmillan, London (2011)
9. Chang, L.V.: Information, education, and health behaviors: evidence from the MMR vaccine autism controversy. Health Econ. **27**(7), 1043–1062 (2018)
10. Chaudhuri, P.: No, US and UK have not selected PM Modi to lead a coronavirus task force. In: AltNews.in (2020)
11. Deodia, A.: Fact check: Delhi Govt advert for riot victims morphed with communal twist. In: India Today (2020)
12. Grinberg, N., Joseph, K., Friedland, L., Swire-Thompson, B., Lazer, D.: Fake news on Twitter during the 2016 U.S. presidential election. Science **363**(6425), 374–378 (2019)
13. Gupta, A., Lamba, H., Kumaraguru, P., Joshi, A.: Faking sandy: characterizing and identifying fake images on Twitter during hurricane sandy. In: Proceedings of the 22nd International Conference on World Wide Web, pp. 729–736 (2013)
14. Hirst, M.: Towards a political economy of fake news. Polit. Econ. Commun. **5**(2), 82–94 (2017)
15. Jolley, D., Douglas, K.M.: The effects of anti-vaccine conspiracy theories on vaccination intentions. PLoS ONE **9**(2), e89177 (2014)
16. Lamble, L.: Killings of police and polio workers halt Pakistan vaccine drive. In: The Guardian (2019)
17. Mrozek-Budzyn, D., Kieltyka, A., Majewska, R.: Lack of association between measles-mumps-rubella vaccination and autism in children: a case-control study. Pediatr. Infect. Dis. J. **29**(5), 397–400 (2010)
18. Newman, N., Fletcher, R., Kalogeropoulos, A., Nielsen, R.K.: Reuters Institute Digital News Report 2019. Technical report, Reuters Institute, Oxford (2019)
19. Outlook: Mob Lynchings Back as Child Lifting Rumours Spread on WhatsApp. Outlook (2019)
20. Press Trust of India: Prince Charles' Office Denies Indian Minister's Ayurveda Cure of COVID-19. In: NDTV (2020)
21. Ravi, S.: Why India Man was Lynched Over Beef Rumours. In: BBC (2015)
22. Read, M.: Donald trump won because of Facebook. In: New York Magazine (2016)
23. Ritchie, H.: Read all about it: The biggest fake news stories of 2016. In: CNBC.com (2016)
24. Rosenberg, M., Confessore, N., Cadwalladr, C.: How trump consultants exploited the Facebook data of millions. In: The New York Times (2018)
25. Sharma, K., Qian, F., Jiang, H., Ruchansky, N., Zhang, M., Liu, Y.: Combating fake news: a survey on identification and mitigation techniques. ACM Trans. Intell. Syst. Technol. **10**(3), 1–42 (2019)
26. Silverman, C., Singer-Vine, J.: Most Americans who see fake news believe it, new survey says. In: BuzzFeed News (2016)

27. Singh, B.: Internet existed in the days of Mahabharata: Tripura CM Biplab Deb. In: The Economic Times (2018)
28. Smith, M.J., Ellenberg, S.S., Bell, L.M., Rubin, D.M.: Media coverage of the measles-mumps-rubella vaccine and autism controversy and its relationship to MMR immunization rates in the United States. Pediatrics **121**(4), e836–e843 (2008)
29. Starbird, K.: Examining the alternative media ecosystem through the production of alternative narratives of mass shooting events on twitter. In: Proceedings of the Eleventh International AAAI Conference on Web and Social Media (ICWSM 2017), pp. 230–239 (2017)
30. Subramanian, S.: Inside the Macedonian fake-news complex. In: Wired Magazine (2017)
31. Vinck, P., Pham, P.N., Bindu, K.K., Bedford, J., Nilles, E.J.: Institutional trust and misinformation in the response to the 2018–19 Ebola outbreak in North Kivu, DR Congo: a population-based survey. Lancet Infect. Dis. **19**(5), 529–536 (2019)
32. Vosoughi, S., Roy, D., Aral, S.: The spread of true and false news online. Science **359**(6380), 1146–1151 (2018)
33. Wardle, C.: Misinformation has created a new world disorder. Sci. Am. **321**(3), 88–93 (2019)
34. Weintraub, K.: "Fake news" web sites may not have a major effect on elections. Sci. Am. (2020). https://www.scientificamerican.com/article/fake-news-web-sites-may-not-have-a-major-effect-on-elections/
35. Zhou, X., Zafarani, R., Shu, K., Liu, H.: Fake news: fundamental theories, detection strategies and challenges. In: Proceedings of the Twelfth ACM International Conference on Web Search and Data Mining (WSDM2019), pp. 836–837 (2019)

Misinformation and the Indian Election: Case Study

Lyric Jain and Anil Bandhakavi

Abstract Diverse demographics, culture, and language; a troubled history of communal violence, polarised politics, and sensationalist media; and a recent explosion in smartphone ownership and Internet access have created a "fake news" crisis in India which threatens both its democratic values and the security of its citizens. One of the unique features of India's digital landscape is the prevalence of closed networks – ideologically homogeneous groups of individuals communicating on private platforms – in which misinformation proliferates. This poses several challenges: the encryption of private messages makes tracking and analysing the spread of information through these channels difficult; accusations of censorship and surveillance can prevent governments from tackling misinformation propagated through private groups; ethical considerations associated with the extraction of data from encrypted, private conversations; highly influential means of disseminating information, with users receptive to messages which fit the common worldview of the group; and speed of proliferation, i.e. information spread to a large user base at the touch of a button. In this chapter, we showcase some of our core technologies in the areas of credibility and veracity assessment. Our key findings during the Indian general election 2019 indicate that individual users play a major role in the solution to fake news and misinformation. Adopting content verification strategies even by 1% of WhatsApp users will help vaccinate India's WhatsApp networks against fake news. Whilst volume must increase, the speed of fact-checks is the vital improvement which the fact-checking industry must undergo – semi-automated fact-checks proved 35 times more effective than traditional fact-checks in fighting fake news on WhatsApp – to debunk rumours early in their propagation path and spread fact-checks to audiences that came across the original piece of problematic content.

Keywords Fake news · Misinformation · Credibility · Fact-checking · WhatsApp · Indian general elections

L. Jain · A. Bandhakavi
Brookfoot Mills, Brookfoot Industrial Estate, Brighouse, UK
e-mail: contact@logically.co.uk

© Springer Nature Switzerland AG 2021   257
Deepak P et al., *Data Science for Fake News*, The Information
Retrieval Series 42, https://doi.org/10.1007/978-3-030-62696-9_13

1   Misinformation and Disinformation in India

1.1   Misinformation and Disinformation in India

The prevalence of misinformation in India is worrying, though not surprising. Everything about the country's make-up has made it fertile ground for misinformation to spread. The hugely diverse demographics, the sheer number of different cultural groups and languages, its troubled history of communal violence, polarised politics and sensational media, and even the geography of the country leave it vulnerable.

A major component of Narendra Modi's 2014 election campaign was a promise to roll out Internet and mobile access across huge swathes of India, and the government was hugely successful in doing so. According to the "Household Survey on India's Citizen Environment & Consumer Economy", in 2016, whilst only 60% of Indian households had access to basic sanitation, 88% owned mobile phones.[1] This, compounded with the global shift to a new digital information ecosystem, has contributed to the fact that India has developed what many are suggesting is a "fake news" crisis.

The effects of misinformation in India should not be underestimated. One need simply look at the widely reported spate of mob vigilante killings that occurred over an 18-month period between mid-2018 and the start of 2019 to see the real effects of rumour and falsehood. Nicknamed the "WhatsApp lynchings", often the killings were a direct result of rumours of child abductions which spread over the messaging platform to rural communities. The victims were mostly strangers, passing through communities and not known to the locals who – spurred on by false rumours – carried out the attacks.

Other instances of mob violence were related to cow vigilantism directed towards Muslims and Dalits. This type of religious and communal violence often follows similar patterns to that directed at presumed child abductors – instigated by false rumours spread over messaging apps and directed towards someone considered "other". In many instances, law enforcement was impotent and unequipped to deal with mobs stirred to violence. According to a report by the Armed Conflict Location & Event Data Project (ACLED), "the increasing number of events involving vigilante violence against presumed criminals is indicative of the perceived lack of law enforcement and lack of patience and trust in India's criminal justice system".[2]

The lack of trust in India's criminal justice system mirrors that of the loss of trust across most areas of the state: traditional journalism, politicians, and government

---

[1]Moonyati Mohd Yatid, "Truth Tampering Through Social Media: Malaysia's Approach in Fighting Disinformation and Misinformation", The Indonesian Journal of Southeast Asian Studies, vol. 2, no. 2, Jan 2019 (pp. 203–230), p. 206.

[2]The Complexity of Disorder in India (p. 3) ACLEDdata, https://www.acleddata.com/2018/08/17/the-complexity-of-disorder-in-india/

institutions. This atmosphere of distrust, anxiety, and division set the scene for the world's largest election and was undeniably capitalised on by political actors who spread their own disinformation to leverage electoral gain.

The organic spread of rumours occurring on messaging apps such as WhatsApp and Telegram mimics the way political disinformation is spread by bad actors on the same platforms and can have similar outcomes. Senior Fellow at the Observer Research Foundation, Maya Mirchandani suggests that "in India, these spaces provide both tacit and overt sanction for rising incidents of majoritarian violence as identity-based, populist politics dominate the country's landscape".[3]

1.2   Closed Networks for Disinformation

One significant challenge for those aiming to combat the spread of misinformation is that many of the more dangerous stories and rumours are spread on social media and, more specifically, on closed networks and messaging apps. The fact that these platforms are the primary tool by which misinformation seems to be spread raises a number of complicated issues.

First and foremost, there are practical hurdles blocking attempts to research and counter the spread of misinformation on platforms such as WhatsApp. Because these messaging apps are mostly encrypted, it is very difficult to assess the proliferation of the information that is circulating inside private groups. Unless we can gain access to groups on these platforms, it is near impossible to see what content is being shared and stop anything considered false or misleading. Issues of censorship and surveillance are also raised at the thought of the government or others gaining access to private chat groups – even if the honest intention is to combat the spread of misinformation.

Another problem for researchers are the ethical considerations of extracting data from encrypted conversations between private individuals. This has already had implications on previous attempts to study misinformation in India and other countries. The Oxford Internet Institute struggled with these very issues during a study conducted on misinformation on WhatsApp in India: "the encrypted nature of the platform (WhatsApp), amorphous structure of public groups and our strict ethical considerations pose significant challenges to joining and extracting data at scale from WhatsApp groups. We note that, our strategy does not ensure that all WhatsApp groups are adequately represented, however, forwarded content circulating within these groups could readily by shared in other private groups that the members belong to and therefore it is possible that the content analysed in this study has been viewed by a much larger network of WhatsApp users".[4]

---

[3]Digital Hatred, Real Violence: Majoritarian Radicalisation and Social Media in India, ORF Occasional Paper, August 2018 (p. 1).

[4]"News and Information over Facebook and WhatsApp during the Indian Election Campaign", COMPROP DATA MEMO 2019.2, Oxford Internet Institute, May 13, 2019, p. 7.

The closed nature of messaging platforms has also been observed to create an environment in which users feel freer to share questionable content as well as for such content to be readily believed. In his study of fake news in Asia, Andy Yee suggests that these apps create "walled gardens" in which the information shared "can especially resonate since receivers are more likely to trust their circles of like-minded people".[5]

These closed networks pose significant problems for anyone wishing to tackle misinformation. Stringent research ethics stunt attempts to analyse the extent of the problem, questions of censorship and surveillance plague attempts at intervention, and the very private and personal nature of messaging groups provides the perfect conditions for misinformation to spread unchecked.

1.3   Scale, Prevalence, and Complexity of the Problem

To suggest that misinformation is a new phenomenon is clearly incorrect. The recent spike in interest in "fake news", inspired by its popularisation and politicisation firstly in the USA and then globally, has, however, sparked a renewal of interest in the phenomena. In addition to this, evidence of a sustained Russian disinformation campaign during the 2016 presidential election – mostly played out over social media platforms – has woken up the international community to the fact that technology has created new avenues for bad actors to exploit and weaponise information.

In a country in which the number of voters with access to a smartphone – and by extension digital messaging apps – has nearly doubled from 21% in 2014 to 39% in 2019, it's easy to see why such focus was paid to digital campaigning techniques by the main political parties. With the ability for a single person to share a message or story with around 1280 different individuals in seconds at almost no cost, WhatsApp and other messaging services became key to the parties' campaign strategies. It was reported that the Bharatiya Janata Party (BJP) had enlisted nearly 900,000 volunteers to carry out their WhatsApp campaign. Other parties followed suit, and although it is impossible to say just how effective digital strategies were in comparison to traditional election campaigning, the scale and prevalence of political messaging over digital platforms cannot be understated.

Political parties were warned in advance of the election by the Electoral Commission and by social media platforms themselves not to misuse social media during the election. At one stage Facebook removed 687 accounts linked to the Indian National Congress (INC) which they believed were being used to spread highly partisan and divisive content.[6] Although Congress claimed that none of their

---

[5] Andy Yee, "Post-Truth Politics and Fake News in Asia", Global Asia, vol. 12, No. 2, 2017 (pp. 66–71), p. 68.

[6] https://www.wsj.com/articles/facebook-removes-hundreds-of-fake-accounts-ahead-of-indian-elections-11554129628

official party accounts were affected, it is clear that Facebook was being used by supporters, if not the party itself, to spread misinformation and politically divisive messages for electoral gain.

Whilst technological advancements have increased the scale and reach of misinformation, and also added to difficulties in tracking and stopping its spread, the phenomenon itself is complex and has deep roots in society. In India, a country with a history of political and religious division, much of the political messaging that is found to be problematic touches on pre-existing cultural divisions. Benson Rajan of Christ University, Bangalore, has shown that the particular brand of Hindu nationalistic misinformation spread by the BJP in the 2019 general election can be found in its nascent stages as far back as 2014. During the 2014 Indian general election – which saw the BJP sweep to a historic victory – there is evidence of divisive and distinctly religious misinformation being spread online.[7] The apparent electoral success of these messages was consequently built on and perfected in time for 2019.

The blurring of political, religious, and cultural messaging adds further layers of complexity to India's misinformation problem. It means that the solution is not just technological and requires us to look past the knee-jerk application of blame that has been evident in recent years. Technology can be exploited to spread divisive messages and false content. However, the readiness for society at large to accept these messages requires a far more nuanced approach. This case is made in a study released by the Berkman Klein Center for Internet & Society at Harvard University in 2018. The authors write, "Specific technologies, under specific institutional and cultural conditions, can certainly contribute to epistemic crisis... WhatsApp in India, suggest(s) that technology, in interaction with particular political-institutional conditions, can become the critical ingredient that tips societies into instability... But we have not seen sufficient evidence to support the proposition that social media, or the internet, or technology itself can be a sufficient cause for democratic destabilization at the national scale... it is only where the underlying institutional and political-cultural fabric is frayed that technology can exacerbate existing problems and dynamics to the point of crisis".[8]

1.4   Early Solutions and Fact-Checking in India

Having looked at the ways in which both the Indian government and institutions such as the electoral commission and the social media giants have tried to address the misinformation crisis without much success, it would be prudent to touch on

---

[7]Benson Rajan, "New Mythologies of Fake News: WhatsApp and Misrepresented Scientific Achievements of Ancient India", Handbook of Research on Deception, Fake News and Misinformation Online, Christ University, India, 2019.

[8]Network Propaganda: Manipulation, Disinformation, and Radicalization in American Politics, Oxford Scholarship Online, October 2018.

other ongoing attempts. Over the last few years, India has seen an explosion in the number of fact-checking operations popping up. Some are small and independent; others are born out of the operations of existing publishing giants.

Whilst the fact-checking industry is growing and becoming more prominent, there are a number of issues that these organisations are encountering. First and foremost, there is a serious lack of funding for the organisations that are appearing in what in many ways is still a cottage industry. Despite the increasing attention being shown to fact-checkers, many still rely on volunteers, handouts, and access to open-source digital tools, which can often be rudimentary.

Some funding has been made available from social media companies, whilst Facebook runs a third-party fact-checking operation which sees select organisations work with the social media giant to verify content on its platforms. This scheme, however, is far from perfect. Firstly, there is an issue of scale. No matter how many organisations Facebook partners with and even with their own content moderation teams, they will never be able to fact-check and debunk every piece of misinformation on their platform. Further to this, the posts that are fact-checked are not removed from the site; they only have their visibility reduced to users. Questions have also been raised over a number of organisations who have partnered with Facebook in India, with claims that they themselves have spread misinformation.[9]

During the election itself, an independent group of journalists, fact-checkers, and technologists worked with WhatsApp to fact-check rumours and posts on the messaging platform. Billed as a "research project", Project Checkpoint set up a WhatsApp hotline to which users could send requests for the team to verify.[10] This project too suffered from issues of scale, and as it was only a research project, intended to build a database of India-specific misinformation, it likely had little tangible effect or outcome.[11,12] Although there have been a number of attempts at finding solutions to the spread of misinformation in India, none yet have appeared to be either successful or sustainable.

2 Logically's Capabilities

2.1 Automation to Augment Value

At the heart of our approach is the principle of extended intelligence (see later). In the current environment of information warfare, bad actors are equipped with

---

[9]https://www.nytimes.com/2019/04/01/technology/india-elections-facebook.html

[10]https://medium.com/@meedan/press-release-new-whatsapp-tip-line-launched-to-understand-and-respond-to-misinformation-during-f4fce616adf4

[11]https://www.checkpoint.pro.to/

[12]https://www.poynter.org/fact-checking/2019/is-whatsapps-new-tip-line-debunking-hoaxes-about-the-indian-election-not-really/

sophisticated tools and bots. By ignoring potential assistive applications of AI technologies, we risk bringing a knife to a gunfight on the information battlefield.

Our view of our role in the fact-checking ecosystem is to bring the industrial revolution to an industry described by many as a cottage industry.

As well as scaling our human fact-checking team and equipping them with cutting-edge dashboards and workflows to improve efficiency in line with a constantly expanding user base, we're also developing pioneering automation technologies. Together, these technologies will stimulate a technological revolution in the fact-checking industry. Just as initially the assembly line and more recently the robotic arm enabled the rapid development of the automobile into the everyday product it is today, Logically's fact-checking dashboards and automated veracity assessment tools will enable the publication of fact-checks on a scale previously unimaginable.

In addition to the scalability of automation, the lower latency of automated fact-checks and verifications makes them orders of magnitude more effective in countering mis/disinformation compared to their traditional counterparts.

This is a vital component of our ambitions in the field, both in terms of our ability to stem the immediate flow of misinformation online and our long-term ambitions to develop new technologies and approaches for a more sustainable and impactful fact-checking industry.

One of the biggest factors in the recent explosion in the volume and impact of misinformation has been its ability to propagate quickly online through social networks. Human fact-checking procedures are time-consuming and expensive, a huge obstacle both in terms of the impact and scale of fact-checking efforts. Whilst Logically's advances in the operational efficiency of human fact-checking processes are a positive step, full automation of fact-checking processes would enable a truly scalable solution to the problem of misinformation, and one which can negate the impact of misinformation by verifying, or debunking, claims in real time.

We continue to develop machine learning and common-sense reasoning technologies that judge incoming new claims from our users against our universe of facts established by our fact-checking teams, single sources of truth, and trust partners. This technology is already in action in the Logically app, where incoming claims follow a hybridised approach which first seeks to automatically verify a claim, relying only on human verification where this isn't possible. The development of this technology to enable full automation is one of our key priorities over the coming months and years and a central component of our mission.

2.2 Credibility vs. Veracity

It's important to distinguish two of Logically's core capabilities: credibility assessment and veracity assessment. Our credibility assessment looks at indicators of content's overall accuracy, without specifically analysing claims contained within.

We draw on content, distribution, and metadata analysis to assess the content in its entirety. Veracity assessment, on the other hand, is concerned with a single statement or claim, conducting primary and secondary research (a "fact-check") to conclusively verify its factual accuracy. Put simply, credibility assessment represents an accurate well-informed guess approach to information verification, whilst veracity assessment enables an unequivocal assessment of a claim's accuracy.

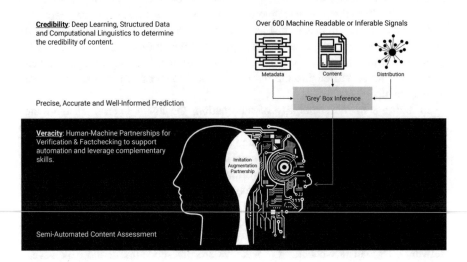

Inevitably, the applications of the two capabilities are therefore quite different, with the most important difference – other than the judgement's certainty – being the speed with which the assessment can be carried out. Whilst Logically's credibility assessment is fully automated, veracity assessment requires human input and a critical mass of location or domain-specific facts to have been established in order to automatically evaluate a claim's accuracy, making a real-time judgement impossible at this stage.

3   Credibility Assessment

3.1   Credibility Assessment

Logically is unique in its combined analysis of network, content, and metadata to reach its conclusions. Our AI leaves no stone unturned, evaluating every possible indicator of the overall article's accuracy, as well as the specific claims contained within the text, to inform more sophisticated conclusions than rival models which rely on just one or two of these analytical channels.

Network Analysis

- Propagation Analysis

  Accurate, misleading, and false information follows different propagation paths online after publication. We analyse the ways in which a story develops and proliferates on social media, looking for indicators which provide insights into the nature of the content itself.
- Bot Detection

  We can identify when a story is being spread by bots rather than humans based on common indicators we've identified. For example, a social media account that engages with and shares a story almost instantaneously is likely run by a bot, since human reaction speeds would be far slower. If a story is being spread by bots, then there is a high probability that its content is inaccurate and being maliciously propagated.
- Nodal Analysis

  We're building methods of detecting specific accounts or networks on social media websites that are more likely to propagate false or misleading content.

Metadata

We analyse the text's contextual surroundings online in order to gain insights relevant to an evaluation of the content itself. This analysis can be broken down into three categories:

- Author level

  Analysis of metadata surrounding a particular author and their biases, perspective, subject interests, and track record of accuracy.
- URL level

  Analysis of metadata surrounding the webpage on which a piece of text is hosted. This will include an analysis of that page's advertisements, tags, and recommended content.
- Domain level

  Publisher-level analysis and domain health analysis. We can access spamblock lists and government databases relating to domain violations to evaluate the reliability of webpages hosted under that domain, as well as using our trust-linking capabilities to evaluate publishers themselves.

Content Analysis

Logically relies on various capabilities to analyse the content of a piece of information with great sophistication and detail. Our AI clusters indicators of misinformation under the headings of Fallacy, Context, Claim, Tone, Bias, or Other and can trace which indicators are identified within each article in order to achieve our goal of explainable AI. Under each category, there are various subcategories which outline even more clearly the precise flaw that is identified in the text. For example, we won't simply classify an article as biased but rather by the type of bias exhibited: either statistical, cognitive, media-based, contextual, prejudicial, or conflict of interest bias. As shown in the diagram (right), we further refine each of these subcategories; contextual bias alone contains 16 sub-classifications. Our AI will quantify the level of each sub-classification contained within an article to inform its overall conclusion as to the article's veracity.

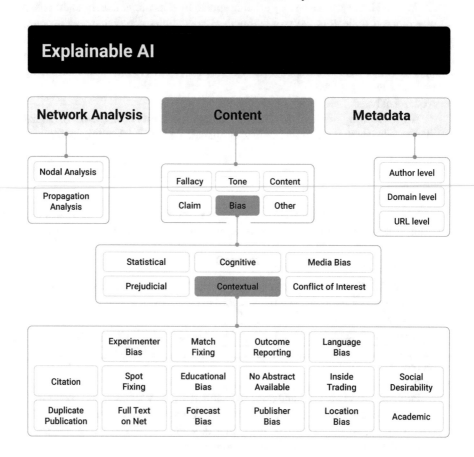

3.2 Credibility Assessment Methodology During Indian Elections

Logically deployed its credibility assessment capabilities for 2 months during the elections, and assessments for each URL were logged both 1 min and 1 h after its publication. The automated assessments were validated by expert annotation and fact-checking teams. Over 100,000 websites from around the world were monitored for misinformation and other forms of problematic content during the month of April. Logically's credibility assessment capabilities were deployed across all content from these websites related to and/or originating within India and made up 3836 websites; 174 of these were from mainstream media sources and 3662 were from fringe outlets.

Findings During Indian Elections

From the period of 1 to 30 April 2019, Logically analysed 944,486 English articles.

We found that 33,897 articles were completely fake. However, the study also highlighted that 85% of English news being reported in the Indian media didn't contain any factual inaccuracies. Perhaps more significantly, the vast majority of the "fake" articles were from fringe websites and not mainstream media.

This might mean Indian media has been vindicated of charges of "fake news" and paid media; however, if the problem of fake news is not taken seriously, then the incremental deterioration of trust in the fourth pillar of democracy will pose a serious threat to publishers and India's netizens. Furthermore, whilst mainstream publishers tended to be factually accurate, the political biases in coverage were evident and in the cases of a handful of publishers were extreme.

The study pointed that during the same period, fake news pieces were shared more than 1 lakh times, hateful articles were shared more than 3 lakh times, and 15 lakh shares were connected to extremely biased stories likely reflecting the sharer's personal opinions on topics. As a result, readers could be entering filter bubbles and echo chambers on their own.

Hateful and junk science articles were shared at significantly higher rates on Facebook (20 and 34 shares per article), and "fake" articles were shared at significantly lower rates (3 per article) suggesting that the platforms' efforts at downranking misinformation and already fact-checked content have been very successful. This further highlights the unique traits of mis/disinformation in India – the most significant platforms for problematic content have been closed networks such as WhatApp, private Facebook groups – more on this in the Veracity section (Figs. 1 and 2).

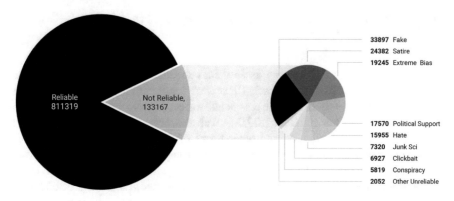

Fig. 1 No. of articles published in April by class

Fig. 2 Shares on Facebook per article by content type

Credibility Assessment: Evaluation

We constantly evaluate and iteratively improve our models, which are tested according to the yield they attain. We benchmark automation performance against that of our expert annotation team, producing an F score which reflects both the recall and precision of our models. We've been delighted with the performance of our models in each of the three areas of analysis we conduct:

- Content analysis: 92%
- Distribution analysis: 93%
- Metadata analysis: 88%

When grouped these deliver a level of performance significantly beyond the current state-of-the-art approaches – over 95%.

Fig. 3 Life cycle of a claim

4 Veracity Assessment

4.1 Methodology: The Life Cycle of a Claim

Logically boasts the world's largest team of dedicated fact-checkers, supported by in-house journalists, innovative technology, and efficient, streamlined processes designed to safeguard the integrity of our fact-checks whilst maximising their efficiency. Whilst we aspire to a fully automated fact-checking solution, we have developed a hybridised process which supports the development of our fact-checking algorithms, enables the incremental adoption of our automated fact-checking technology as it matures, and enables efficient and high-quality fact-checking in the meantime (Fig. 3).

1. Incoming Claim

    Users can submit suspect claims for fact-checking that they encounter within the Logically app or from third-party publishers by sharing the article with Logically or pasting the URL into the app. Once we have the article, our claim detection technology gets to work by extracting the factual claims within, enabling the user to select the one they'd like to be verified and submit it to our fact-checkers. Users can also enter raw text or paste a message/post from another platform.

    Claims then appear on the bespoke dashboard developed to help our team progress claims through our process efficiently, maximising our chances of preventing false claims from spreading.

2. Automated Fact-Check

    Incoming claims are first checked against our single source of truth databases through a vectorised analysis of the incoming claim against related claims. This represents an initial processing stage of a larger scalable automated fact-checking solution which will become more effective as our universe of known and inferred "facts" is expanded through human fact-checking and sourcing from single sources of truth databases such as government datasets.

    Additionally, users are able to determine if images are manipulated via the Logically app. The analysis of the image is returned to the user within seconds with colour-coded highlights on all areas the models believe likely to be manipulated. Users can forward the same image for fact-checking if they would like the content of the image itself verified.

3. Random Sample Verification

Logically conducts random verifications of claims fact-checked through our automation capabilities to ensure that any errors are detected and used to develop the algorithms which conduct the fact-check.

4. Human Fact-Check

All claims sent to our human fact-checking team are subject to the following processes:

- *Selection and Prioritisation*

  We classify all incoming claims according to our taxonomy of claim types. This helps us quickly identify unverifiable claims which don't require human attention and prioritise claims according to their relevance, virality, and estimated verification time.

- *Assignment*

  A moderator or supervisor will assign an incoming claim to one of the team based on the complexity of the claim and the level of research required for verification.

- *Research and Conclusion*

  Our fact-checkers are trained to follow carefully constructed processes according to the type of claim they're verifying, which can be either statistical or text-based. We provide full justifications of our judgement and links to relevant primary sources used to conduct the fact-check.

- *Moderation*

  All completed fact-checks must be signed off by a moderator before publication to ensure that the judgement is correct and the justification convincing and substantiated. Fact-checkers can escalate difficult claims to moderators and supervisors as needed, should they be unable to verify its accuracy.

5. Judgement

Finally, we publish our verdict, which will be either True, Partially True, False, Misleading, or Unverifiable. Logically's verdicts have been carefully designed to encourage sharing and maximise virality; the judgement and claim are featured on an eye-catching image designed to look great on social media platforms. Through these viral verdicts, we hope to maximise the impact of each fact-check and help ensure that any false claims are debunked as broadly as possible.

4.2 *Methodology During Indian Elections*

Logically's verification work during the 2019 Lok Sabha was conducted by our 27-person team of fact-checkers, moderators, and supervisors. The team worked full-time in order to check as many requests as possible, putting in countless hours of overtime to make sure that we could verify as much as possible.

In keeping with the mission of Logically, in helping users to access better quality information and engage in constructive, empathetic debates around current affairs, our fact-checking operation prioritises user requests for verification. This means that the majority of claims that we check have been submitted by individuals through the Logically app. Our fact-checking and editorial teams will also sometimes find claims that they'd like to check, if they feel that the issue is particularly important or lacks adequate coverage by other organisations.

Once the claim has been researched and our team reaches a verdict, a Safe to Share image is created. This has been designed to include an image, easy-to-read verdict, quick summary of the fact-check, and then a longer explanation of how it was conducted. This format was inspired by the success of Verificado, a pop-up fact-checking operation during the Mexican general election in 2018, which adopted innovative storytelling techniques in order to optimise its work for social media audiences. Our own Safe to Share images have been developed specifically with WhatsApp in mind, reflecting the critical role that the platform plays within the Indian information ecosystem and its recent struggles with "fake news".

Findings During Indian Elections

Our team of fact-checkers has been trained using industry-leading standards and methodology from organisations including Google, Poynter, First Draft News, and the International Fact-Checking Network.

With turnaround time a key performance metric, given the fast-paced nature of the news cycles, our 27-person team of fact-checkers have continued to improve in efficiency since the commencement of the project in January 2019. After an initial month of training, the team to date have completed approximately 6000 fact-checks on everything from current affairs to sports and celebrities, defence, economics, politics, and science. Each member of our team has gone from producing approximately one final fact-check a day to producing six. Complete oversight from our moderators, as well as a rigorous system of checks and balances and emphasis on excellent sourcing of evidence, ensures that we produce high-quality fact-checks as quickly as possible, to stop misinformation before it spreads (Fig. 4).

During the initial months before the election, the improvements in productivity from updated dashboards and tools take time to materialise – following a typical learning curve that appears to stagnate at the 120 per day level once the election is in full swing. This level of output suggests maximum feasible productivity given the current state of supporting technologies has been achieved (Fig. 5).

During the initial weeks of deployment, the automation capabilities answered just 20% of incoming claims; however, they significantly improved over time, and in the month of May once a substantial database of facts was established and a few optimisations were made to the automation systems, they dealt with 40%–70% of all incoming requests (Fig. 6).

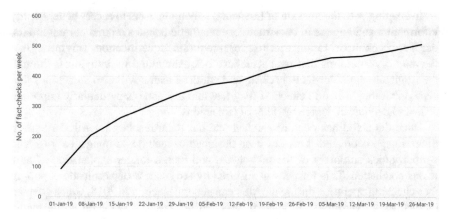

Fig. 4 Weekly fact-checking output – pre-election

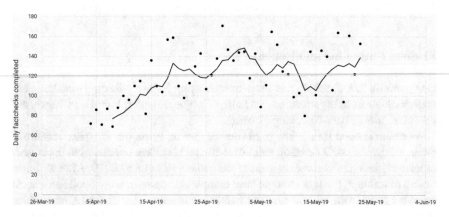

Fig. 5 Daily human fact-checks during elections

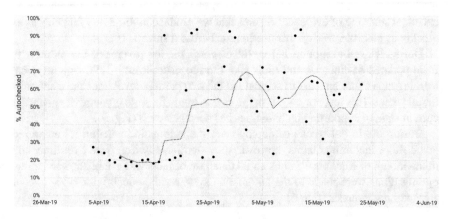

Fig. 6 Percentage of user requests responded to with autochecks

Perhaps more importantly, the automation systems conducted 90% of fact-checks on high traffic days. These occurrences highlighted by the chart occur on days where over 500 geographically and topically concentrated requests were received.

Evaluation

The objectives of our fact-checking operation in India were ambitious, with a very short turnaround time between implementation and the commencement of the 2019 Lok Sabha elections. A user-centric model of verification, in which we prioritised user requests for clarification or fact-checks, meant that our team was reliant on the Logically app at the same time as it was being launched.

At particular points during the project, these factors acted as temporary constraints on our fact-checking output. A tight timeline posed some challenges as we assembled and trained the team, meaning that there was less room for error and more pressure to learn and adapt to a challenging role quickly. Technical issues with the app in its beta phase also imposed some structural limitations upon the team's operations.

However given the demanding nature of the project, the team rose quickly to these challenges, rapidly adjusting to political developments in India and incorporating these insights into their work. The result was an acceleration in fact-checking output to approximately 600 fact-checks per month in the first 3 months.

As we begin to expand our fact-checking operations to include claims from the UK, this rate continues to improve, and our team of fact-checkers, with the support of complete editorial oversight from Logically London, is quickly developing a thorough understanding of the UK political and information environment. We hope that these skills can be implemented effectively in time for the 2020 US presidential elections, as well as any impromptu elections in the UK.

5   WhatsApp Solution for a Sharing Nation

Logically's solution to the problem of misinformation in India seeks to harness the rapid digitalisation of India's information ecosystem; India has become a sharing nation, an interconnected web of information disseminators pursuing individual goals ranging from humour to ideological conversion. These conditions have greatly exacerbated the problem of misinformation, and whether we like it or not, they're the conditions in which we must find a solution to that same problem.

India has undergone something of a digital revolution in recent years, with consumer technology growing at a remarkable rate. Smartphone ownership has risen from 33% to 40% since 2017,[13,14] reflecting a broader and even more significant explosion in remote data usage. In 2017, 108 petabytes of data was transmitted online. This figure is expected to grow by 490% by 2022, with the nation consuming 646 petabytes.[15]

Unsurprisingly, social media companies have benefited from the nation's growing connectivity, with the nation's active social media users growing by 24% from 2018 to 2019, reflecting a consistent trend which is expected to continue in the coming years.[16] Such growth in misinformation-friendly platforms has had inevitable consequences on the information ecosystem in India. WhatsApp groups have emerged around ideologically homogeneous groups dedicated to sharing information which supports their worldview. These groups, often with hundreds of members, become echo chambers congenial to the spread of misinformation: an ideologically blinkered network of potential propagators ready to embrace content supporting their common worldview.

The role of social media in spreading misinformation is well established; the engagement-oriented content discovery algorithms incentivise sensationalism, over-simplification, and gossip rather than detailed and balanced journalism. This was prevalent in the Indian election, which featured a proportion of polarising political news topped only by the US election in 2016. More than a quarter of the content shared by the election's victorious party, the Bharatiya Janata Party (BJP), was "junk news" of some form.[17]

The combined efforts of the academic, technology, and media industries have yet to find a solution to misinformation capable of functioning in this environment. Attempts have thus far had one thing in common though: they fought, rather than embraced, the nation's orientation towards shareable online content. Whether encouraging user verification and responsible dissemination of information or imposing those responsibilities on the platforms themselves, solutions have always been reactive, seeking to counteract the virality phenomenon.

Logically has flipped this approach on its head, creating a simple yet innovative solution to misinformation which, based on early data collected during the 2019 Indian election, is highly effective. Our role in combating misinformation is engineered to harness the same benefits of the modern information ecosystem which enables misinformation's spread in the first place, achieving high virality and engagement through the format and strategic publishing of our fact-checking output. We produce shareable graphics featuring a headline image, attention-grabbing

[13]"Digital 2019: India", We are Social and Hootsuite. 31st January 2019.

[14]"Digital 2017: India", We are Social and Hootsuite. 1st February 2017.

[15]https://qz.com/india/1483368/indias-smartphone-internet-usage-will-surge-by-2022-cisco-says/

[16]"Digital 2019: India", We are Social and Hootsuite. 31st January 2019.

[17]Vidya Narayanan, Bence Kollanyi, Ruchi Hajela, Ankita Barthwal, Nahema Marchal, Phillip N. Howard; "News and Information over Facebook and WhatsApp during the Indian Election Campaign", Data Memo 2019.2, Oxford, UK: Project on Computational Propaganda.

judgement, and a concise justification, providing all the details required by the IFCN's fact-checking regulations.

We've also entered into strategic partnerships which will maximise the exposure and virality of our fact-checks, extending their reach beyond in-app verification for Logically users. Deployment on third-party apps (most notably in the context of India, WhatsApp) provides a means of accessing Logically's verification capabilities through immensely popular platforms and will enable us to upscale our fact-checking operations and underpin an even more ambitious solution to misinformation in India. Moreover, we'll be able to access the ideologically affiliated groups which play such an important role in the national discourse, debunking claims and stifling the radicalisation that such polarised and homogenous communities could foster.

Logically is committed to fighting the issue of misinformation in India going forward, and data collected around the 2019 election supports its theory that India is and will remain a sharing nation, and organisations dedicated to addressing the growing misinformation crisis must find solutions within this environment.

5.1 Long-Standing Questions

As suggested previously, WhatsApp is likely to be the most significant avenue by which misinformation is circulated in India – significant not just because of the frequency in which it is used but also because of its encrypted nature, also previously touched upon. These two features of WhatsApp usage in India also pose two distinct yet closely linked challenges:

1. How can we get access to the kind of information being shared in order to spot problematic content, whilst respecting the privacy of users and without compromising encryption?
2. How can we find a way of disseminating credible information, verifications, and debunks back to the same audiences who are receiving this content in the first place?

5.2 Related Work

Previous work has been done in attempts to address misinformation during elections, specifically with messaging platforms such as WhatsApp. Arguably the most successful of these projects took place during the elections in Mexico in 2017, which turned into an unprecedented fact-checking collaboration. In all, the collaboration involved over 100 journalists representing about 60 different media partners. The project named Verificado – Spanish for "Verify" – looked into ways that they could most effectively spread fact-checks and debunks online and how they could ensure that this content reached those most affected by misinformation.[18]

One aspect of Verificado's project was to set up a public WhatsApp account to which members of the public could send requests for the verification of individual pieces of content. Once received by the Verificado team, the content was verified and graphics were made up to illustrate the team's verdict of that claim or piece of content, along with the official Verificado logo and a timestamp. This was then published on various platforms and sent to other organisations involved in the collaboration who in turn published it on their own platforms.

By the end of the election in Mexico, the Verificado brand had amassed a widespread following among voters. The official website had received over 5 million views, and the team had published around 400 fact-checking notes and around 50 videos. These videos were also posted on numerous platforms and websites, some receiving more than a million views each. Most importantly, when we consider transferring similar methods of combating misinformation to an Indian context, the WhatsApp group had garnered more than 10,000 contacts.

5.3 Exposing Misinformation on Closed Networks

Individuals on closed networks such as WhatsApp are exposed to disinformation as a result of organised propaganda campaigns and inaccurate viral stories. As these networks tend to be encrypted, a way to obtain access to content whilst respecting platform policy and user privacy would be to encourage whistleblowing on these networks. By encouraging the individuals themselves to share content they have received, fact-checkers and other organisations gain instant access to problematic content which would otherwise go unchecked. The question therefore is: How do we motivate this sort of whistleblowing? (Fig. 7).

There are a number of obvious motivations for individuals to share content from their closed network. Concern for their communities and networks, sparked by seeing content that they are suspicious of, is probably most likely to instigate a whistleblowing response – either this or out of a sense of activism spurred on by seeing negative information targeted at.

These motivations are also broad enough within news consuming audiences as the primary psychologically gratifying motivations for news consumption are information foraging, socialising, and status-seeking (Fig. 8).

By establishing a network of motivated and active "forwarders" who are consistently exposed to mis/disinformation, we can gain access to highly shared questionable content. In the 50 days between 4 April 2019 and 23 May 2019, 11,560

---

[18]https://verificado.mx/

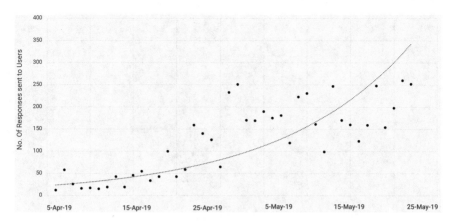

Fig. 7 User requests answered – without key dates

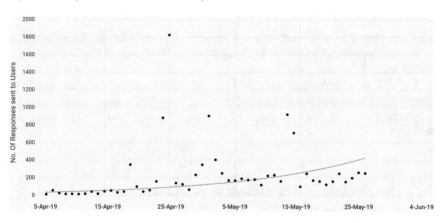

Fig. 8 User requests answered – including key dates

unique user requests were received. The level of activity during non-key dates (key dates are days of and the day before an election stage) shows an exponential rise over the 50-day period implying user satisfaction and improving operational capacity.

By looking at the key dates, it's clear that these are days of heightened traffic with 1823 requests answered on 23 April.

Users are more likely to expose content if they are confident they will receive a response – an outright debunk, verification, or even a statement acknowledging the nuances and complexity of a particular message.

There are persona-wide differences in the sensitivity to response time. However, the general trend holds. If a fact-check is responded to any later than 30 min, over 50% of users are unlikely to expose questionable content.

The immediacy of response appears to help in exposing novel content. Only 793 of all received requests were related to third-party fact-checks indicating that this methodology uncovers new information previously hidden from fact-checkers and

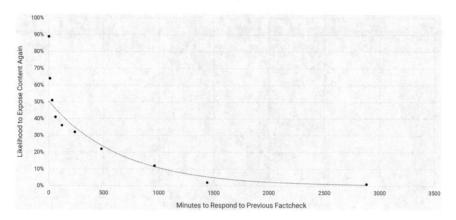

Fig. 9 Impact of response time on verification frequency

other activists. This phenomenon wasn't just to niche content but also some of the most popular claims. Only 9 of the 25 most popular requests we received were also fact-checked by a third party (Fig. 9).

~~We will be continuing with this body of work focusing on more precisely~~ quantifying the critical mass of active forwarders required in our network to expose a statistically significant percentage of all malicious and false content spread on WhatsApp.

5.4   Disseminating Verifications to Audiences Exposed to Mis/Disinformation

The response to each request we received was in the form of a sharable image, similar to the concept used by Verificado in Mexico.

Out of the 11,560 user requests answered, 3690 were shared with WhatsApp users. The sharing rate was particularly high during key election dates (Fig. 10).

Each response was accompanied by a unique trackable link, and unique visits to this URL were logged. The chart below shows the number of unique visits per shared response (Fig. 11).

On average, the link accompanying the responses received eight visits. However, the sharable image is designed to be self-sufficient, and most viewers are unlikely to click on the link. The click-through rate on such messages is expected to be between 10% and 30% (Source: Gupshup). Using this assumption, we can project the following reach for our responses.

The share-worthiness of response appears to be highly sensitive to the response time. Automated fact-checks (less than 2 min) receive 11 clicks during their lifetime

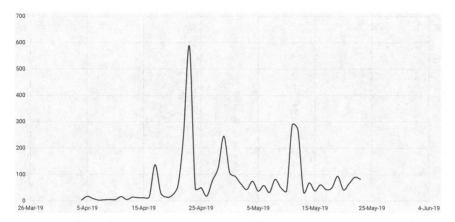

Fig. 10 Number of responses shared on WhatsApp per day

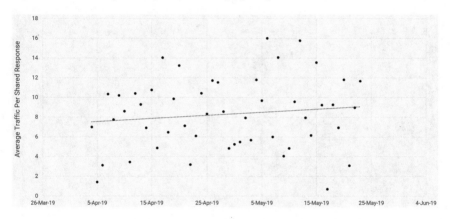

Fig. 11 Unique clicks per shared response fact-checks and verifications

but are outperformed by human fact-checks – these receive 14 clicks during their lifetime (Figs. 12 and 13).

The additional benefit of quick response times would be to slow down or altogether prevent forwarding chain reactions. Based on user surveys, users are almost certain to share verifications of a claim if they discovered it in a group. Based on the link tracking data, our earlier hypothesis that a verification can be forwarded repeatedly has been validated. Furthermore, in a quick turnaround scenario, fact-checks would be delivered to multiple groups that have only recently received the initial problematic message – thereby acting as a deterrent to anyone considering forwarding the original message. There is some evidence to suggest that individuals within these groups share the fact-checks more broadly; however, the precise individual and group dynamics resulting from a user sharing a fact-check in their conversation remain unclear.

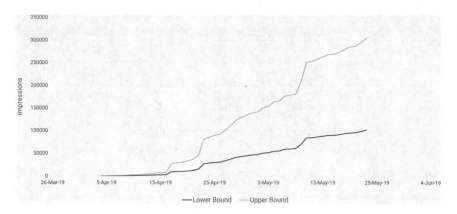

Fig. 12 Est. cumulative impressions on WhatsApp

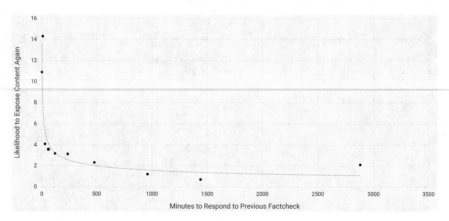

Fig. 13 Impact of response time on expected lifetime impressions

The average active user of Logically's verification services was able to spread verifications of content to 10–30 WhatsApp users. By scaling these shallow networks nationally to 1–3 million engaged whistleblower/shares, this approach may be able to reach all users affected by mis/disinformation on WhatsApp in India. This approach need not be limited to WhatsApp and in principle can apply to other closed networks such as Facebook Groups, Messenger, and Telegram.

An additional hypothesis that can be proposed on the basis of some of these findings is that repeated exposure to quick verifications of content from a single source would build resilience within that group and the routes travelled by both the original content and its corresponding verification. In addition to this, we will be continuing with this body of work focusing on gamification strategies that can be leveraged to help incentivise greater sharing rates, and quantifying the critical mass of active forwarders required in our network to disseminate verifications and debunks to a statistically significant portion of audiences initially exposed to the original messages.

STS, Data Science, and Fake News: Questions and Challenges

Renny Thomas

Abstract Fake news has become the norm in our times. With the coming of social media platforms, where anyone can write/post on any issues without any regulations, sometimes based on what they read from various platforms and sometimes based on what they are asked to believe in by political ideologies, fake news has become a phenomenon that requires serious academic investigation as it has dangerous consequences in society. This chapter attempts to argue for a possible and productive conversation between science and technology studies (STS) and data science to talk about the politics of fake news production. It argues that STS can work as a close ally of data science to bring in questions of power and politics associated with fake news, and its methods can be used in data science to make it more socially relevant.

Keywords STS · Truth · Power · Knowledge · Objectivity

In addition to describing, representing, or theorizing data science, STS researchers have an opportunity to shape its rollout, whether informing design, doing it, or something else. And since the activities of data science will undoubtedly continue to overflow any technical definition, STS too will be entangled with the rollout of data science and its consequences.—David Ribes[1]

---

[1] David Ribes [1] "STS, Meet Data Science, Once Again", *Science, Technology, & Human Values* 44 (3), p.535.

R. Thomas
Department of Humanities and Social Sciences, Indian Institute of Science Education and Research (IISER), Bhopal, India

© Springer Nature Switzerland AG 2021
Deepak P et al., *Data Science for Fake News*, The Information
Retrieval Series 42, https://doi.org/10.1007/978-3-030-62696-9_14

1   Introduction

Fake news has become the norm in our times. With the coming of social media platforms, where anyone can write/post on any issues without any regulations, sometimes based on what they read from various platforms and sometimes based on what they are asked to believe in by political ideologies, fake news has become a phenomenon that requires serious academic investigation as it has dangerous consequences in society. Science and technology studies (STS) can help data science to deal with fake news using its methodological and conceptual tools.

STS as a discipline can play an important role because as an intellectual exercise it was meant to study and critically engage with notions of truth and power, especially the idea of a *singular truth*. STS can help data scientists to go beyond the binary of truth and post-truth and will allow scientists to think about many truths that exist.

Fake news is not free from power. On the one hand, data science enables us to deal with fake news using various methods and techniques, and on the other hand, STS can make sure that data surveillance and data exploitation are not happening by making data science more accountable.

STS as a discipline has always been critical of what is called "objective" scientific data and information, which comes out of laboratories, devoid of politics and power. Precisely because of this reason, STS is also equipped enough to deal with fake news. Before we criticize fake news and the various problems associated with it, we should remember that fake news and scientific information can't be seen as oppositional categories. This is not to say that both are the same. Here I argue, in order to engage with and criticize fake news, we also need to engage and criticize the very idea of singular truth and scientific information. Fake news, like scientific information, is also about power and ideology. This particular position makes STS the most eligible discipline to engage with fake news. Because if we are going to look for "scientific information" as an alternative to fake news, we also have to make sure that we question and challenge the power associated with science and scientific information.

2   Truth, Power, and Knowledge

Though, as argued earlier, as a discipline STS helps us in criticizing truth and power, it is important to argue that STS is also a discipline that challenges fundamentalism of all kinds, be it scientific or religious, scientific reductionism or relativism. As STS scholar Banu Subramaniam very aptly argued, "Given this history, it is not surprising that feminist studies, postcolonial studies, and STS have been at the forefront of not only critiquing the powerful institutions of science and religion and their perversions of truth, but have also been at the forefront of moving us away from the totalizing language of 'objective truth' and 'objectivity.' These are the fields that understand not only the dangers of truth but also the dangers of RELATIVISM – in both poles, the strictures of power conquer, colonize, and corrupt completely" [2, p. 212]. When science and scientific truth are controlled by scientists and the state, the stakeholders who manufacture fake news come from different parts of the world, and therefore the (mis)information that they create has no accountability, and that makes the situation more complicated. It is called fake news because it is difficult to trace the source of the news/information.

When we think of fake news in that sense, we need to go beyond the binary of fake–real and try to think about the possibility of finding fake in "real news" as well. That is to argue how "the real" science and "scientific information" also have aspects of unreal/non-real, as they are not beyond the control of the state and experts. Very often we see this when we think about the statistics of various events. Often we hear different versions of statistics from different stakeholders, be it of COVID or any other events. How do we make sure that a particular statistics given by a country/stakeholder is correct? Here we can use theories and methods of STS and ask the question: can we really believe the numbers provided by a state or a laboratory? This is where STS plays a role by being part of data science and provokes data science to rethink the binary of fake and real, truth and post-truth, allowing us to look for fake in real news as well, because the very history of science and scientific information has to do with power and domination. Using this formulation, we can study fake news and its source, foundations, politics, and power.

Increasingly, we see that right-wing governments across the world use fake news as a weapon and strategy; they spread news, create news about various issues from politics to knowledge, and present it as "truth," It is important to observe here that the consumers of fake news are not just "villagers" and the "uneducated" but also the educated ones, with technical degrees in engineering and sciences. Of course, one is not saying that the educated ones are already always matured enough to differentiate between fake news and truth. Ideologically and politically driven, they manufacture a truth that is convenient to them and use social media, especially Twitter, to spread that "information" as knowledge. The consumers of this particular truth conveniently do not think about the validity of the news. They just follow what is given to them as "truth."

Here with the help of STS, one can ask the nature of political power and fake news. We need to ask questions about fake news not just from authoritarian and autocratic political regimes. We should also be able to study the production of fake news in democratic regimes. We see that in many democratic countries, the media that criticizes the ruling government gets labelled as "fake news." Fake news here is a tool for the powerful to label the side that opposes them.

3   Truth Versus Post-truth

We very often hear that we live in a post-truth society and that there is nothing called real truth/reality. Though STS as an intellectual exercise helped us in talking about the power and authority of the so-called experts, and the politics of scientific expertise, STS also helps us in understanding how the very construction of numbers and statistics is not free from power, and hence fake news becomes a natural choice for STS scholars to engage with. STS as a discipline, for example, can help data scientists to understand the politics behind the denial of climate change and can be used to understand the larger global network of politicians, capitalists, and other interest groups who construct the narrative and spread the fake news that climate change is not a reality [3].

Data science is a field slowly coming in many universities, including India. It is important that these institutions include STS as part of their training and make STS scholars part of their exercise. With the help of other scientists, STS scholars can make data science more accountable and make fake news a matter of public discussion.

It is difficult to differentiate between fake news and scientific information in our times, as very often fake news is presented as "information" by the powerful regimes. STS scholarship can help data scientists to deal with questions that are not necessarily seen as part of "science": politics of knowledge production. Science and technology studies as a field can help data science to be more democratic, and, more importantly, it is necessary that the data science collective include STS scholars as part of the exercise as they can help methodologically and conceptually in dealing with the politics of fake news production. Therefore, asking questions of power and authority that STS did historically by studying the power of scientists and laboratories can be used in studying fake news as well.

One of the important sites where fake news spreads like a virus is Twitter, and also on many other social media, and WhatsApp. Data science can take the help of STS and anthropology methodologically. Doing an ethnography of fake news on Twitter or WhatsApp will help us understand the nature of fake news and the politics associated with fake news production and consumption. STS scholars had been doing ethnographies of lab and other sites of knowledge production. The same method can be used to study various sites of fake news production. A rich ethnography of fake news by studying Twitter or WhatsApp can inform us the political economy of fake news.

To conclude, as fake news increases, we need innovative ways to deal with it, and that's what data science is promising: to help us identify fake news and its producers. STS can work as a close ally of data science or as part of data science to bring in questions of power and politics associated with fake news, and its methods can be used in data science to make it more socially relevant. Data science clearly can't work like a laboratory science. It needs the public, and STS will help data science practitioners to engage with the public using qualitative methods. Studies demonstrate that data science and STS can have a very productive relationship

[1, 4], and there is no reason to believe that STS will not be a significant ally for data science. With the conceptual tool of STS, we can understand the larger network of fake news, different stakeholders, sites of fake news production, and the politics of these sites. As sociologist and STS scholar David Ribes, who worked extensively with data scientists,[2] argued, "In addition to describing, representing, or theorizing data science, STS researchers have an opportunity to shape its rollout, whether informing design, doing it, or something else. And since the activities of data science will undoubtedly continue to overflow any technical definition, STS too will be entangled with the rollout of data science and its consequences" [1, p. 535]. In future, there will be more conversation between STS and data science. It is inevitable.

References

1. Ribes, D.: STS, meet data science, once again. Sci. Technol. Human Values. **44**(3), 514–539 (2019)
2. Subramaniam, B.: Holy Science: The Biopolitics of Hindu Nationalism. Orient Blackswan, New Delhi (2019)
3. Kofman, A.: Bruno Latour: the post-truth philosopher, mounts a defense of science. The New York Times Magazine, October 28, 2018. Accessed 18 May 2020
4. Kolkman, D.: STS and data science: making a data scientist? EASST Rev. **35**(4) (2016)

---

[2]Trained in sociology and STS, Ribes currently works as an Associate Professor at the Department of Human Centered Design & Engineering, University of Washington, USA.

Linguistic Approaches to Fake News Detection

Jane Lugea

Abstract To date, there is no comprehensive linguistic description of fake news. This chapter surveys a range of fake news detection research, focusing specifically on that which adopts a linguistic approach as a whole or as part of an integrated approach. Areas where linguistics can support fake news characterisation and detection are identified, namely, in the adoption of more systematic data selection procedures as found in corpus linguistics, in the recognition of fake news as a probabilistic outcome in classification techniques, and in the proposal for integrating linguistics in hybrid approaches to fake news detection. Drawing on the research of linguist Douglas Biber, it is suggested that fake news detection might operate along dimensions of extracted linguistic features.

Keywords Linguistics · Fake news · Biber · Register · Deception · Detection

1 Introduction

We know that fake news is a problem. What we don't know is how to tackle it. This chapter discusses the role of linguistics in research on fake news detection to date and makes suggestions for its inclusion in future interdisciplinary studies.

Yet how can the academic community solve the problem of fake news detection if we cannot always detect it ourselves? Research shows that people are not very skilled at detecting when others are deceiving them through language [1]. Several scholars [2, 3] have commented that fake news detection is complicated by the fact that digital news is media rich, using images, video, audio, hyperlinks, and embedded content, as well as text. Parikh and Atrey [3] suggest that "readers" of digital news do not engage with the textual content to any great extent, citing a news story stating that "70% of Facebook users only read the headline of science stories

J. Lugea
School of Arts, English and Language, Queen's University Belfast, Belfast, Northern Ireland, UK
e-mail: j.lugea@qub.ac.uk

© Springer Nature Switzerland AG 2021
Deepak P et al., *Data Science for Fake News*, The Information
Retrieval Series 42, https://doi.org/10.1007/978-3-030-62696-9_15

before commenting" [4]. One might assume that this would force us to question the utility of linguistics at all in fake news detection, but think again. Ironically, their citation is a satirical news story, which, beyond the headline, contains only filler text. Its creators wanted to see how true their false claim was, and, indeed, the empty headline has been shared 192,000 times and, now it seems, in an academic publication.

This example illustrates several points about the usefulness of linguistics as part of an interdisciplinary approach to fake news detection. The remainder of this chapter is structured around these points. If there is one thing that the COVID-19 pandemic has taught us, it is that "the science" is only as good as the premise or assumptions on which it is based. Therefore, the automation of fake news detection depends on the quality of the input data and its classification system. After discussing definitions of fake news (Sect. 1.1), Sect. 1.2 describes how, as a discipline, linguistics is dedicated to carefully describing, labelling, categorising, and classifying linguistic data, beginning with a sketch of the sub-disciplines involved. Although linguists have yet to comprehensively describe the language of fake news, "true" news articles have specific features that are well accounted for in linguistic description (Sect. 1.3) and could serve as a springboard for fake news detection. Moreover, there is a substantial body of linguistic research on deception, outlined in Sect. 1.4. Section 1.5 outlines the significance of context in shaping texts and their features and in language-sensitive research. Linguistics and, arguably, the humanities and social sciences more widely have much to contribute in describing, characterising, and contextualising fake news.

Section 2 outlines linguistic approaches adopted in fake news detection research thus far, beginning with "bag of words" and Linguistic Inquiry and Word Count approaches; these methods are compared with the similar frequency-based corpus approach in linguistics (Sect. 2.1). Research in stylometry and computational linguistics has investigated readability and punctuation (Sect. 2.2) and deep syntax (Sect. 2.3). To a lesser, but no less significant, extent, computational linguists have investigated rhetorical structure and discourse-level features of fake news texts (Sect. 2.4). In the concluding remarks (Sect. 3), I elaborate on Shu et al.'s [5] proposed framework for the detection of fake news on social media, by suggesting where linguistics might support the data science approach to this problem in the future.

1.1 Defining Fake News

Defining fake news is not straightforward, but it is an important exercise because the definition adopted shapes the characterisation of the phenomenon, the selection or design of the training dataset, and, ultimately, the detector's accuracy on novel data. In scholarly definitions of fake news, the common thread is news which is "intentionally deceptive" [5–10]. A broad definition of fake news includes satire, parody, propaganda, advertising [11], and hoaxes [8, 12].

Several studies surveyed here include satire in fake news [8, 10, 12]. In attempting to differentiate between satire and fake news, some suggest that satire does not intend to make readers believe in its truth value [10, 13]. However, ascertaining authorial intention is methodologically challenging, and some satirical news does deceive its readers (see Sect. 1.0). Satire is a substantially different genre [14] with, from a discursive perspective, ostensibly different authorial intentions and the communicative purpose of entertaining rather than misinforming [15].

Another challenge in defining (and identifying) fake news is its potential similarity with "real" news, which might be characterised as fact-based (although see Sect. 1.3). Guo and Vargo [10] acknowledge the difficulty in distinguishing between fake news and fact-based media, especially when the latter is partisan (e.g. Fox News). Fact-based news may also use a sensationalised style, especially in tabloid or magazine formats translated to online platforms. The content of fact-based news could also contain some misleading or false content, mixed in with truth [5]. Therefore, there is a grey area between true and false that needs to be considered [16], and some fake news detection research is laudable for its adoption of a graded notion of truthfulness [12].

The term "fake news" is further confounded by those, like former President Trump and some followers, who attribute it to credible news organisations in order to delegitimise their content. The proliferation of such false claims makes it difficult to base fake news detectors on user responses online. Because the content of fake news is often controversial and its style sensational, it can generate many comments and shares, fuelling its virality; false information spreads six times faster than truth on Twitter [17]. Its virality may, in turn, lend it credibility [16, 18] and even influence the agenda or tone of fact-based news [10, 19].

An approach to fake news that is sensitive to its language will necessarily need a narrow definition, as the linguistic characteristics of each of the text types discussed above vary. For such purposes, this chapter adopts the "narrow" definition of fake news as "a news article that is intentionally and verifiably false" [5], although the studies surveyed here adopt a range of definitions.

English is the most widely used language on the Internet [20], so it is no surprise that the vast majority of scholarship on fake news investigates English fake news texts. Therefore, the studies referred to in this chapter are investigations into fake news in English, except for one which examines fake news in Polish [9]. It must be clear, however, that there are very few linguistic universals, so what may characterise fake news in English may not hold true for fake news in other natural languages.

1.2 Linguistics, Sub-disciplines, and Methods

Because reference will be made to linguistic sub-disciplines and methods throughout this chapter, it is useful to sketch the field first. The discipline of linguistics is just as rich and varied as the phenomena it aims to describe; it encompasses the languages of the world, language variation and change, and patterns in sound or syntax and investigates how we create, understand, and reproduce meanings in cultural contexts. Language operates on several different levels which work – for the most part – simultaneously in use:

(i) *Phonetics and phonology:* deals with patterns in speech sounds
(ii) *Morphology and syntax:* deals with the structure of words, phrases, and sentences
(iii) *Lexico-semantics:* deals with the meaning of words and phrases
(iv) *Pragmatics:* deals with implied meaning, which depends on shared contextual or cultural knowledge
(v) *Discourse analysis:* considers texts as whole artefacts, as well as their content and their communicative contexts

These are the levels on which languages are used and studied, so they form the main sub-disciplines of linguistics as well. Whilst deception can be a feature of written or spoken language (see Sect. 1.4 and [21]), fake news is generally in written form and can therefore be studied using the sub-disciplines (ii–v). Section 2 outlines approaches to fake news detection which draw on these linguistic levels. However, the pragmatic level remains to be understood by machines, as communication at this level demands a wide range of knowledge, particular to specific contexts and cultures. For example, if someone in the same room remarks, "It is very cold in here", you may understand this as a request to close the window. Such indirectness requires shared knowledge of the context and a human to search for implications of the statement. The politeness strategy may be culturally specific and not shared by all humans. As a sub-discipline of linguistics, pragmatics offers frameworks for conceptualising the kinds of knowledge necessary for communication and how that knowledge is put to use.

In their research, linguists can examine a range of linguistic levels (i–v) at once or adopt an approach not simply defined by the levels of language but by the kinds of texts and contexts they are interested in (e.g. see the remainder of Sect. 1). Linguistics describes social behaviour and interaction, as well as considers how humans create and interpret meanings, and so straddles the social sciences and humanities to varying degrees. Research in linguistics can be purely theoretical or may draw on data to advance its theories, adopting an empirical approach which may be qualitative or quantitative. The latter necessitates larger quantities of language data, which are called corpora (from the Latin word for "body") and are studied in corpus linguistics. These datasets of natural language are compiled with careful attention to sampling methods [22], making them particularly useful for language-sensitive research. Stylometrics is another sub-discipline which uses computational and quantitative approaches to identify linguistic markers of style, leading to, for example, authorship attribution or forensic insights. Computational linguistics aims to understand language from a computational perspective, using scientific and engineering approaches to create ways of processing and producing language [23]. The remainder of this chapter refers to fake news detection research from all of these relevant fields.

1.3   News in Linguistics

The underlying linguistic characteristics of fake news have not yet been fully understood [5], yet there is a wealth of linguistic scholarship on news discourse more generally. As mentioned in Sect. 1.1, "discourse" refers to the study of entire texts in their contexts but also to the use of language as a social practice, enacting ideologies and representing the world. If we want to characterise the language of fake news, we might first consider what is the language of "news", what kinds of texts count as "news", and how they are linguistically constructed.

Critical linguistics and (critical) discourse analysis are two sub-disciplines which have advanced a great deal of scholarship on news discourse. Print news discourse is formulaic in structure (headline, lead, main events, etc.) [24, 25] and traditionally places what the journalist considers to be the most "newsworthy" information at the top of the piece [26]. The media's selection of events that seem to be newsworthy content is subjective but follows patterns that have been identified as "news values", such as negative and timely events [27]. Once events are selected for representation in news and mediated through language, they are no longer objective facts but linguistic representations, which can vary depending on the perspective adopted by the writer. Linguistics studies the choices made by reporters, such as naming strategies (e.g. freedom fighter or terrorist?) and how syntax can be used to, for instance, attribute agency and cast participants in the roles of perpetrators or victims. The linguistic choices can be related to ideological or rhetorical aims of the news provider. Consider the difference between broadsheet journalism and the red-top tabloid press; sociolinguistic research has shown that they use language differently [26, 28] due to differences in "audience design" [26]. The careful analysis of news language has been extended to other semiotic modes, such as the visual, in multimodal discourse analysis of news, and online news [29, 30], scholarship on which may prove relevant for developing visual-based fake news detection methods [29, 30].

The mainstream press has translated some generic practices to their online news platforms. However, news is "mediated" by the communicative form (e.g. newspaper, television, online), meaning it is shaped by the constraints and affordances of the particular medium. For example, although online headlines can emulate the sensational headlines of tabloid print journalism through "clickbait", they may adopt more informative language to consider the demands of search engines. The wording of a headline, like other elements of news language, can be shaped by ideological perspective as well as communicative goals and context, a point elaborated in Sect. 1.4. With most people now consuming their news online, linguists must keep up with the ways news discourse is evolving in the new settings. Because of the proliferation of user-generated content, news is no longer the product of traditional news institutions only. Studies have shown that fake news has the power to influence fact-based news, including the issues reported in partisan and emerging media [19] and the emotional tone of reporting about Trump during the 2016 US election [10]. Therefore, creating a distinction between the language of fake news and "real" news is not straightforward, and a continuum or probabilistic method may be more suitable than one that is binary.

1.4 Deception in Linguistics

Online deception is usually targeted at individuals, whereas fake news is disseminated on a greater scale through social media [31]. Despite these differences, in the absence of a comprehensive linguistic description of fake news, it is worth referring to the substantial body of research on the language of deception more generally. Some of this research deals with spoken lies, but the literature on deceptive written texts has more relevance to fake news. The basic premise of deception research is that certain behaviours, known as "leakage cues" [7, 32, 33], can be correlated with deception, despite – or perhaps because of – the author's attempt at manipulation. Research demonstrates that there are differences in the ways that truth-tellers and deceivers use language [34], not simply at the level of content but at the level of style [35]. In other words, deception can be identified not by *what* is said but *how*.

Research has been carried out into the automatic detection of deception in online reviews [33, 34], online advertising [36], online dating [37], and crowdfunding platforms [38]. Several studies indicate that negative sentiment is more frequently found in language used to deceive [35, 39], a feature that appears to also apply to fake news [9, 13]. However, rather than representing a "leakage cue" of deception generally, this may be due to deceivers exaggerating the sentiment they want to convey, as in fake negative online reviews [33]. Pinning down the precise linguistic characteristics of deceptive texts is difficult, because it varies depending on the nature of deception carried out in the data, as well as other textual and contextual variables. For example, Newman et al. [36] find that when people lie about their personal opinions, they use fewer first- and second-person pronouns ("I", "you"), while Rashkin et al. [12] discover that these personal pronouns are more common in fake news. Topic-specific studies can have limited generalisability in the detection of deception and, as argued in the next section, in the detection of fake news [7].

1.5 Different Texts and Contexts

The text is the basic unit of analysis in linguistics. This section considers texts of different lengths and varieties and the significance of their context in shaping their content.

In the fields of pragmatics, truth-conditional semantics, and the philosophy of language, individual utterances have long been studied for their truth value, resulting in a rich body of theoretical literature [40] which has not yet been operationalised in fake news detection (although see [41]). This linguistic approach to truth at the level of the sentence would be applicable to fake news in the form of tweets, although automating the detection might prove difficult when so much of what happens at the semantic and pragmatic levels is detectable by human knowledge (Sect. 1.2). In the practice of fake news detection, expert journalists on the online service PolitiFact evaluate the truthfulness of individual statements from (US-based) public officials, using a gradable scale from "true" to "pants-on-fire-false". Yet manual fact-checking by experts is time-consuming and intellectually demanding, even when the text is brief [5]. It remains to be seen how we can approach the truth value of texts beyond sentence length, never mind many texts in a dataset. The advantage of analysing fake news in texts of longer length is the presence of more textual features which might serve as "leakage cues" (Sect. 1.4).

In a language-sensitive approach to deception or fake news detection, it is essential to consider the topic and domain of the texts in the dataset. This is because "almost any kind of text has its own characteristic linguistic features" and these depend on its "register", "genre", and "style" – key terms in linguistics, described in a seminal monograph [42]. "Register" refers to the communicative purposes and situational context of a text, although it is often over-simplified to refer to formal and informal situational contexts. A celebrity gossip blog, which aims to entertain as well as inform, will use linguistic features that serve these particular communicative functions. On the other hand, news articles found on the website of a traditional print-based broadsheet newspaper (e.g. *The Guardian*) use very different features because this is a different register. "Genre" refers to how a text's structure is shaped according to the variety of text it is, and the interplay between domain and structure. Structural constraints result in linguistic differences between online genres, such as tweets and websites. Linguistic "style" arises as a result of patterned choices that language users make, which can be idiosyncratic (my conversation style may differ from yours), rhetorical (I may have affective or persuasive aims), or aesthetic (artful use of language). Research in stylometry adopts computational methods to characterise styles (e.g. of authors, or deception) and better detect them in novel data.

From surveying the literature on fake news detection, it appears that some of the most successful studies make good attempts at controlling the data for topic and/or domain (although see Castelo et al. [43] who avoid the problem of topic variation in fake news). However, many studies fail to consider register differences in the training and/or the test data, which may have adverse consequences on accuracy rates. For example, an influential study examined the difference between real and deceptive news stories [8], but their dataset comprised news stories read aloud on an entertainment radio show, from which listeners had to guess the fake. They quite rightly assumed that, by having the same expert writers, the real and fake news stories in their dataset would be comparable, but with hindsight they realised that news stories written for entertainment purposes will be qualitatively different in terms of their stylistic and pragmatic features. Another study uses two distinct datasets to train their fake news detector [6], the first comprising online fake news articles from Kaggle [44] and the second made up of "real" news from online versions of *The New York Times* and *The Guardian*. Their research is sensitive to differences in topic but overlooks the fact that the sources of their "real" news are high-brow "serious" broadsheet journalism, which adopts different linguistic features from the sources of fake news (Sect. 1.3). Therefore, their detector may well be leveraging deep neural networks in fake versus "broadsheet" style journalism and may not be very accurate when faced with real news that is more tabloid in linguistic style (Sect. 1.3). These examples highlight the importance of considering the register in selecting data for fake news studies.

A study by Pérez-Rosas et al. makes laudable efforts to create training datasets with fake and real news from a wide variety of domains (sport, business, entertainment, politics, technology, education, and celebrity news) [33]. Whilst the majority of their "real" news comes from online news sources, "fake" versions of real news are crowd-sourced through Amazon Turk for their fake news dataset. However, the Amazon Turk workers produced the fake "fake news" in very different contexts, for different purposes, and with their own styles. From a linguistic perspective, crowd-sourcing data may replicate the genre "online news", but it ignores the vital role of register and style in shaping textual features.

2 Linguistic Approaches to Fake News Detection

Language-based approaches are adopted in 34% of fake news detection studies [16]. This section summarises some of this scholarship, organised according to the kind of language approach adopted (although some studies use several simultaneously).

2.1 Bag of Words and LIWC

The "bag-of-words" (BoW) method of representing texts "regards each word as a single, equally significant unit [whereby] individual words or 'n-grams' (multiword) frequencies are aggregated and analysed to reveal cues of deception" [7]. These authors go on to explain how n-grams can then be further analysed by tagging them according to their grammatical or lexico-semantic function and that, despite the simplicity of the method, it overlooks contextual or ambiguous meanings (see also [43]). Arising from the BoW method of data representation is the use of word/category frequencies in pre-coded examples to train "classifiers" to predict instances of future deception.

BoW is one of the analytical methods Dey et al. [13] use on two datasets they constructed and classified using K-nearest neighbours (KNN): one of credible articles and another of malicious articles. Their dataset of malicious articles includes Donald Trump's tweets about his rival for presidency, Hillary Clinton; using BoW they identify an abundance of superlatives (e.g. "big") and abusive terms (e.g. "crooked", "phony") in Trump's tweets about Clinton, concluding that his tweets are therefore "polarised" and "subjective" in contrast with "legitimate" tweets. The style and veracity of Trump's tweets is a topic deserving of a literature review all of its own [44–49], albeit outside the scope of a computer science paper. However, to include Trump's tweets in a dataset of "malicious articles" compromises the research design in two ways. First, tweets are a different genre to articles (Sect. 1.5), so the dataset is made up of incomparable data. Second, to select tweets in which Trump attacks a rival for the malicious dataset and then interpret the language used therein as malicious means that the research is circular.

Rashkin et al. [12] examined the "lexical resources" in a dataset of trusted and fake news articles drawn from various online news platforms. This included the Linguistic Inquiry and Word Count (LIWC), a sentiment lexicon which searches for words that indicate strong or weak subjectivity, and hedging, which indicates uncertainty or obfuscation. They compiled their own list of terms that indicate dramatisation (e.g. comparatives, superlatives, action adverbs, modal adverbs, and manner adverbs). They found that words of exaggeration were used more in fake news, whereas words used to offer concrete figures – comparatives, money, and numbers – appear more often in truthful news. They found that trusted news uses more reporting verbs (e.g. said, announced), indicating more frequent citation of sources, which contrasts with Dalecki et al.'s findings [51], a difference which may derive from the different data; Dalecki et al. examined print news fabricated by journalists cutting professional corners, whereas Rashkin et al.'s study is focused on fake news online. The model created by Rashkin et al.'s study had a success rate of 65%, which, although better than chance, may have been improved if their datasets were more carefully compiled and comparable, an issue also present in other studies [6, 9, 13, 31].

Despite the methodological issues with Pérez Rosas et al.'s datasets (Sect. 1.4), their use of the LIWC tool reveals interesting findings [33]. They found that legitimate content often includes words describing cognitive processes (such as insight and differentiation), as well as more function words (e.g. pronouns "he", "she"), negations, and expressions of relativity. These features might hint at more carefully considered content in "real" news. Conversely, fake news uses more social and positive words, expresses more certainty, and focuses on present and future actions. Such features might be related to over-statement and a rhetorical attempt at persuading rather than informing the reader. Moreover, the authors of fake news use more adverbs, verbs, and punctuation characters than the authors of legitimate news. Their domain-sensitive research also reveals differences between legitimate and fake content in celebrity news. Specifically, "legitimate" celebrity news seems to use more first-person pronouns, references to time, and positive emotion words, which correlates with previous work on deception detection [52]. On the other hand, fake content in this domain has a predominant use of second-person pronouns ("he", "she") and negative emotion words and focuses on the present. Note that a higher frequency of second-person pronouns is associated with fake celebrity news but with "real" news in other domains. This demonstrates the significance of news domain as a variable in characterising and detecting fake news.

The BoW method shares much in common with corpus linguistics (Sect. 1.2), including the way in which data can be preprocessed, for example, by tagging parts of speech. However, corpus linguistics has a long tradition of careful and systematic corpus design which, in the spirit of social science research, aims for representative and balanced sampling [22, 51–54]. Corpus research often begins quantitatively, and once statistical patterns in the data are identified, empirically driven decisions about which data merits further qualitative analysis can be made, addressing the over-simplicity raised by Conroy et al. [7]. Moreover, a corpus can be contrasted with a comparable reference corpus to produce a list of "keywords" (i.e. words or phrases that are statistically over- or under-used), helping to reveal what is lexically characteristic of a dataset. The issues raised in this section could be addressed by adopting more careful corpus design methods, lending improvements to the data selection procedures, the data analysis, and therefore the success of computational fake news detection. As Shu et al. note: "A promising direction is to create a comprehensive and large-scale fake news benchmark dataset, which can be used by researchers to facilitate further research in this area" [5]. I propose that corpus linguistic design methods are invaluable in achieving this aim.

2.2 Readability and Punctuation

Features of the linguistic content can be extracted and measured for their "read-ability" according to pre-existing metrics, such as Flesch–Kincaid, Flesch Reading Ease, or Gunning Fog. Readability refers to the complexity of the textual structure and the relative ease or difficulty of reading the text. Measures include the number of characters, complex words, words per sentence or paragraph, etc.

Pérez Rosas et al. use readability and punctuation alongside the LIWC method for an integrated approach [33]. Another study used the Flesch Reading Ease Score measure of readability to analyse a dataset of newspaper articles, some of which were known to have been invented by journalists and others "true" articles from the same journalists [50]. Their findings showed that the deceptive stories were significantly more readable than the true ones (a finding corroborated in relation to Polish fake news [9]) and used more direct quotes. It is suggested that deceptive news represents a "simpler world" through more simple language and that direct quotes lend credence to the deception. The researchers overlook the fact that news stories are only ever a representation so may contain degrees of truth/deception [12]. Although their findings pertain to print journalism – and a particular kind of unprofessional deception – they may be investigated further in digital fake news. The number of direct quotes in the corpus of online news, for example, can easily be extracted using punctuation features or reporting verbs (e.g. "announced") and phrases (e.g. "according to").

2.3 Deep Syntax

Beyond the level of word use, punctuation, and readability measure, the language of fake news can be investigated at the syntactic level. Using context-free grammar (CFG), sentences in the data are transformed into a parse tree to describe syntax structure (e.g. noun and verb phrases), which are then rewritten by their syntactic constituent parts. The rewritten syntax can then be contrasted with known structure or patterns of lies, leading to differentiation between fake and true content. CFG-based features have been shown as useful in deception detection research [55] and are adopted in some fake news detection studies [31, 56]. Third-party tools (such as the Stanford parser) can be used in the automation. Syntax analysis alone may not be enough to detect fake news [7], but it can work well when integrated with linguistic or other detection methods [31].

2.4 Rhetorical Structure and Discourse Analysis

Whilst a range of fake news detection studies have examined the role of lexico-semantics and syntax, the discursive and pragmatic features of potentially deceptive texts have been overlooked, with the work of Victoria Rubin and colleagues being a notable exception. Rubin et al. argue convincingly that "[s]ince news verification is an overall discourse level decision – is the news fabricated or not? – it is reasonable to consider discourse/pragmatic features in each news piece" [8]. Based on earlier research in detection deception [57], Rubin et al. apply rhetorical structure analysis and the vector space model (VSM) to a dataset of news stories [8]. This method involves manually identifying the rhetorical function of segments of textual data, and they note the difficulty of achieving inter-coder agreement, perhaps based on the subjectivity involved and problems with the complex classification system. Nonetheless, their findings showed the VSM model was able to correctly assess 63% of the stories as true/false. The rhetorical relations "disjunction", "restatement", "purpose", and "solutionhood" pointed to true stories, whilst "condition" pointed to deception. Whilst the problem with using fictional news stories as a dataset was critiqued earlier, the researchers do recognise the difficulty of obtaining a reliable and representative dataset, a challenge linguistics can help with (Sect. 2.1).

3 Conclusions

Throughout this chapter, it has been emphasised that there is a tenuous distinction between "fake" and "true", since fake news may commonly mix true statements with false claims, "true" news is representational, and some "true" news sources are more sensational than others. Therefore, it is suggested that fake news detection

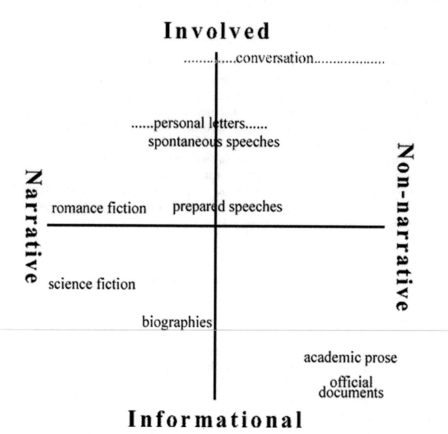

Fig. 1 Example of register dimension (from Biber [58])

research abandons a binary value (fake/not fake) and instead attempts to predict the likelihood of fake news [5]. The linguist Douglas Biber has carried out extensive research on corpora of the English language and, based on linguistic features of a text, is capable of recognising the register to which it belongs, tracing data on scalar continuums or "dimensions" (Fig. 1).

Adapting this approach to the identification, characterisation, and detection of fake news presents a promising line of enquiry in future research.

Several scholars distinguish between the linguistic approach to fake news detection, outlined in this chapter, and the "network" approach [7, 31], which can involve metadata or fact-checking. The meta-features extracted might include the links or comments on a website, or features from different modalities, such as a webpage's visual components. Although these approaches are discussed separately, there may also be a role for linguistics in analysing meta-data, not least in analysing textual comments in response to (fake) news and/or in applying the frameworks in multimodal discourse analysis for categorising images with some semiotic rigour. Fact-checking entails identifying fake news by examining social network activity

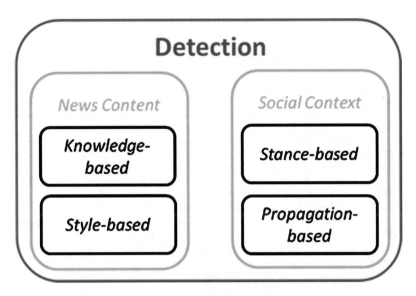

Fig. 2 Data mining framework for fake news detection (adapted from Shu et al. [5])

(e.g. identifying tweets which question or discredit a news article's truthfulness) or contrasting content with external sources (e.g. FakeCheck.org or Snopes.com). Linguistics has limited use for fact-checking because it relies on the availability and reliability of extra-textual information. Conversely, the use of textual analysis in fake news detection has the benefit of being limited to text-intrinsic features. It seems that a hybrid of linguistic and network approaches would address the fact that fake news can be detected through intrinsic and extrinsic features.

Based on an extensive review of the literature from a data mining perspective, Shu et al. [5] propose a framework for the detection of fake news on social media (Fig. 2). This chapter concludes by considering where linguistics might support the data science approach to this problem.

Shu et al. propose that in fake news detection, research can focus on "news content" or the "social context" of data, a distinction which has some overlap with the linguistic and network approaches outlined above. Although linguistics has limited relevance to fact-checking ("knowledge-based" analysis of "news content"), it has everything to do with style, as discussed throughout this chapter. If online news is evaluated according to its "social context", linguistics has limited relevance to "propagation-based" models which examine the interrelation of relevant social media posts. One exception might be if the method involves tracing similar linguistic structures across posts or pages. Linguistics has a lot to input to stance-based models whereby reactions to (fake) news material are utilised in its detection. Characterising such reactions and opinions is another of the myriad ways in which linguistics can support the automated detection of fake news.

References

1. Bond, C.F., BM, D.P.: Accuracy of deception judgments. Pers. Soc. Psychol. Rev. **10**(3), 214–234 (2006)
2. Chen, Y., Conroy, N.J., Rubin, V.L.: News in an online world: The need for automatic crap detector. Proc. Assoc. Inf. Sci. Technol. **52**(1), 1–4 (2015)
3. Parikh, P., Atray, P.K.: Media-rich fake news detection: a survey. In: IEEE Conference on Multimedia Information Processing and Retrieval. IEEE, Maimi, FL (2018)
4. Science Post Team. Study: 70% of Facebook users only read the headline of science stories before commenting. The Science Post, 2018. [Online]. Available: http://thesciencepost.com/study-70-of-facebook-commenters-only-read-the-headline/. Accessed 19 August 2020
5. Shu, S.K., Silva, A., Wang, S., Tang, J., Lui, H.: Fake news detection on social media: a data mining perspective. ACM SIGKDD Explorations Newsletter. **19**(1), 22–36 (2017)
6. O'Brien, N., Latessa, S., Evanglopoulos, G., Bois, X.: The language of fake news opening the black-box of deep learning based detectors. In: 32nd Conference on Neural Information Processing Systems (NIPS 2018). IEEE, Montreal, Canada (2018)
7. Conroy, N.K., Rubin, V.L., Chen, Y.: Automatic deception detection: methods for finding fake news. ASIST. **52**, 1–4 (2015)
8. Rubin, V.L., Conroy, N.J., Chen, Y.: Towards news verification: deception detection methods for news discourse. In: Proceedings of the Hawaii International Conference on System Sciences (HICSS48) Symposium on Rapid Screening Technologies, Deception Detection and Credibility Assessment Symposium. Grand Hatt, Kauai (2015)
9. Marquardt, D.: Linguistic indicators in the identification of fake news. Mediatization Stud. **3**, 95–114 (2019)
10. Guo, L., Vargo, C.: Fake news and emerging online media ecosystem: an integrated intermedia agenda-setting analysis of the 2016 US presidential election. Commun. Res. **47**(2), 178–200 (2020)
11. Tandoc, E.C., Lim, Z.W., Ling, R.: Defining "fake news": a typology of scholarly definitions. Dig. Journal. **6**, 137–153 (2018)
12. Rashkin, H., Choi, E., Jang, J.Y., Volkova, S., Choi, Y.: Truth of varying shades: analyzing language in fake news and political fact-checking. In: Proceedings of the 2017 Conference on Empirical Methods in Natural Language Processing. Association for Computational Linguistics, Copenhagen, Denmark (2017)
13. Dey, A., Rafi, R.Z., Parash, S.H., Arko, S.K., Chakrabarty, A.: Fake news pattern recognition using linguistic analysis. In: 2018 Joint 7th International Conference on Informatics, Electronics & Vision (ICIEV) and 2018 2nd International Conference on Imaging, Vision & Pattern Recognition (icIVPR). IEEE, Kitakyushu, Japan (2018)
14. Paskin, D.: Real or fake news: who knows? J. Soc. Media Soc. **7**(2), 252–273 (2018)
15. Simpson, P.: On the Discourse of Satire: Towards a Stylistic Model of Satirical Humour. John Benjamins, Amsterdam and New York (2003)
16. De Beer, D., Matthee, M.: Approaches to identify fake news: A systematic literature review. In: International Conference on Integrated Science ICIS 2020: Integrated Science in Digital Age 2020. Springer, Cham (2021)
17. Vosoughi, S., Roy, D., Aral, S.: The spread of true and false news online. Science. **359**, 1146–1151 (2018)
18. Albright, J.: Welcome to the era of fake news. Media Commun. **5**(2), 87–89 (2017)
19. Vargo, C., Luo, L., Amazeen, M.A.: The agenda-setting power of fake news: a big data analysis of the online media landscape from 2014 to 2016. New Media Soc. **20**, 2028–2049 (2018)
20. I. W. Stats. (2020). https://www.internetworldstats.com/stats7.htm. Accessed 18 August 2020
21. Enos, F.: Detecting deception in Speech, PhDd thesis (2009). http://www1.cs.columbia.edu/~frank/enos_thesis.pdf. Accessed 31 August 2020
22. Wynne, M.: Developing Linguistic Corpora: A Guide to Good Practice. Oxbow Books/Arts and Humanities Data Service, Oxford (2005)

23. Schubert, L. Computational linguistics (2020). https://plato.stanford.edu/archives/spr2020/entries/computational-linguistics/. Accessed August 2020
24. van Dijk, T.: Studies in the Pragmatics of Discourse. Mouton de Gruyter, Berlin (1981)
25. van Dijk, T.: News as Discourse. Lawrence Erlbaum Associates, Mahwah, NJ (1988)
26. Bell, A.: The Language of News Media. Blackwell, Oxford (1998)
27. Bednarek, M., Caple, C.: News Discourse. Continuum, London and New York (2012)
28. Jucker, A.: Social Stylistics: Syntactic Variation in British Newspapers. Mouton de Gruyter, Berlin and New York (1992)
29. Caple, H., Knox, J.S.: A framework for the multimodal analysis of online news galleries: what makes a "good" picture gallery? Soc. Semiotics. **25**(3), 292–231 (2015)
30. Chovanec, J.: Multimodal storytelling in the news: Sequenced images as ideological scripts of othering. Discourse Context Media. **28**, 8–18 (2019)
31. Jin, Z., Cao, J., Zhang, Y., Zhou, J., Tian, Q.: Novel visual and statistical image features for microblogs news verication. IEEE Trans. Multimedia. **19**(3), 598–608 (2017)
32. Gupta, A., Lamba, H., Kumaraguru, P., Joshi, A.: Faking sandy: characterizing and identifying fake images on twitter during hurricane sandy. In: WWW 2013 Companion. ACM, Rio de Janeiro, Brazil (2013)
33. Pérez-Rosas, V., Kleinberg, B., Lefevre, A., Mihalcea, R.: Automatic fake news detection. In: Proceedings of the 27th International Conference on Computational Linguistics. Association for Computational Linguistics, Santa Fe (2018)
34. Ali, M., Levine, T.R.: The language of truthful & deceptive denials & confessions. Commun. Rep. **21**, 82–91 (2008)
35. Ott, M., Cardie, C., Hancock, J.: Negative deceptive opinion spam. In: Proceedings of NAACLHLT. Springer, Atlanta (2013)
36. Newman, M.L., Pennebaker, J.W., Berry, D.S., Richards, J.M.: Lying words: predicting deception from linguistic styles. Pers. Psychol. Bull. **29**(5), 665–675 (2003)
37. Feng, V., Hirst, G.: Detecting deceptive opinions with profile compatibility. In: Proceedings of the Sixth International Joint Conference on Natural Language Processing. Asian Federation of Natural Language Processing, Nagoya, Japan (2013)
38. Zhang, L., Guan, Y.: Detecting click fraud in pay-per-click streams of online advertising networks. In: The 28th International Conference on Distributed Computing Systems. ICDCS, Beijing (2008)
39. Toma, C., Hancock, J.: Reading between the lines: linguistic cues to deception in online dating profiles. In: Proceedings of the 2010 ACM conference on Computer supported cooperative work. ACM, New York (2010)
40. Shafqat, W., Lee, S., Malik, S., Kim, H.: The language of deceivers: Linguistic features of crowdfunding scams. In: Proceedings of the 25th International Conference Companion on World Wide Web. International World Wide Web Conferences Steering Committee, Geneva (2016)
41. Hancock, J., Toma, C., Ellison, N.: Lying in online data profiles. In: CHI 2007. ACM, San Jose, CA (April 28–May 3, 2007)
42. Biber, D., Conrad, S.: Register, Genre and Style. Cambridge University Press, Cambridge and New York (2019)
43. Castelo, S., Almeida, T., Elghafari, A., Santos, A., Pham, K., Nakamura, E., Freire, J.: A topic-agnostic approach for identifying fake news pages. In: Companion Proceedings of the 2019 World Wide Web Conference. ACM, New York (2019)
44. Kaggle. Getting real about fake news (2016). https://www.kaggle.com/mrisdal/fake-news. Accessed 21 August 2020
45. Potthast, M., Kiesel, J., Reinartz, K., Bevendorff, J., Stein, B.: A stylometric inquiry into hyperpartisan and fake news. Comput. Lang. (2017). https://doi.org/10.18653/v1/P18-1022
46. Kreis, R.: The "tweet politics" of President Trump. J. Lang. Polit. **16**(4), 607–618 (2017)
47. Ross, A.S., Rivers, D.J.: Discursive deflection: accusation of "fake news" and the spread of mis- and disinformation in the tweets of President Trump. Soc. Media Soc. **4**(2), 1–12 (2018)

48. Stolee, G., Caton, S.: Twitter, Trump, and the base: a shift to a new form of Presidential talk. Signs Soc. **6**(1), 147–165 (2018)
49. Lockhart, M.: President Donald Trump and his political discourse: ramifications of rhetoric via Twitter. Routledge, London and New York (2018)
50. Clarke, I., Grieve, J.: Stylistic variation on the Donald Trump Twitter account: a linguistic analysis of tweets posted between 2009 and 2018. PLoS One. **14**(9) (2019)
51. Dalecki, L., Lasorsa, D.L., Lewis, S.C.: The news readability problem. Journal. Pract. **3**(1), 1–12 (2009)
52. Pérez-Rosas, V., Mihalcea, R.: Cross-cultural deception detection. In: Proceedings of the 52nd Annual Meeting of the Association for Computational Linguistics. Association for Computational Linguistics, Baltimore, MA (2014)
53. Biber, D.: Representativeness in corpus design. Literary Linguistic Comput. **8**(4), 243–257 (1993)
54. Herring, S.: Computer-mediated discourse analysis: an approach to researching online behavior. In: Designing for Virtual Communities in the Service of Learning, pp. 338–376. Cambridge University Press, Cambridge (2004)
55. Beißwenger, M., Storrer, A.: Corpora of computer-mediated communication. In: Corpus Linguistics. An International Handbook, vol. 1, pp. 292–308. Mouton de Gruyter, Berlin (2008)
56. Weisser, M.: Practical Corpus Linguistics: An Introduction to Corpus-Based Language Analysis. Wiley Blackwell, Chichester (2016)
57. Feng, S., Banerjee, R., Choi, Y.: Syntactic stylometry for deception detection. In: Proceedings of the 50th Annual Meeting of the Association for Computational Linguistics. Association for Computational Linguistics, Jeju Island (2012)
58. Biber, D.: Variation Across Speech and Writing. Cambridge University Press, Cambridge (1988)